IN PURSUIT OF THE GENE

# In Pursuit of the Gene

## ~ FROM DARWIN TO DNA ~

James Schwartz

HARVARD UNIVERSITY PRESS
Cambridge, Massachusetts, and London, England — 2008

*To my wife, Ann Hochschild, whose unwavering belief in this book sustained me through the many years it took to write. Her selfless devotion to excellence, her clear mind, and the pleasure she takes in getting things right are an unending source of wonder to me. I dedicate this book to her in acknowledgment of my gratitude and love.*



Printed in the United States of America

Library of Congress Cataloging-in-Publication Data

Schwartz, James, 1955–
In pursuit of the gene : from Darwin to DNA / James Schwartz.
p. cm.
Includes bibliographical references and index.
ISBN-13: 978-0-674-02670-4 (cloth : alk. paper)
ISBN-10: 0-674-02670-5 (cloth : alk. paper)
1. Genetics—History. 2. Genes—History. I. Title.
[DNLM: 1. Genetics—history. 2. DNA—history. QU 11.1 S399i 2008]
QH428.S24 2008
576.509—dc22 2007034830

# CONTENTS

# ILLUSTRATIONS

# PREFACE

— SCIENCE IS FUNDAMENTALLY DIFFERENT from art, it is often claimed, because the sensibility of the scientist isn't involved. Had X not made the discovery, Y would eventually have gotten to the same place. Made into a caricature, this is the view of scientists in white coats engaged in a selfless quest for truth. At the opposite extreme is the postmodernist idea that science is a social construct—that no independent, objective reality exists outside of one's perceptions. While both views are clearly absurd, I confess that I am more sympathetic with the former than the latter. The consequences of denying the existence of an objective scientific reality were dramatically illustrated in the Soviet Union in the period between 1933 and 1953 when the ideas about genetics that had been built up over the previous century were undermined and then outlawed, plunging Soviet agriculture into free fall.

While it may not matter in the long run how a scientific idea came into being, nonetheless learning about the thoughts, feelings, and interactions of the people who worked on it can illuminate the science in unexpected ways. This may be particularly true of the study of heredity, which is both the most mathematical and abstract of the life sciences and at the same time the most personal. Intimately connected with the origin of statistics, the early study of inheritance also touched on age-old philosophical questions about free will and determinism, the relationship of parent and child, and the extent to which human beings can be reduced to the sum of their parts. Fueled by their concern with these elemental themes, I believe, the early geneticists were a particularly passionate group, pathologically competitive in some instances and utterly selfless in others, prone to intense loyalties as well as overwhelming hatreds, singularly idealistic and ruthlessly pragmatic.

For these reasons I have chosen to take a biographical approach to the history of genetics, and my research was greatly facilitated by the fact that many of the players who made the key discoveries were skilled, prolific letter writers. Darwin's epistolary output was staggering, as was that of his cousin Francis Galton, who was in many ways the pioneer of the modern theory of heredity. Similarly, Galton's disciples Frank Weldon and Karl Pearson, and their nemesis William Bateson, all of whom played leading parts in the battle over Mendelism, were devoted correspondents. Most of the papers of Gregor Mendel himself, who managed to divine the essence of modern genetics in 1865 working alone in a monastery in Moravian Silesia, were destroyed, but enough survived to give a sense of the man behind the scientist. The sad story of Hugo de Vries, who played a key role in the restitution of Mendel's work in 1900, can be pieced together from De Vries's correspondence with Darwin, Bateson, and others. The standard view of the post-1900 development of genetics, which largely took place in the lab of Thomas Hunt Morgan at Columbia, was shaped by Morgan's favorite student, A. H. Sturtevant, in his 1965 *History of Genetics,* but an account entirely at odds with the one in that book later came to light in a few surviving letters on the topic by Morgan as well as in the remembrances of several other geneticists who were members of the fly room. This alternative view is given full expression in the correspondence of Morgan's most celebrated student, Hermann J. Muller, which is held at the Lilly Library at Indiana University, and in a previously unknown collection of 550 letters donated by Muller's son to the Cold Spring Harbor Laboratory Archives.

From these letters and other primary source material, it is possible to trace the idea of the gene as it left one consciousness and entered another. Of course many people were thinking about the same thing, and it is necessary to ignore a host of offshoots in the service of following the main thread. Sometimes ideas were passed from one scientist to another just as the originator intended. Other times, it was a terrible confusion in one person's mind that resulted in a brilliant illumination in the next. Still other times, those who found the correct path did so entirely by accident after setting off on the wrong one or, conversely, began with a brilliant flash of insight and proceeded to wander deep into the wilderness.

The story of the gene begins with Darwin's arcane and deeply flawed "provisional hypothesis of pangenesis," which reflected his recent conversion to another highly problematic theory, that of Lamarckian inheritance. Despite its many problems, Darwin's theory posited the existence of microscopic hereditary particles that would play a crucial role in the development of the modern ideas about the gene. No sooner had Darwin proposed his model than his cousin Francis Galton commandeered it to support a theory of inherited intelligence that was itself based on a host of erroneous assumptions and ad hoc arguments. Nonetheless, Galton's hereditary theory was much closer to the truth than that of his far more emotionally balanced, reasonable, and steadfast cousin.

The Dutch botanist Hugo de Vries also tried to enlist Darwin's theory of pangenesis in the service of his own theory, hoping to displace Darwin as the central figure in modern biology. In the course of his work De Vries stumbled on Mendelism, but he failed to properly appreciate that he had uncovered the discovery of the century and instead spent the remainder of his life devaluing Mendel's achievement and promoting his own erroneous theory, the mutation theory, in its place. It was the English zoologist William Bateson who took up the cause of Mendelism, driven to a feverish pitch in its defense by his desire to expose the folly of his once best friend Frank Weldon, whom Bateson felt had betrayed both him and the cause of truth.

In 1909, Morgan, who stood opposed to all three of the major developments in contemporary biology—Darwinism, the chromosome theory, and Mendelism—began to study the common fruit fly. The following year, despite his doubts about both Mendel's factors and the importance of the chromosomes, Morgan discovered the first definitive proof that a Mendelian factor was associated with a particular chromosome—the sex-linked X chromosome. But Morgan's continued equivocation, about Darwinism and the implications of the chromosome theory, led to a falling out with Muller, who had already begun to see his way toward a new synthesis of Mendel and Darwin. Ten years later, in 1926, Muller proved that X-rays could be used to artificially induce mutations in flies, a discovery that would ultimately win him the Nobel Prize. Though Morgan and Sturtevant partially succeeded in tainting Muller's reputation, Muller would go on to develop the modern theory of the gene, which served as the foundation for modern molecular biology.

In my view, discussing the history of genetics without describing actual experiments would be like analyzing the history of art without looking at any of the paintings. Accordingly, although this book is intended for a lay audience, I have described many of the key experiments in considerable detail. Some research papers read like finished masterpieces and others like works in progress. Many of the papers of the frenetic Morgan, for example, were vague and unformed. Often in a muddle over the meaning of his discoveries, Morgan was nonetheless a good enough scientist to recognize something new and important when he stumbled on it. At the other extreme was Mendel, whose paradigm-shifting paper "Experiments on Plant Hybrids" was an exquisitely polished argument, supported by a vast array of data that confirmed his model with astonishing precision.

— THE LONGER I STUDIED this material, the more the history of the developing concept of the gene began to coalesce into a coherent narrative. While I tried to emphasize the continuities and recurrent patterns, I was also acutely aware of the need to remain faithful to the facts. As a result, I excluded observations that seemed to fit with my sense of the characters but could not be independently verified, and I provided documentation for each of the more controversial claims in this book. Further danger lurked in the accounts of the scientists themselves, who often attempted to shape the history to their own ends, as when Sturtevant wrote in his biographical memoir of Morgan for the National Academy of Sciences that Muller was often in and out of the fly room but never actually had a desk there. But to a remarkable degree, I think, it was possible to get beyond the distortions. One thing that helps is the system itself—no matter what prejudices a scientist might bring to his work, each new idea, in order to survive, must be tested in controlled experiments and must hold up to the scrutiny of other scientists. As the scientific facts ultimately sort themselves out, I believe, so do the facts about the scientists who produced them, and little by little it is possible to move closer to an undistorted view of the history as it unfolded.

**Charles Darwin (top) and Francis Galton.**

From Karl Pearson, *The Life of Francis Galton,* vol. 1 (Cambridge: Cambridge University Press, 1914), after p. 56.

CHAPTER 1

# Viva Pangenesis

— IN THE WINTER OF 1868, nine years after the publication of his *On the Origin of Species,* Charles Darwin published his two-volume, nearly thousand-page *Variation of Animals and Plants under Domestication.* Among other things, *Variation* was a review of the multitude of observations concerning the phenomenon of inheritance. In fact, *Variation* was so crammed with detail that Darwin's publisher elected to use two fonts, a large one for the most important material and a smaller one for fine points recommended only for the serious naturalist. In a letter to his friend Joseph Hooker on how to approach his new work, Darwin advised, "Skip the whole of Vol. 1, except the first chapter (and that only to be skimmed) and skip largely the 2nd volume; and you will say it is a very good book."[1] The penultimate chapter, however, was not to be missed, for this chapter contained an exposition of his new theory of pangenesis, an idea more dear to Darwin than any save his theory of natural selection.

The "provisional hypothesis of pangenesis," as he called his new theory, was Darwin's heroic attempt to make a coherent whole out of the chaotic mass of information pertaining to inheritance. Composition of the chapter had been a terrible ordeal. Even in the best of times Darwin suffered from a variety of physical and psychological maladies, but his troubles reached such a pitch while writing about pangenesis that he sought out Dr. John Chapman, former editor of the *Westminster Review,* expert in psychological medicine, and consultant to many famous Victorian neurotics. Under Chapman's care, Darwin was treated with ice packs on the spine for ninety-minute sessions, three times a day. At first the treatment appeared to be helping, but Darwin soon concluded that "ice to the spine did nothing."[2] Suffering greatly, he nonetheless managed to complete his forty-page chapter.

From the start, the theory of pangenesis met with a chilly reception. Emma, Darwin's devoutly religious wife, considered the name "wicked" for its suggestion of pantheism.[3] To his skeptical friend Hooker, Darwin had written, "It is a comfort to me to think that you will be surely haunted on your death-bed for not honouring the great god Pan." Thomas Huxley, an ardent defender of the faith who had suffered the humiliation of being an early and vituperative foe of natural selection, wrote Darwin of his fear that "somebody rummaging among your papers half a century hence will find Pangenesis and say, *See this wonderful anticipation of our modern theories, and that stupid ass Huxley prevented his publishing them.*"[4]

The fundamental building blocks of the new theory were microscopic hereditary particles that Darwin called "gemmules." As he envisioned it, each cell constantly shed gemmules, at each stage of development. These gemmules circulated throughout the body and eventually found their way to the reproductive cells, where they could be assembled into germ cells, the sperm and egg cells, capable of giving rise to new organisms. "Speaking strictly," Darwin explained, "it is not the reproductive elements nor the buds, which generate new organisms, but the cells themselves throughout the body."[5]

Darwin's attachment to pangenesis stemmed largely from the fact that it solved a central unsolved problem in his theory of natural selection. In Darwin's model, the slow and relentless action of natural selection drove evolutionary change, sifting out the most advantageous from a large number of *small* differences among organisms. The problem arose from Darwin's working assumption that the child was a blend, in not necessarily equal measure, of each of his parent's traits. For under the so-called *blending hypothesis,* children were more or less averages of their parents' traits—for example, the mating of tall and short adults would result in children of average height—and, in a matter of very few generations, all the diversity in a population would disappear. Without a new source of variation, evolution would soon grind to a halt.

Pangenesis came to the rescue by providing a ready source of heritable variation. The idea was that the environmentally induced changes in organisms—"acquired traits"—would be transmitted to their offspring via the gemmules. The modifications in cells and the organs made up from them were presumed to give rise to *modified* gemmules. The modified gemmules, in turn, would give rise to modified cells in the body of the offspring. Because any new feature of an organism acquired due to a change in its habits or conditions of life was passed on to its children, acquired traits provided a wellspring of diversity.

The doctrine of the heritability of acquired characteristics had a long and checkered history before Darwin. In fact, in 350 B.C., in his treatise "On the Generation of Animals," Aristotle had already argued against a pangenesis-like model and the idea that children inherited their parent's scars. In the second volume of his 1796 treatise *Zoonomia,* Erasmus Darwin, Charles Darwin's grandfather, suggested that changes in parents due to their activities could be passed on to their children. The idea that acquired traits could be passed on soon came to be known as Lamarckian inheritance after the French biologist Jean-Baptiste Lamarck, who promoted it with single-minded intensity.

Although Charles Darwin had initially been an ardent anti-Lamarckian, characterizing "adaptations from the slow willing of animals" as "Lamarck nonsense," he did a complete turnaround after the publication of the *Origin* and began to collect evidence for the heritability of mutilations and many other acquired characteristics.[6] He also found it "highly probable" that acquired mental habits were passed on, as "gemmules derived from modified nerve-cells are transmitted to the offspring."[7] No doubt Darwin took solace in the thought that he had passed on to his children not only his frail constitution but his highly virtuous mental habits. Despite his legendary reputation for balanced, evenhanded weighing of evidence, Darwin grew so attached to the Lamarckian principle that he permitted himself to be carried away: "It is probable," he wrote in *Variation*, "that hardly a change of any kind affects parents without some kind of mark being left on the germ."[8]

— FRANCIS GALTON, who was Charles Darwin's half first cousin, read *Variation* with great excitement, scribbling furiously in the margins as he read. Just as it had ten years earlier when he first read the *Origin*, Darwin's work took hold of Galton and changed the course of his life.[9] The three chapters on inheritance fascinated him most of all. For many years now, Galton had been developing his own ideas about inheritance, which he had first set forth in a long article for *Macmillan's*, a high-brow literary magazine, in the summer of 1865.[10] While Darwin had carefully skirted the case of man in presenting the thesis of descent from a common ancestor, hoping to avoid the additional controversy that was sure to result by insisting on man's animal nature, his less diplomatic cousin plunged right into the heart of the controversy.

Contrary to the widespread belief that each human soul is created by God, Galton insisted that mental as well as physical traits are inherited. Furthermore, he insisted, just as it is possible by careful selection to produce a superior breed of cattle or horse, it ought to be possible to produce a superior race of men by judicious marriages during several consecutive generations.[11] Eighteen years later, he would introduce the term *eugenics*, from the Greek *eugene*, for "good in stock," and devote his considerable energy to laying the groundwork for a national eugenics program.[12] His first goal, however, was to demonstrate that human intelligence and character are inherited to the same degree as any physical feature in humans or animals. To prove his point, Galton designed a simple statistical study that purported to show that relatives of eminent people are far more likely to be eminent than randomly chosen members of the educated public, and thus that exceptional talent could be passed from one generation to the next. In this, his first treatment, he quickly glossed over his erroneous assumption that eminence accurately reflected natural talent, a fallacy that Galton would commit again and again in future years.

Although his study was fundamentally flawed, Galton made clever use of the resources at hand. For data, Galton used a number of well-known biographical sources, including a compilation of four centuries of "original minds": a list of biographical sketches called "Men of Our Time" that was a Who's Who of England, the Continent, and America. In addition, he consulted a well-known *Dictionary of Painters,* another of musicians, a guide called *Lives of Chancellors,* and the list of top-ranking classics students from successive classes at Cambridge. In each case he calculated the percentage of eminent relatives among the eminent men. For example, he found that of the 391 eminent painters who appeared in the dictionary, 65 were closely related to another painter in the same book, and likewise, 1 in 12 of his "distinguished" literary men had a father, son, or brother who was also distinguished.[13] Overall, the probability of having a near male relative in a given field was 1 in 6. In contrast, he estimated that the probability was 1 in 3,000 that a randomly chosen educated man would achieve eminence. Thus there was a 500-fold greater probability of eminence among relatives of the eminent.

The effect was impressive. Galton considered it unmistakable evidence of the enormous effect of hereditary influence, confidently asserting, "The overwhelming force of a statistical fact like this renders counter-arguments of no substantial effect."[14] Yet on the very next page he succinctly states the obvious counterargument: "When a parent has achieved great eminence, his son will be placed in a more favourable position for advancement, than if he had been the son of an ordinary person," a view to which Galton gave no credence.

On the eve of his forty-fourth birthday, shortly after the appearance of his *Macmillan's* article, Galton suffered a debilitating nervous breakdown. Ordinarily gregarious, Galton was forced to withdraw from London society, and for months at a stretch he and his wife traveled in Italy and Switzerland. But even at the peak of his psychological distress, Galton began work on his next project, *Hereditary Genius,* a book-length treatment of the material he'd begun to explore in his magazine piece.[15]

While Galton went to great lengths to impart to *Hereditary Genius* an air of scientific neutrality and objectivity, the subject matter had a deeply personal significance for him. From an early age, Galton's parents had encouraged him in the belief that he was a child prodigy and regularly trotted him out to recite long passages of memorized poetry for visitors. By the end of high school, Galton revered above all else mathematical talent and those who possessed it. Helped by the timely intervention of his cousin Charles, Galton persuaded his father to allow him to abandon the study of medicine in order to pursue a mathematics degree at Cambridge, where he hoped to join the ranks of his intellectual heroes.

In an unmistakably autobiographical passage, Galton traced the trajectory of the gifted child in the opening chapter of *Hereditary Genius.* "The eager boy," he wrote, "when he first goes to school, and confronts intellectual difficulties, is astonished by his progress."[16] Judging by his growing capacity, he becomes convinced that he will be among those "who have left their mark upon the history of the world." Slowly, the sober facts of the world begin to assert themselves. "The years go by; he competes in the examinations of school and college, over and over again with his fellows, and soon finds his place among them." Eventually he comes to understand the reality, "unless he is incurably blinded by self-conceit." With this knowledge: "He is no longer tormented into hopeless efforts by the fallacious promptings of overweening vanity, but he limits his undertakings to matters below the level of his reach, and finds true moral repose in an honest conviction that he is engaged in as much good work as his nature has rendered him capable of performing."[17]

But unlike his eager boy who found repose in a life of diminished expectations, Galton was a man obsessed, collecting data on men and their male ancestors from all walks of life—from English judges to musicians and painters to "Wrestlers of the north country." As they had in his youth, mathematicians held a special fascination for him. "There hardly be a surer evidence of the enormous difference between the intellectual capacity of men," he explained near the beginning of his new book, "than the prodigious differences in the numbers of marks obtained by those who gain mathematical honours at Cambridge."[18]

Some years thirty years earlier, when he had been embarking on his university career, Galton had set his sights on achieving the august rank of "Cambridge Wrangler," the highest honor available to a mathematics student at Cambridge. Based on the results of a grueling forty-five-hour exam, the top forty finishers were granted the honorary title of Wrangler, and of those forty, the highest scorer was made Senior Wrangler. But the teenage Galton was already aware that such high ambitions contained a lurking danger, writing home of the case of a fellow student who had "worked himself almost to madness, but was quite unable to succeed on account of his natural powers; poor fellow he did the best thing that he could do though."[19]

Attending lectures for two hours daily and reading for ten and a half, he proudly wrote home, "I verily believe that I never worked so hard before."[20] However, by the end of November the strain had taken its toll. He was confined to bed for twelve straight days, putting "a pro tempore dead stop to Maths."[21] Worried about falling behind, Galton planned to make up for lost time over the Christmas holiday. However, he hadn't kept to his plan, instead spending several weeks visiting with his siblings and finishing off with a trip to Paris. Although his conscience seemed to be bothering him, he continued to ignore his maths when he returned to Cambridge, instead diverting his energies toward designs for a pick-proof lock, a more-accurate balance, and a rotary steam engine.[22] Galton received a disappointing "third class" on his exams.

In the fall of his second year Galton signed up to study with the revered Cambridge math coach William Hopkins, who had tutored some of the finest scientific minds of the age, including James Clerk Maxwell and Lord Kelvin. Hopkins was "in no way Donnish," a reinvigorated Galton wrote home. "He rattles on at a splendid pace and makes mathematics anything but a dry subject by entering thoroughly into its metaphysics. I never enjoyed anything so much before."[23] That winter Galton was acutely aware of the performance of the seniors on their exam. "Cayley is first and Simpson 2nd wrangler, these were far superior to the rest," he reported to his father.[24] Not only the fact that the top two performers outshone the others but also the precise spread of the scores fascinated Galton. "Hopkins told me today that Simpson was 1000 marks ahead of the 3rd wrangler and the getting of 500 marks only entitles a man to be a wrangler."[25] In the spring of his second year, he was forced to endure his most significant trial to date, the traditional "Little Go" exam. Galton and two of his fellow Hopkins students scored in the second class while seven other Hopkins students were in the first.[26]

Despite the fact that Hopkins praised his performance and was quite encouraging about his prospects, Galton was devastated. As a result of his test results, he resolved to forego the traditional spring break and stay behind in Cambridge to prepare for the end-of-year Scholarship exam. Three days into the vacation period, he had a sudden change of heart. As there was no chance of winning, there was no point in even attending the exam, he wrote his father. "I should not know my standing in Maths in the college from the result of this examination any better than I do now," he added, listing the other Hopkins students who had outperformed him. "I would, however, have possibly tried my best," he explained, "but my head is rather bad from having overworked myself."[27]

A summer break spent in Scotland with his classmates did not seem to improve his mental state. In August he wrote, "I have been able to do but little reading since I have been here and altogether am very low about myself."[28] In early November of his third year, he complained, "My head is very uncertain so that I can scarcely read at all."[29] In the same despairing letter, he reported on the failures of half a dozen of his fellow classmates to survive the mathematics program. His assessment of the competitive exam system overall was that "the satisfaction enjoyed by the gainers is very far from counterbalancing the pain it produces among the others."

By the end of the month, Galton's symptoms had become so acute that it was impossible for him to continue on with his math degree. "I find myself quite unable to do anything in reading," he now wrote, "for by really deep attention to Maths I can bring on my usual dizziness etc. almost immediately though generally I feel much better than I used to do. Palpitations of the heart have lately come on when I read more than I ought to do which I am rather glad of than otherwise, as it saves my head."[30] His condition was so severe that he was forced to take the term off and return home. Fifty years later, he still vividly recalled the trauma of these student days: "I suffered from intermittent pulse and a variety of brain symptoms of an alarming kind. A mill seemed to be working inside my head; I could not banish obsessing ideas; at times I could hardly read a book, and found it painful even to look at a printed page."[31]

The symptoms of his second breakdown were very much like those he had experienced twenty years earlier. In his autobiography, he recalled that he was obsessed by small problems day and night, and tried vainly to "think them out." In the winter of 1865–66, on the eve of his forty-fourth birthday, his mother wrote imploring him, "Give yourself a complete year's rest and perhaps it may be the saving of your life."[32] Similarly his older sister Bessie advised: "Your heredity will also be better by returning to it with a fresh eye and refreshed head." Six months later he precipitously withdrew from the August meeting of the British Association of Science, where he was scheduled to deliver an address, and the following year he resigned as general secretary because he could not tolerate the "sustained racket" of the meetings. "It would have been playing with death had I continued to hold it," he wrote in his memoirs.[33] "Those who have not suffered from mental breakdown can hardly realize the incapacity it causes," he explained. A *sprained brain* was like a *sprained joint*, "after recovery seems to others to be complete, there remains for a long time an impossibility of performing certain minor actions without pain and serious mischief, mental on the one hand, physical on the other."

Returning to the obsession of his undergraduate years, Galton had in the course of his research for *Hereditary Genius* managed to procure a list of the scores of the high finishers for two separate years from a Cambridge examiner. "The precise number of marks obtained by the senior wrangler in the more remarkable of these two years was 7,634, by the second wrangler in the same year, 4,123; and by the lowest man in the list of honours, only 237," he reported. With loving attention, he teased out what he considered to be a subtle injustice the test did to the best and the brightest. For although he considered the scores to be strictly proportional to mathematical talent at the low end, he speculated that the scores of the most gifted, who would be unable to write quickly enough to capture the torrential flow of their thoughts, were likely to be an underestimate of their true abilities. On the other hand, he marveled, "They find their way at once to the root of the difficulty in the problems that are set, and with a few clean, apposite, powerful strokes, they can overthrow it."[34]

Despite the terrible ordeal he'd suffered at Cambridge, Galton now proposed to classify the entire nation based on mental aptitude. In order to accomplish his goal, he intended to make use of the then still rarified knowledge of the normal curve, classifying people according to their degree of deviation from the population average. He had got the idea of applying the normal curve to analyze human phenomena from the Belgian mathematician Adolphe Quetelet, whose book *Letters in Probabilities* was published in English in 1849. In *Hereditary Genius,* Galton presented Quetelet's analysis of the chest measurements and heights of 5,783 Scottish soldiers and more than 100,000 French conscripts, a strikingly beautiful demonstration of the regularity of variation in a set of data. According to the "law of deviation from an average," as Galton referred to the theory, the various size classes would tend to be distributed like a bell curve, where the frequency of the most common size class (the average) was proportional to the area under the apex of the bell and the frequency of the exceptionally tall and short classes to the area near the tails. "The marvelous accordance between fact and theory must strike the most unpracticed eye," Galton delighted.[35] Quetelet's successful application of the law of averages suggested other measurements to which the law ought apply:

> Now if this be the case with stature, then it will be true as regards every other physical feature—as circumference of head, size of brain, weight of grey matter, number of brain fibres, &c.: and thence, *by a step on which no physiologist will hesitate,* as regards mental capacity.[36]

Of course, he knew it was a leap from the tangible world of brains and brain fibers to the abstraction of mental capacity, and yet he did not hesitate to draw conclusions about intelligence. "Analogy clearly shows there must be a fairly constant average mental capacity in the inhabitants of the British isles, and that the deviations from that average—upwards toward genius, and downwards towards stupidity—must follow the law that governs deviations from all true averages."[37]

But he could do better than analogy, Galton claimed; he would show directly that mental aptitude was normally distributed by examining test scores. As it turned out, the Cambridge math exam scores didn't follow a normal distribution, but he'd found a civil service exam given by the Royal

Military College at Sandhurst that did. "There is, therefore, little room for doubt," he asserted, "if everybody in England had to work up some subject and then to pass before examiners who employed similar figures of merit, that their marks would be found to range, according to the law of deviation from an average, just as rigorously as the heights of French conscripts, or the circumferences of the chests of French soldiers."[38]

Although it was true that certain tests could be found to give a normal distribution of scores, the equation of test scores with mental capacity remained open to question. Nonetheless Galton forged ahead, using the assumption of normality to divide the population into 14 groups, separated by equal increments of "natural aptitude." Arrayed to the right of center were classes A, B, C, D, E, F, and G of increasing aptitude. To the left were the mentally challenged, fittingly denoted in lowercase, a, b, c, d, e, f, and g. The size of each class could be predicted by using the same table Quetelet had used to find his theoretical values for height. Galton arranged the scale so that the top two classes, F and G, combined would contain 1 in 4,000 men, which was his best estimate of the fraction of all educated men who achieved eminence. He stretched for new poetic heights in order to explain the extraordinary thing it was to be an F or a G: "On the most brilliant of starlight nights there are never so many as 4,000 stars visible to the naked eye at the same time; yet we feel it to be an extraordinary distinction to a star to be accounted as the brightest in the sky."[39]

Despite the fact that he still had no solid basis for believing that intelligence is analogous to height and chest size, Galton did not hesitate to apply Quetelet's reasoning to the mental realm. While reading Darwin's theory of pangenesis in the winter of 1868–69, Galton suddenly saw how Darwin's hereditary particles could be used to argue that intelligence, or any other mental or physical trait, for that matter, would be normally distributed. As he expressed it in a hastily composed concluding chapter of *Hereditary Genius*, "It gives a key that unlocks every one of the hitherto unopened barriers to our comprehension of [inheritance]."[40]

In the simplest case, in which a trait was determined by two forms of a particular gemmule that existed in equal numbers in each of the parents, one would expect the trait to be normally distributed. If, for example, intelligence depended on the inheritance of two forms of a gemmule, one that produced brilliant nerve cells and the other for dull ones, and the two forms existed in equal numbers in both parents, then one would see a bell-shaped distribution of intelligence among the offspring. The classes F and G would correspond to those lucky few who had chanced to inherit a collection consisting of a vast excess of gemmules for superior neurons; the majority of the population in the middle ranges, in the classes e, d, c, b, a, A, B, C, D and E, inherited a nearer to equal mixture of superior and dull gemmules, and the unfortunate classes f and g, the so-called "idiots" and "imbeciles," were the recipients of a nearly uniform collection of inferior gemmules.[41]

With the particulate theory of inheritance provided by pangenesis, not only did Galton have a mechanism to explain the normal distribution of intelligence, but he could also use the distribution of intelligence in one generation to predict the distribution in the next one. Four pages from the end of *Hereditary Genius,* he spelled out how a breeding program would work: "Not only might the average qualities of the descendants of groups A and B, A and C, A and D, and every other combination be predicted, but also the numbers of them who deviate in various proportions from those averages."[42] In the eleventh hour, it appeared, Galton had hit upon something big—a unifying principle that lent an air of scientific credibility to his theory of hereditary genius and his hopes for improving the human race.

While Darwin's pangenesis seemed to give a tremendous boost to Galton's theory of hereditary genius, there was one big problem with it from his point of view. As Darwin conceived of them, the particles were extremely malleable. If the environment modified the hereditary material, then identifying the talented and encouraging them to breed with each other could hardly be more effective than spreading the advantages of good nutrition, housing, sanitation, and education. However, if he could throw out the inheritance of acquired traits and keep the particles, Galton could have his cake and eat it too.

— SHORTLY AFTER PUBLISHING *Hereditary Genius,* Galton wrote Darwin to propose that they conduct a joint experiment to test the theory of pangenesis. Galton did not mention his doubts about the Lamarckian implications of pangenesis, writing instead, "I want to make some peculiar experiments that have occurred to me in breeding animals and want to procure a few couples of marked and assured breeds."[43] As an afterthought, seemingly intended to pique Darwin's interest, he added, "Pray excuse my troubling you; the interest of the proposed experiment—for it really is a curious one—must be my justification."

Galton's experiment was designed to test for the presence of circulating particles that carried hereditary information. *Circulation* of the particles was the key to the inheritance of acquired traits, for circulating particles were required to transmit changes in one part of the body (neurons, for example) to distant reproductive cells. The ubiquitous gemmules ought certainly be found in the blood of mammals, Galton reasoned, and thus it ought to be possible to transfer the gemmules from one animal to another by performing a blood transfusion. In particular, Galton proposed to transfuse blood from mongrel rabbits into pure breeds. If the theory were correct, one would expect pure-bred rabbits containing mongrel gemmules to give rise to mongrel progeny. A positive finding would go a long way toward establishing the existence of circulating gemmules and their role in inheritance. A negative finding, on the other hand, would constitute a near fatal blow to pangenesis. For if blood did not contain gemmules, then it seemed reasonable to conclude that cells of the body were not actively shedding them, and then there was no longer a mechanism by which to transmit acquired traits.

Darwin's assistance to Galton would be invaluable: Not only was Darwin a compendium of information about the various breeds of rabbits (and virtually any other animal or plant), but he had a network of relationships with animal breeders throughout England. However, it wasn't Darwin's connections or his vast knowledge of the natural world that would be most helpful to Galton. Galton was lifted up by his connection to the great man himself. The fact that Darwin had been intoxicated by his initial reading of *Hereditary Genius* meant the world to him. Forsaking his usual reserve, Darwin had written on December 23, 1869: "I have only read 50 pages of your Book, but I must exhale myself, else something will go wrong in my inside. I do not think I ever in all my life read anything more interesting and original."[44] In a swoon, Galton had written back the following day: "There is no one in the world whose approbation in these matters can have the same weight as yours."[45] Twenty years later, on the occasion of being awarded the gold medal by the Royal Society, Galton still put Darwin before anyone: "I valued his encouragement and approbation more, perhaps, than that of the whole world besides."[46]

For his part, Darwin leaped at the opportunity to test his hypothesis, unaware of the perils this posed to his beloved pangenesis. With Darwin's help, Galton had no difficulty procuring several different rabbit breeds and enlisting an expert from the Zoological Society to help him with the transfusions into the jugular veins of his rabbits. In March 1870, Galton reported on the first experiment, which had ended in premature death of the litter but contained a "hopeful" case. One of the offspring of a pair of true-breeding silver greys transfused with the blood of a common yellow showed a head much lighter than its siblings. "The head was certainly . . . irregularly coloured, being especially darker about the muzzle, but I did not and do not care to build anything about such vague facts," Galton wrote Darwin, expressing the cautious optimism of an impartial experimentalist. "Quite sick with expected hope and doubt," he added, concealing, perhaps even from himself, the true nature of his hope and doubt.[47]

Two days later Galton reported another sign that could be construed as evidence for mongrelization, but immediately pointed out how it might not in fact mean anything at all. Consciously or not, Galton was toying with Darwin. It appears, though, that Darwin was not so easily put off the track. Several days after Galton's letter, Mrs. Darwin wrote to her daughter Henrietta with news of her father's collaboration: "F. Galton's experiments about rabbits (viz. injecting black rabbit's blood into grey and vice versa) are failing, which is a dreadful disappointment to them both." At the end of the month Galton sent another positive message, proclaiming, "Better news—decidedly better," and again, eight days later, he reversed himself. In May, Galton wrote yet again of "Good rabbit news!"—a white forefoot on one of the progeny of a transfused doe.[48] By June this latest evidence had begun to show signs of unraveling.

Still playing the part of loyal disciple, Galton next suggested the possibility that the gemmules might reside in the insoluble fibrin that was eliminated from the transfused blood. To test the possibility that the gemmules had been eliminated from the defibrinized blood, Galton turned to cross-circulation, a method in which the donor and recipient, the mongrel and his pure-bred cousin, were mechanically connected by means of a shunt between their respective carotid arteries. This technique ensured that whole blood, in its natural state, could be used for the transfusions. As Galton had expected, the transfusion of whole blood also appeared to lack the power to communicate hereditary gemmules.

Galton's record of his elegant experiment appeared in *Proceedings of the Royal Society* on March 30, 1871. In the article, he testified one final time to his "great hopes that Pangenesis would turn out to be true," but the evidence against pangenesis was overwhelming. The *Royal Society* article presented the definitive results of an experiment in which three silver-grey bucks and four silver-grey does were cross-circulated with the blood of an equal number of commoner bucks and does. This transfused stock gave rise to 88 rabbits in thirteen litters, and in none of the offspring was there even the slightest hint of mongrelization. "The conclusion from this large series of experiments is not to be avoided," he wrote with authority, "that the doctrine of pangenesis, pure and simple, as I have interpreted it, is incorrect."[49]

Despite the fact that the experiments had been pointing in this direction all along, the Royal Society paper took Darwin completely by surprise. In an uncharacteristic fit of pique, he fired off an angry letter to *Nature*. "I have not said one word about the blood, or about any fluid proper to any circulating system," he objected indignantly. "It is, indeed, obvious that the presence of gemmules in the blood can form no necessary part of my hypothesis; for I refer in illustration of it to the lowest animals, such as the Protozoa, which do not possess blood or any vessels; and I refer to plants in which the fluid, when present in the vessels, cannot be considered as true blood."[50] Gemmules might well move from tissue to tissue in some medium other than blood, argued Darwin, and, "though [Galton] deserved the highest credit for ingenuity and perseverance," Darwin was not prepared to concede defeat. "It does not appear to me that Pangenesis has, as yet received its death blow," he wrote in his concluding sentence, "though, from presenting so many vulnerable points its life is always in jeopardy; and this is my excuse for having said a few words in its defense."

Darwin's *Nature* letter was somewhat disingenuous in that it provided no hint that he had been privy to the design and execution of the rabbit experiments from the beginning. There was only one cryptic reference to the fact that he and his cousin had been collaborating for more than a year: "When I first heard of Mr. Galton's experiments," he wrote, "I did not sufficiently reflect on the subject and saw not the difficulty of believing in the presence of gemmules in the blood."

Darwin's account was also subtly self-serving in another way. While he had ascribed to insufficient reflection his failure to challenge the idea that gemmules were present in the blood, he had in fact changed his mind after the hypothesis had been disproved. As he apparently recognized, it was bad science to undertake an experiment to test a hypothesis and, finding the hypothesis disproved, then to claim that the hypothesis was no longer relevant. So when he came to rewrite his pangenesis chapter for the second edition of *Variation*, published in 1875, Darwin recanted, admitting that he "should certainly have expected that gemmules would have been present in the blood, but this is no necessary part of the hypothesis, which manifestly applies to plants and the lowest animals."[51]

The morning he received his copy of *Nature* containing Darwin's letter, Galton immediately set to work to repair the rift. "My Dear Darwin," he wrote in great haste so as not to miss the morning mail, "I am grieved beyond measure to learn that I have misrepresented your doctrine, and the only consolation I can feel is that your letter to 'Nature' may place that doctrine in a clearer light and attract more attention to it."[52] But as solicitous as he was, he was not prepared to back down. The following week he wrote Darwin again, this time announcing his intention to reply to his *Nature* letter in the following issue. He was confident, he insisted, that Darwin would find the "half-plaintive" end of his letter amusing.

Although Galton genuinely revered his cousin, who was by now undoubtedly the most famous scientist in England and perhaps the world, he did not accord Darwin's scientific opinions any special status. In fact, Darwin's defense of pangenesis did not appear to influence Galton's thinking on the subject at all. Galton's "apology" in the May 4 issue of *Nature* was actually a spirited defense of his interpretation and not the least bit apologetic. It had been misleading of Darwin to describe his gemmules as "circulating freely" and to refer to "the steady circulation of fluids," Galton pointed out, as these phrases strongly encouraged one to think of blood. "Blood has an undoubted claim to be called a circulating fluid, and when that phrase is used, blood is always meant." Furthermore the process of particles moving through solids, as the gemmules were now supposed to move through the body's tissues, was properly spoken of as "dispersion," not "diffusion," which applied, he said, more properly to the "movement in or with fluids."[53] These were all changes Darwin saw fit to include in the later revised edition of the pangenesis chapter.

But the key paragraph of the letter consisted of an allegory that would have been nearly indecipherable to anyone but Darwin. In the passage, Galton relates the interaction between a wise tribal leader and his followers. The scene is imagined to take place at the dawn of the invention of language when "some wise ape-like animal . . . first thought of imitating the growl of a beast of prey so as to indicate to his fellow-monkeys the nature of expected danger." Halfway through the narrative, Galton switches to the first person, a change in point of view that would not be lost on Darwin. It is "as if I had been assisting at such a scene," Galton wrote. "As if, having heard my trusted leader utter a cry, not particularly articulated, but to my ears more like that of a hyena than any other animal." Seeing as no one had stirred, Galton takes it upon himself to investigate, "down a path of which I had happily caught sight, into the plain below, followed by the approving nods and kindly grunts of my wise and most respected chief." Upon returning with the happy news that it was a false alarm and there was no hyena in the plain, he is told that he has misinterpreted his master's signal, that it was not a hyena in the plain, but rather a leopard in the trees that posed the threat. "Well, my labor has not been in vain," Galton concludes on an upbeat note, "it is something to have established the fact that there are no hyenas in the plain, and I think I see my way to a good position for a look out for leopards among the branches of the trees. In the meantime, Viva Pangenesis," wrote Galton, acknowledging for the first time that it was the death of pangenesis as Darwin had conceived it that was at stake.

Although pangenesis had escaped the fatal blow, it had been badly wounded. It soon received another damaging blow from Dr. Lionel Beale, a well-respected scientist, doctor, and inventor. In another letter to the editor of *Nature,* Beale praised Galton's "numerous," "well-devised," "difficult," "laborious," and "honest" tests of Darwin's hypothesis. "If such a well-thought out and well-executed series of experiments had no effect," Beale wrote, "I do not believe it possible to obtain a series of experimental results which would lead the supporters of Pangenesis to abandon the hypothesis," he wrote. Pangenesis was, in Beale's estimation, based "upon the fictions of the fancy."[54]

Ever mindful of Darwin and manipulating events to his advantage, Galton hastened to disassociate himself from the critical letter and commiserated with Darwin over the rough handling he'd had, assuring him that he did not "in any way share the animus of the letter."[55] The content of Beale's letter, with which Galton was in perfect accord, was not mentioned. But Galton showed a new self-assurance, deliberately dispelling any hope that he might be won back to the cause. "My new experiments are not hopeful—alas!" he wrote in a postscript. "I hope Pangenesis will get well discussed now."

It is a testament to Darwin's forgiving and genial character and Galton's deep attachment to him that their conflict over pangenesis did not cause a permanent rupture in their personal relationship. In fact, both men continued on as if nothing much had happened. Despite the fact that they now both agreed that the gemmules weren't likely in the blood, and that perhaps they ought not to have been looking for them there in the first place, the blood transfusion experiments continued. It wasn't Galton's style to force a confrontation. Instead, he gradually and quietly withdrew. He and Louisa traveled that summer, as they did every summer, and the rabbits were entrusted to Darwin's care. In the fall, he wrote to thank Darwin for taking over the care of the rabbits and supervising the experiments and promised to resume his cross-circulation experiments that October. But October seems to have come and gone without word from Galton. By late November he was quite abashed: "I am truly ashamed to have trespassed so long on your kindness, in keeping the rabbits, but until now, owing to a variety of causes . . . I could not ask for them back."[56]

In December 1871 the cross-circulation experiments were resumed, and rabbits were now routinely being shuttled back and forth from the Darwin estate in Down to University College in London where the surgery was performed. Darwin was now a full partner in the breeding program, urging Galton to take rabbit progeny off his hands in order to make room for more breeders. In February Galton made a half-hearted request for another mating pair, and further expressed his appreciation to Darwin for his efforts, which were now far outstripping his own. "I feel perfectly ashamed to apply again to you in my recurring rabbit difficulty," Galton wrote in May, informing Darwin that he and his wife were off on their annual extended European holiday and that he did not trust his servants to carry on the experiments in his absence. "Is it possible that any of your men could take charge of them and let them breed?" he inquired, offering to pay "even a large sum—many times the cost of their maintenance—to any man who would really attend to them."[57] It was an elaborate display of Victorian graciousness, for Galton clearly had no further interest in the rabbits and was only continuing on with the experiments for Darwin's sake, instinctively understanding that Darwin needed more time to part with his beloved pangenesis.

Darwin wrote the very next day to say that he would be happy to keep the rabbits and breed from them and that he'd commissioned his footman to do the work. "I have said that you would give him a present, and make it worth his while: and that of course adds to the expense that you will be put, and I have thought that you would prefer doing this to letting me do so, as I am most perfectly willing to do."[58] If Darwin finally understood that Galton had been less than candid in his dealings with him, Galton seemed more than willing to pay the price. The following day Galton wrote back with all the particulars concerning the rabbits. In what appeared to be a casual postscript, he mentioned his plan to read "a little paper" at the Royal Society on "Blood-relationship."[59] Even the title was rich with resonance, as it was blood that, metaphorically at least, went back generations and bound kin together, and blood that Darwin now insisted had no necessary part in his gemmule theory.

The paper, published in *Proceedings of the Royal Society* on June 13, 1872, enunciated a new anti-Lamarckian theory of heredity that was meant to displace pangenesis. In this paper, Galton proposed that an individual is composed of two distinct parts, a "patent" part and a "latent" part. The patent part consists of the visible traits of an individual, which correspond to "latent elements." In addition, Galton speculated, there are purely "latent" elements that do not manifest themselves in the individual but whose existence is demonstrated by "the well-known circumstance that an individual may transmit to his descendants ancestral qualities that he does not himself possess."[60] Although the manifest expression of an element might be altered in the course of adult life, the "latent equivalent" did not change. The transmission of characters from parent to child involved the transfer of latent elements in a process that he compared to drawing balls from an urn. "The true hereditary link connects," Galton concluded, "not the parent with the offspring, but the primary elements of the two, such as they existed in the newly impregnated ova, when they were respectively developed."[61] This view, which would come to be known as the "continuity of the germ line," would form the basis for the modern theory of heredity.[62]

By the winter of 1872, nearly three years after Galton had first proposed the rabbit experiments, even the infinitely patient Darwin was showing signs of fatigue. Writing to Galton he ruefully reported that the last lot of rabbits bred "perfectly true in character," and that he was now ready to breed one more generation. "If the next one is as true as all the others, it seems to me quite superfluous to go on trying."[63] At age 63, Darwin seemed reluctant to waste any more time on pangenesis: "My career is so nearly closed, that I do not think it worth while [to continue with the investigation]. What little more I can do, shall be chiefly new work." This must have been welcome news to Galton. As far as he was concerned the case against Darwin's fluidly changeable gemmules was closed, and the rabbit experiments had provided confirmation of what he'd believed all along, that heredity was immutable and the primary determinant of human nature.

**Portrait of Galton as a young man.**

From Karl Pearson, *The Life of Francis Galton*, vol. 1 (Cambridge: Cambridge University Press, 1914), after p. 92.

CHAPTER 2

# Reversion to the Mean

IMMEDIATELY AFTER he'd completed "On Blood-Relationship," Galton turned to eugenics and his passion for race improvement. Rather than risk his ideas for a human breeding program being lost in the backwater of a scientific journal, he chose to publish in *Fraser's Magazine,* a widely read literary magazine that published the likes of Thackeray and Carlyle. The article was simply entitled "Hereditary Improvement."[1] "Whoever has spent a winter at the health-resorts of the South of France," he began, "must have been appalled at witnessing the number of their fellow-countrymen who are afflicted with wretched constitutions, while that of the sickly children, narrow-chested men, and fragile, delicate women who remain at home, is utterly disproportionate to the sickly and misshapen contingent of the stock of any of our breeds of domestic animals."[2] The appalling physical condition of his fellow countrymen pointed to an even deeper problem, as Galton saw it: the dearth of intellectual talent among Britain's leaders.

Fortified by his new theory of heredity, Galton was now prepared to set forth a specific agenda to improve the situation. On the one hand, the state would encourage intermarriage and reproduction of the gifted. On the other, it would discourage, or outlaw, if necessary, procreation of the untalented. As a first step, government would create a national register, a "golden book" as he called it, of the naturally gifted. This would be part of a broad effort to collect and publish the pedigrees of gifted families, fully illustrated with measurements and photographs. He was anxious to assure skeptics that there would be "no coercion in the matter of marriage" and, in fact, "no direct steps *at first* beyond simple enquiry." All boys at school would be tested, examined, and classified as to their natural physical and mental gifts, and the most promising registered. Two generations hence, he anticipated, the lists would contain 1 percent of the population, an estimated 300,000 "magnificent specimens of manhood and womanhood." Like the residents of Garrison Keillor's Lake Wobegon, the children of the gifted would be above average, "on the whole, better in every respect than the children of other people—stronger, healthier, brighter, more honest and more pleasant."[3]

The idea was "not to begin by breaking up old feelings of social status," he hastened to reassure his highly privileged readership, but rather "to build up a caste *within* each of the groups into which rank, wealth, and pursuits already divide society." Ultimately, the upper classes would be divided according to ability, and those with natural gifts would be given special privileges. The gifted would soon begin to prefer their own company, for not only would "they succeed better by themselves," but "their own society would be by far the more cultured and pleasant."[4] The inevitable result would be the formation of a caste of the naturally gifted, and it was his hope that the "sentiment of caste" would encourage them to intermarry.

When the gifted were firmly ensconced in their seats of power, nothing would prevent them from treating their inferiors with "all kindness, so long as they maintained celibacy." However, he warned, if undesirables "continued to procreate children, inferior in moral, intellectual and physical qualities, it is easy to believe the time may come when such persons would be considered as enemies to the State." State control of breeding was a price well worth paying, Galton argued. He was confident that the goal of creating a "high race" out of the "present mongrel mass of mankind" could be achieved. "Marvelous effects might be produced in five generations," he predicted.[5]

Among the fast-growing number of his illustrious friends and connections, Galton continued to value Charles Darwin's opinion above all others, and Darwin had been quick to respond to Galton's "Hereditary Improvement." On the idea of the "spontaneous formation of castes of the naturally gifted" and the intermarriage of their members, he remained skeptical. In fact, the extended Darwin family was just such an elite, socializing and breeding together over many generations, and Darwin lived in perpetual fear that his first-cousin marriage to Emma Wedgwood was the cause of the many health problems his children suffered. In addition, Darwin saw great difficulty in choosing individuals for such a register of the elite. He even questioned whether such superior individuals actually existed. "How few are above mediocrity in health, strength, morals and intellect," wrote the sober-minded evolutionist, clearly failing to resonate to the splendor of Galton's magnificent specimens of manhood and womanhood. Though Darwin expressed regret that he saw "so much difficulty" with the scheme, he granted, "The object seems a grand one." Darwin encouraged Galton to put aside the practical eugenics program and instead proceed with his scientific investigation of the nature of inheritance, reminding him that this too was "part of your plan."[6]

In closing, Darwin chastised Galton for falling prey to the most basic misunderstanding of Darwinian theory. In his great enthusiasm for racial improvement, Galton had written that "the life of the individual is treated as of absolutely no importance, while the race is treated as everything,"[7] forgetting the most fundamental of Darwinian principles, that selection operated on the level of the individual. "Would it not be truer to say that nature cares only for the superior individuals and then makes her new and better races," Darwin politely inquired.[8] "But we ought both to shudder using so freely the word 'nature' after what De Candolle has said," he added, making reference to a new book by the Swiss botanist, Alphonse de Candolle. This seemingly innocuous comment would have stung more than all the other criticism, for De Candolle's *Histoire de science and des savants depuis deux siècles* contained a sweeping critique of Galton's hereditarian interpretation of mental ability,[9] and Galton had been stewing over the book for weeks.

Not only had De Candolle challenged Galton's views, but he criticized the quality of Galton's study in *Hereditary Genius.* "The very title and the first sentence of his book shows that he regards heredity as the dominant cause," De Candolle had written in his *Histoire.* "I do not see, however that he has given proof of it nor that he has scrutinized the question in a specialized enough manner." Furthermore, he argued that his own approach, which had resulted in a very different conclusion, was superior. "I employed completer biographical documents, drawn from French, English, and German works," he asserted. "I thus flatter myself to have penetrated farther into the heart of the question . . . I obviously have a larger and solider base than that of M. Galton."[10]

Rather than use biographical dictionaries to show that eminence ran in families, as Galton had in *Hereditary Genius,* De Candolle had set out to estimate the percentage of eminent scientists produced by different countries as measured by their representation among the foreign associates of the leading German, French, and English scientific societies. In particular, he had looked at the lists of foreign membership of the Royal Society of London, the Academy of Sciences of Paris, and the Academy of Berlin in the years 1750, 1789, 1829, and 1865. As De Candolle pointed out, because foreign associates were more stringently selected than natives and their choice was less likely to be influenced by internal politics, on average they would tend to be more genuinely qualified.

It was immediately and glaringly obvious that countries did not contribute to the pool of eminent scientists according to their relative sizes. In fact, De Candolle's native Switzerland had been the leading contributor for nearly two hundred years, contributing roughly twelve times the fraction of members to the Royal Society as would be expected on the basis of its tiny population (8.2 percent of the foreign members were Swiss, while Switzerland accounted for only 0.7 percent of the population of Europe). In contrast, Russia and Poland combined accounted for nearly 20 percent of the population of Europe while contributing only 6 percent of the foreign members to the Royal Society, fewer than a third the number that would have been expected on the basis of population size alone.[11]

On the face of it, De Candolle's data could be interpreted to lend support for the nature argument—the Swiss were born smarter than Russians and Poles. But De Candolle made use of the fact that he had data from four separate years spread out over a two-hundred-year period to show that the relative productivity of a country changed over time. For example, De Candolle's data unambiguously established that Sweden and Holland, which had outshone Germany and England in the eighteenth century, fell far behind in the nineteenth century. These findings cast serious doubt on the idea that inborn qualities of a population dictated the number of eminent scientists a country would produce. The differences between countries were due, not to hereditary factors, De Candolle suggested, but rather to "other causes."

The elucidation of these "other causes" formed the substance of De Candolle's book. All in all, he had a list of eighteen environmental factors, or *causes favorables,* as he termed the influences that promoted the scientific development of a country.[12] Wealth was the primary factor—richer countries with better-endowed libraries, museums, and laboratories leading poorer ones. He also found that religious tolerance played a significant part in a country's success, as did the extent to which a country's native language had eclipsed Latin as the official language of written scientific communication. Farther down on the list was the effect of climate, cooler countries producing a disproportionate number of eminent scientists.

Most irksome of all to Galton was De Candolle's insistence that scientific disposition was not inborn in the individual, but carefully nurtured by family and public institutions. De Candolle had made brilliant use of his data to infer the importance of family influence. For example, he had noticed that the number of eminent scientists per family varied from country to country: in Switzerland, for example, the majority of families of eminent scientists tended to give rise to several eminent sons, while in Italy and France it was hardly ever the case that there was more than one illustrious son per family. De Candolle suggested that the difference was to be accounted for by the fact that in Italy and France it was customary for the children of the well-to-do to be sent to boarding schools where they were removed from family influence, while in Switzerland it was customary for children to live at home until the age of 18 or 20. Therefore, De Candolle concluded, "education, in each family, by example and advice given, must have a stronger influence than heredity on the special career of the young scientists."[13]

To strengthen his argument, De Candolle had also made clever use of his data on the occupations of fathers of eminent men. Initially he had been greatly surprised by the discovery that eminent pastors produced a large number of eminent scientific sons—more so, in fact than eminent doctors, pharmacists, and other fathers in science-related fields. On strictly hereditary grounds, one would have expected scientists to produce more scientist sons than nonscientists. But the explanation in this case, De Candolle suggested, was to be found in the quality of the family life of the pastor, the absence of distractions, and "a more regular watchfulness of the father."[14]

Three weeks after receiving a copy of *Histoire* from the author, Galton wrote to De Candolle that he had "read and re-read" his book, "with care and with great instruction to myself."[15] While he congratulated him on "the happy idea of accepting the nominations of the French Academy and similar bodies as reliable diplomas of scientific eminence, and on thus obtaining a solid basis for your reasoning," Galton also expressed his dismay at what he considered to be De Candolle's unfair criticism of his work. But rather than challenge De Candolle's arguments on their own merits, Galton had felt obliged to appeal to his illustrious family connections: "I feel the injustice you have done me strongly, and one reason that I did not write earlier was that I might first hear the independent verdict of some scientific man who had read both books. This I have now done, having seen Mr. Darwin whose opinion confirms mine in every particular."[16]

Apparently, Galton was unaware of the fact that De Candolle and Darwin were also friends, and just as he had after reading Galton's *Hereditary Genius,* Darwin had written De Candolle a congratulatory note after reading his *Histoire* in which he expressed his unreserved admiration for the work.

In fact, except for the substitution of the phrase "I have hardly ever read" for "I do not think I ever in all my life read," Darwin used exactly the same words to praise De Candolle's book as he had used to praise Galton's.[17]

While Darwin's appreciation for De Candolle's work did not equate with an acceptance of De Candolle's environmentalist interpretation of human mental ability, it was also true that Darwin had been only a partial convert to Galton's point of view. Galton had confirmed his belief in the importance of inborn differences in intellect between men, but Darwin still maintained that "zeal and hard work" constituted an "an eminently important difference" between them, as he explained in his letter about *Hereditary Genius.*

Although Galton undoubtedly sensed the power in De Candolle's method and perhaps even appreciated that it was in several key respects superior to his own, he refused to be influenced by De Candolle's conclusions: "You say and you imply that my views on hereditary genius are wrong and that you are going to correct them; well, I read on, and find to my astonishment that so far from correcting them you re-enunciate them. I am perfectly unable to discover on what particulars, speaking broadly, your conclusions have invalidated mine."[18] Galton then proceeded to list what he believed were their points of agreement. Above all, they agreed on the issue of race. While neither Galton nor De Candolle went quite so far as "polygenists," who claimed that blacks were a different species, like Galton, De Candolle never doubted that blacks were intellectually inferior.

The surprising discovery that De Candolle shared his belief in the superiority of Caucasians greatly pleased Galton. Indeed, it was remarkable that De Candolle, who had made a brilliant case for the importance of environmental factors in promoting or inhibiting the production of white European scientists, was unable to see that the same logic might be applied to nonwhites. In his private notebook, Galton wrote, "Quote the original—a complete concession," referring to the passage spelling out De Candolle's view on the inferiority of nonwhites.[19] Writing for public consumption in the widely read *Fortnightly Review* (edited by George Lewes, husband of George Eliot), Galton gloated: "I propose to consider M. De Candolle as having been my ally against his will, notwithstanding all he may have said to the contrary."[20] For all his bluster, Galton was nevertheless sufficiently impressed—or worried—by De Candolle's work that he immediately began his own full-scale investigation that would result in his second scientific book, *English Men of Science: Their Nature and Nurture.*[21] The phrase "nature and nurture" not only had a nice ring, it also ensured that Galton would not be criticized again for one-sidedness, at least in his choice of titles.

In Galton's estimation, not only was De Candolle's argument weakened by his glaring inconsistency regarding race, but the book suffered for its failure to penetrate deeply enough into the personal lives of the eminent scientists he studied. As Galton expressed it in his 1873 review, "the author, however, continually trespassed on hereditary questions, without, as it appears to me, any adequate basis of fact, since he has collected next to nothing about the relatives of the people upon whom all his statistics are founded."[22] To remedy this perceived fault, Galton took as the subjects of *English Men of Science* living members of the Royal Society of London, each of whom was required to fill out a gargantuan questionnaire designed to elicit a vast array of information about the man as well as his near relatives. Of course it helped that many of the subjects were also his close friends and associates, including Darwin, Hooker, Huxley, Herbert Spencer, Clerk Maxwell, and George Stokes.[23] "The inquiry is a complicated one at the best," he insisted in his book. "It is advantageous not to complicate it further by dealing with notabilities whose histories are seldom autobiographical, never complete and not always very accurate; and who lived under the varied and imperfectly appreciated conditions of European life, in several countries, at numerous periods during many different centuries."[24]

But it was precisely these "complications" that De Candolle had so cleverly exploited. The stratification of his data into different subgroups, into *several countries, at numerous periods during many different centuries,* made it possible for him to draw his novel conclusions. Although Galton criticized De Candolle's "very imperfect biographical evidence,"[25] his great accomplishment was to find a way of drawing conclusions from data without relying on subjective reports. Indeed, this was one of the principal virtues of De Candolle's method and one of the most glaring weaknesses of Galton's. As it had in *Hereditary Genius,* Galton's use of selected quotations from his subjects had an arbitrariness to it that seemed out of place in an allegedly scientific study. Furthermore, even if one accepted the validity of the self-reports of his eminent men of science, the snippets of direct quotation gleaned from the questionnaires consistently pointed to the great importance that they placed on the influence of family, teachers, and friends in the development of their careers.

— AT THE SAME TIME that he was conducting the study of his colleagues at the Royal Society, Galton was also tinkering with a new invention. Like his grandfather Erasmus, he had always shown a knack for mechanical invention. This new invention, a device he called a *quincunx* after the staggered rows in which fruit trees were commonly planted, was perhaps his most remarkable as well as his least practical. He had commissioned the London scientific instrument makers, Tisley and Spiller, to build it in 1873, and he unveiled it at a public lecture given at the Royal Institute in the winter of 1874.[26]

The quincunx consisted of a vertical board with rows of staggered pegs attached to it. The board was enclosed in a glass case with a funnel at the top, which served as a feed for metal balls (in the form of shot). As the metal balls weaved their way down through the grid of pegs, they were deflected right or left with equal probability, making their way to a series of vertical compartments along the bottom, which prevented the shot from rolling off.

The beauty of the device was that it gave a real-time illustration of the mathematical principle underlying bell-shaped (or normal) distributions—the "law of frequency of errors." As the mathematician Abraham DeMoivre had first proved in his now-famous 1733 paper, any set of measurements, each of which was determined by a large number of independent influences of roughly similar magnitude that added to and subtracted from a final outcome with equal probability, would tend to be distributed in the shape of a bell curve.[27] This phenomenon was perfectly illustrated by the metal shot as it fell through the rows of the quincunx and moved right or left with equal probability at each peg regardless of what had happened in the row above. As predicted by the law, Galton demonstrated that the shot

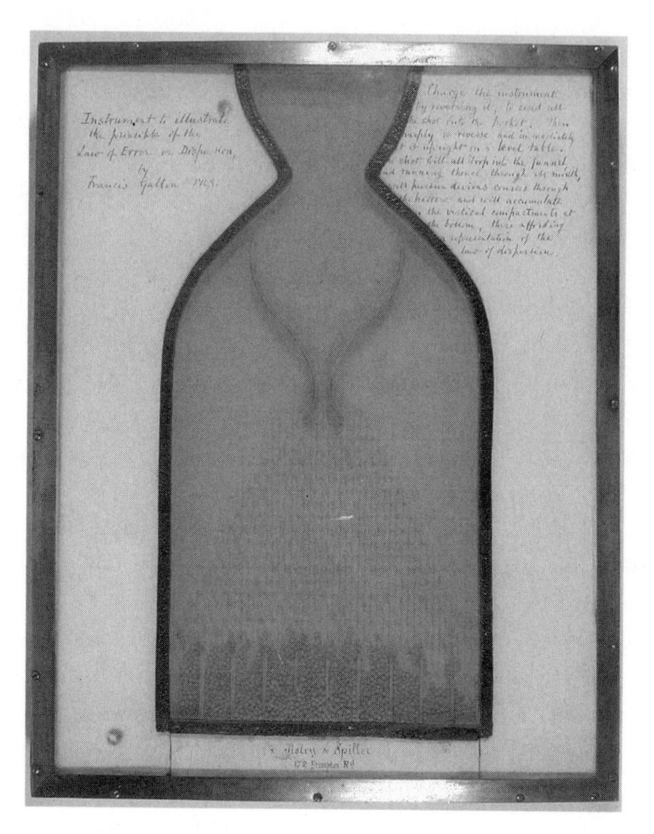


**Galton's first quincunx.** Copyright: The Galton Collection, University College London, image Galt068.

piled up near the center (the leftward and rightward motions tending to cancel each other out), and that fewer pellets made it out toward the edges. The greater the number of rows (independent influences), the more closely the heap of shot resembled the idealized bell curve. It was an inspired idea.

As applied to the problem of heredity, the quincunx illustrated the inheritance of a trait that was determined by a large number of independent factors. Say, for example, there were ten factors that influenced the weight of a seed, and that each of the factors came in either a plus or a minus form, which added to or subtracted from, respectively, average seed weight. The weight would depend on the ratio of plus to minus factors. As-

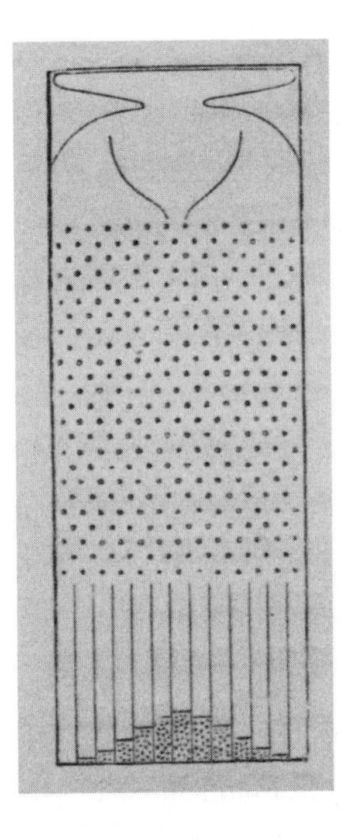

**Galton's schematic representation of the quincunx.** From Francis Galton, *Natural Inheritance* (London: Macmillan and Co., 1889), 63.

suming that each factor had an effect of roughly the same magnitude (and that the factors were inherited independently), the trait would be normally distributed.

With a new confidence in his mastery of the law of frequency of errors, Galton was primed to commence his first quantitative study of inheritance. The idea to use plants came to him while helping Darwin in 1874 with a statistical analysis of the heights of self-fertilizing versus interbred plants.[28] Darwin suggested Galton experiment on sweet peas, and Darwin's friend Joseph Hooker, who was head of Kew Gardens, concurred.[29] Sweet pea plants offered several advantages as a system to study: the plants were largely self-fertilizing, which made analysis of the inheritance results less complicated. Furthermore sweet peas, unlike the common edible pea, were of nearly uniform size in the pod, and they were relatively easy to weigh. Unknown to Darwin, Hooker, and Galton, a brilliant Augustinian monk, studying inheritance in his monastery in Brünn, had also recently completed a study of inheritance in peas, in his case the common garden pea, but Gregor Mendel's remarkable account of this study would remain unread on the shelves of both the Royal Society and the Linnaean Society for another twenty-six years.

Mendel had chosen to analyze the inheritance of single contrasting characters, such as yellow or green peas, or smooth or wrinkled ones. But Galton was interested in a different kind of inheritance, the inheritance of continuously varying traits. According to his cousin Charles's theory, these traits were the raw material on which natural selection acted. The particular trait that Galton settled on was seed size. Specifically, he wanted to see if plants that produced exceptionally large or small peas would give rise, in turn, to plants that also produced above- or below-average-size peas. Darwin was fascinated by botany and natural history in general, but Galton's interest was purely theoretical. Years later he wrote, "It was anthropological evidence that I desired, caring only for the seeds as means of throwing light on heredity in Man."[30]

Galton's first step was to weigh thousands of peas and sort them into seven different weight groups, evenly spread out from lightest to heaviest. After a failed attempt to grow the plants in the summer of 1874, Galton enlisted the support of Darwin and eight other friends who lived throughout England.[31] In the spring of 1875 he sent out nine sets of seeds for cultivation. Each set consisted of seven packets, one for each weight class, made up of ten identical peas of each weight class. Along with the seeds, he included detailed instructions for their planting: Seven beds were to be prepared, one for each size class, and seeds were to be planted in ten evenly spaced holes of 1-inch depth. After they had born fruit, the plants were to be harvested and returned to Galton for analysis.[32]

Seven of the nine crops were successful, which provided an ample supply of progeny seeds from each of the seven parental size classes. In a working notebook Galton made a graph of parental pea size versus average daughter pea size. The seven parental pea sizes were arranged in increasing order from left to right along the base, and the vertical scale gave the daughter pea sizes. Above each of the seven parental pea sizes, he placed a mark at the height corresponding to the size of the average daughter pea

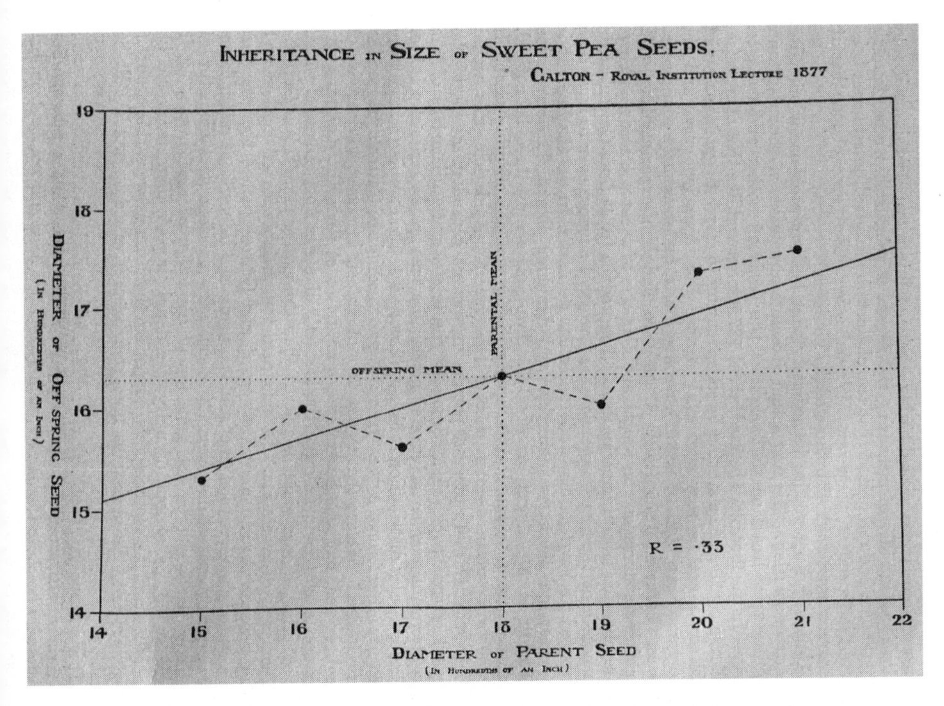


**Galton's original slide with "regression line" added by Karl Pearson in 1930.**

From Karl Pearson, *The Life of Galton,* vol. 3A (Cambridge: Cambridge University Press, 1930), 4.

size. For example, the point in the upper right-hand corner of the graph (see the illustration) corresponded to the average size of the progeny produced by the largest parental peas, and the point in the lower left-hand corner showed the average size of the progeny peas produced by the smallest seeds.[33]

Galton found that the larger classes of peas gave rise to daughters that were, on average, larger than the population mean, but not as large as their parents, and, likewise, the smaller classes gave rise to small peas, but not as small as the parental class from which they had been harvested. This observation would be Galton's central theme for the better part of the next decade. He called the phenomenon "reversion to the mean" and was convinced that he had discovered a fundamental law of hereditary transmission, "a simple and far-reaching law," as he later described it.[34]

The phenomenon of reversion to the mean raised an interesting paradox, which Galton illustrated by considering the case of gigantism. The principle of reversion implied that exceptionally tall people would tend to have somewhat shorter children, and therefore the expectation would be that the relative number of "giants" ought to decrease over time. However, it was clear that the fraction of giants in successive generations remained more or less constant. Two winters after he'd first begun analyzing his pea data, Galton believed he had successfully resolved the apparent paradox. In his grand style, Galton introduced his new theory in a well-publicized lecture delivered in the theater of the Royal Institute in February 1877. By tradition, the men in the audience wore evening dress and the women were done up in their finest gowns.

Preparation of the lecture, subsequently published in three parts in *Nature,* had required painstaking work. As early as the spring of 1876, Galton had sent the manuscript to Darwin's son George for his comments on a preliminary draft. Its unprepossessing title was "Experiments with Plants and the Causes of Statistical Uniformity in Successive Generations." George, the most mathematically inclined of Darwin's sons, had found the manuscript too obscure and advised a total rewrite, a setback Galton took in stride: "These confounded law of error ideas, which in themselves are so simple and clear but to express which no proper language exists and which lies so completely out of the everyday lines of thought are very baffling to deal with and to present, but I don't despair yet."[35]

By January 1877, Galton completed a new, totally revised draft. But the revision still did not earn George's approval. "I am aghast at the trouble my unlucky memoir gives," Galton apologized, promising to undertake yet a third rewrite.[36] The final version of the paper presented a highly contrived model that attempted to reconcile the fact that children tended to revert to the mean with the fact that variability was not actually lost over successive generations.[37] Galton did not even mention the pea data. As he explained a decade later, this data had been so beset by inconsistencies and irregularities that publishing it would have been an invitation to question his model.

Rather than risk creating doubts about the validity of the reversion theory, Galton decided to collect more data, this time primarily from human subjects. Human stature was a nearly ideal subject for a hereditary study. Not only was it normally distributed, as Quetelet had demonstrated, but it was nearly constant over the span of a lifetime and simple to measure. Furthermore, as height played no essential role in survival, natural selection was not likely to be involved in the survival rate of different height classes and thereby skew the inheritance pattern from one generation to the next.

For nearly a decade, Galton was stymied by his inability to get a large enough sample size to do a statistically meaningful study of stature. Not for the first time, Galton's family fortune came to the rescue. Offering generous prizes with a total value of 500 pounds to those subjects who did the best job in completing family histories, he soon had his data set.[38] The prizes were offered in 1883. By 1884 he had collected records of 930 adult children of 205 couples.[39]

With the addition of the human data, Galton introduced his revamped theory of inheritance in three separate publications that appeared in 1885 and 1886. In the eight years since his Royal Institute lecture, Galton had achieved a new clarity about the nature of reversion. His first graph showing the degree of similarity between parent and offspring in peas had not made it into print. This time around the graph of parental versus offspring height took center stage. As in the unpublished graph of the pea data, parental heights were arranged in increasing order along the horizontal axis (see the graph). Above each of the parental height classes (called the "mid-parent value," denoting the average of the two parents' heights), he drew an open circle at the level corresponding to the average height of their adult children (the circles lying along the line labeled "Children"). Just as in the case of the peas, children deviated less from the average than their parents did. Tall parents have tall, but not as tall, children. Short parents produce short, but not as short, adult children.

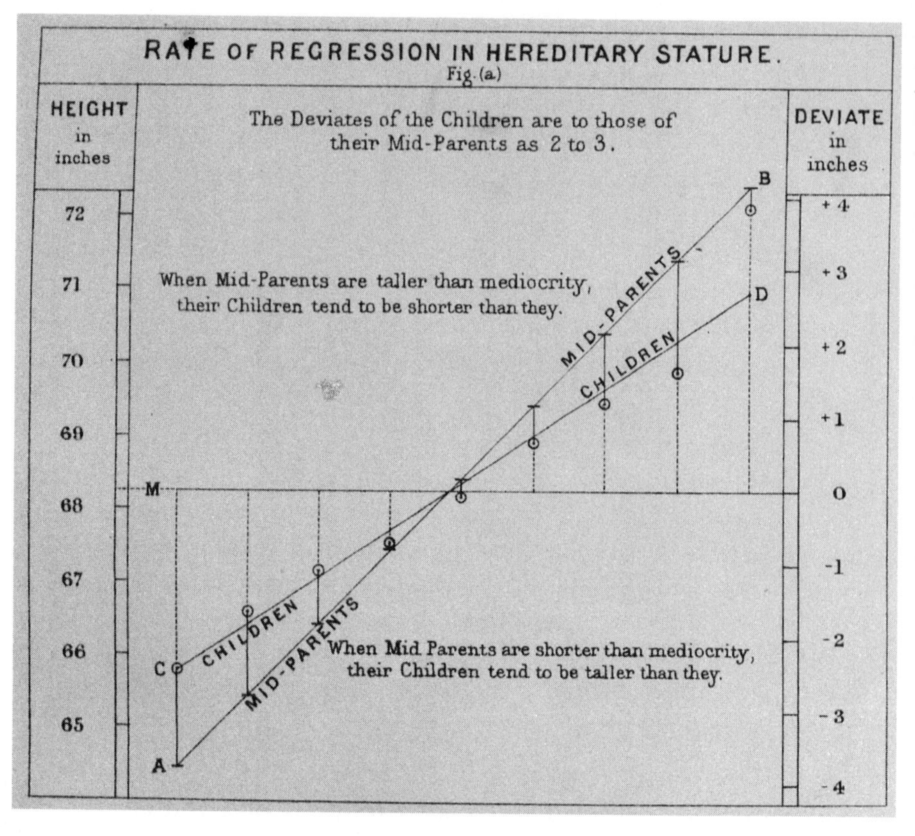


**Rate of regression in hereditary stature.**

From Francis Galton, "Regression towards Mediocrity in Hereditary Stature," *Anthropological Miscellanea: Journal of the Anthropological Institute of Great Britain and Ireland* 15 (1886): 248.

The new conceptual breakthrough had been to draw the straight line that best approximated the data points. This came to be known as the *regression line.* The slope of this line—the line on the graph labeled "Children"—measures how closely children resemble their parents. It is now called the *regression coefficient.* A line inclined at 45 degrees results if the adult children are on average identical to their parents. This is the line (with a slope of 1) labeled "Mid-parents." The more closely progeny resemble the population mean and the less closely they resemble their own parents, the flatter the

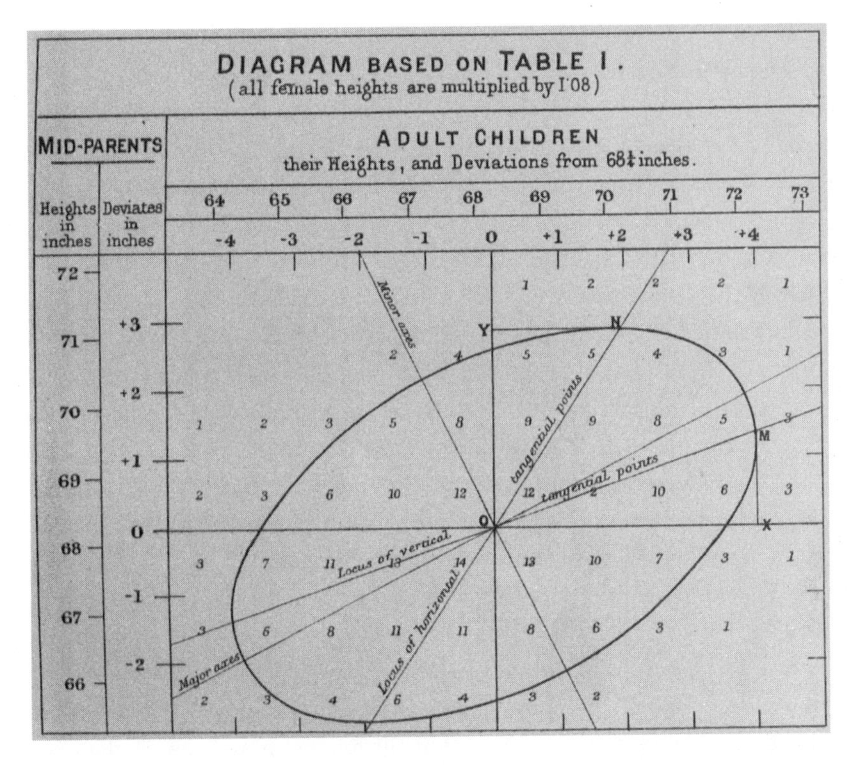


**Ellipse.**

From Francis Galton, "Regression towards Mediocrity in Hereditary Stature," *Anthropological Miscellanea: Journal of the Anthropological Institute of Great Britain and Ireland* 15 (1886): 248.

regression line. An absolutely flat line (with a slope of zero) best approximates the random scatter of points that would result if the heights of parents and their adult children were uncorrelated.

In his 1908 autobiography, Galton recalled the discovery of the phenomenon of regression. Waiting in a railroad station near Ramsgate, he had been poring over a plot of his data on human heights. The data were arranged in the same kind of diagram he'd been staring at now for more than a decade—a grid of squares with mid-parental deviations displayed along the side and adult children deviations along the top. In each square, he'd recorded the number of adult children of the height class, indicated by the scale along the top, born to parents of the deviation, indicated along the side (see the diagram).[40]

As he sat waiting for the train, Galton noticed that numbers of the same value formed a set of ellipses, all sharing the same center, and all of the same basic shape. He described the moment in *Memories:* "The cases were too few for certainty, but my eye, being accustomed to such things, satisfied me that I was approaching the solution. More careful drawing strongly corroborated the first impression."[41] In addition to the hidden pattern of nested ellipses, he picked out two critical lines—the lines he labeled "locus of horizontal tangential points" and "locus of vertical tangential points" on the diagram. The geometry of the picture made it clear to him that the first of these lines was the by-then familiar regression of the adult child on mid-parent height. It seemed to be a nearly miraculous confirmation of more than a decade's work. When he returned home, he made a more careful graph and confirmed his first impression.

The entire afternoon must have had a strange déjà vu quality about it. The last time he'd given serious attention to conic sections (ellipses) was during the fall of his freshman year at Cambridge in 1839. At that time, he had written home to his father and sister that he had never worked as hard. However, conic sections had been difficult for him even then. He had not managed to solve a single one of the practice problems provided by his math tutor.[42]

Forty-six years later, he went to the library of the Royal Institute in London and ferreted out a book on analytic geometry. While struggling with the equations for the conic sections, he was overcome by a familiar sense of despair. As luck would have it, James Dewar, a Scottish chemist, happened to be working in the library that day. "Why do you bother over this?" Dewar asked upon learning of the Galton's difficulties. "My brother-in-law, J. Hamilton Dickson of Peterhouse, loves problems and wants new ones. Send it to him."[43] Dewar's timely intervention seemed to release Galton from his self-inflicted torture. He took up Dewar's suggestion and presented his data on stature to Dickson as an abstract problem in probability theory. Dickson promptly returned a solution, including the exact formulas of the nested ellipses and their tangent lines. Although Galton himself had already found the regression lines simply by staring at the data of the 928 offspring and their parents, it was Dickson's mathematical derivation of them that seemed to thrill him. One can almost hear the awestruck 17-year-old university math student in the 63-year-old Galton's paper for the *Journal of the Anthropological Institute:*

> I may be permitted to say that I never felt such a glow of loyalty and respect towards the sovereignty and magnificent sway of mathematical analysis as when his answer reached me, confirming, by purely mathematical reasoning, my various and laborious statistical conclusions with far more minuteness than I had dared to hope, for the original data ran somewhat roughly, and I had to smooth them with tender caution.[44]

Experimental validation of the theory of regression provided by the data, combined with Dickson's independent proof of the theory, gave Galton a new confidence in his theory of inheritance. Revisiting the meaning of the sweet pea experiments, he now explained: "I was then blind to what I now perceive to be the simple explanation of the phenomenon, so I thought it better to say no more upon the subject until I should obtain independent evidence."[45] Over the intervening decade, he had become convinced that children who deviated from an "ideal form" would tend to regress toward it. The more powerful the pull of the ideal form, the stronger the regression. Bolstered by this mystical idea, Galton now believed that the phenomenon of regression to the mean was "a perfectly reasonable law which might have been deductively foreseen."[46] The transmission of a trait was a balance between regression toward the center—the attraction toward an ideal ancestral type—and dispersion about the mean, the consequence of the random assortment of hereditary factors. The two tendencies, he wrote, "necessarily neutralize one another, and fall into a state of stable equilibrium."[47]

If the equilibrium were to be maintained, it followed that more exceptional variants had to be less likely to become established in a population. The data on human heights proved the point. He had shown, for example, that while more than half of the sons of men of average height (5 feet, 8½ inches) were taller than their fathers, fewer than 7 in a 1,000 sons of exceptionally tall men (6 feet, 5 inches) exceeded their fathers in height. For Galton it was a foregone conclusion that the same was true for mental and other traits:

> The more exceptional the amount of the gift, the more exceptional will be the good fortune of a parent who has a son who equals, and still more if he has a son who overpasses him in that respect. The law is even-handed; it levies the same heavy succession-tax on the transmission of badness as well as of goodness. If it discourages the extravagant expectations of gifted parents that their children will inherit all their powers, it no less discountenances extravagant fears that they will inherit all their weaknesses and diseases.[48]

The possibility of measuring the resemblance between parents and offspring in any quantifiable trait was one of the great insights of Galton's life. By 1890 he realized that the regression coefficient was a special case of a more general phenomenon, *correlation,* as he dubbed it: the degree of similarity between any two sets of data. While the concept of correlation stood the test of time and quickly became one of the cornerstones of the new science of statistics, Galton had misinterpreted the biological significance of regression. His confusion led him to far-reaching claims about heredity and evolution that would not hold up.[49]

In particular, Galton used the observed regression in human stature to argue that evolutionary change could not possibly be the slow and continuous process that Darwin had envisioned. The "slight and uncertain selective influences," he argued, were no match for the blind force of regression. Where Darwin had gone wrong, according to Galton, was in insisting that small continuous variants, which he called "variations," provided the raw fodder for natural selection. The key to understanding evolution as well as to engineering successful race improvement, Galton argued, was the appearance of "sports," individuals who possessed new traits that were qualitatively different from the traits possessed by their parents. "I feel the greatest difficulty in accounting for the establishment of a new breed," he explained, "unless there has been one or more abrupt changes of type." Something about the nature of sports, he conjectured, allowed them to escape regression's relentless pull: "By refusing to blend freely with other forms the most peculiar 'sports' admit of being transmitted almost in their entirety with no less frequency than if they were not exceptional."[50] The goal of eugenicists, it followed, was to encourage the breeding and intermarriage of exceptionally able sports.

It was nearly twenty years since the publication of *Hereditary Genius*, where he had first suggested the possibility that evolution might proceed in a discontinuous fashion, comparing the process of evolution to the movements of a stone consisting of a "vast number of natural facets." If a large enough force were applied, the stone might be coaxed to dislodge from one edge and "tumble over into a new position of stability."[51] In his 1889 *Natural Inheritance*, he once again returned to the metaphor of the polygonal slab, but now he had an elaborate machinery of statistics and a theory of inheritance to support it. In the continuing debate over the efficacy of Darwin's natural selection, Galton's ideas about discontinuous evolution would, for better and for worse, have a profound influence on the next generation.

**W. F. R. Weldon.**

From Karl Pearson, "Walter Frank Raphael Weldon, 1860–1906," *Biometrika* 5 (1906): 1.

CHAPTER 3

# Galton's Disciples

— IN MARCH 1890, Galton was asked to review a zoology paper with a statistical slant for the *Proceedings of the Royal Society*. The paper was a radical departure for its 30-year-old author, W. F. R. Weldon, who was smitten by Galton's new book, *Natural Inheritance*. Up until that point, Weldon, like many zoologists of his generation, had believed it was his lifework to carry the torch of Darwinism forward by providing evidence for natural selection in the comparative anatomy and embryology of animals. But in Galton's statistical methods he saw the possibility for a new approach to the study of evolution, and as a first step he had set out to quantify the degree of variation in a population of common shrimp collected during a year's leave from Cambridge at the new Marine Biological Laboratory at Plymouth.[1] With the help of his wife, Weldon painstakingly measured four different organs in shrimp, and, using the methods pioneered by Quetelet and Galton in studies of human characteristics, fit the results to a normal curve.

In its application of the new statistical methods to the realm of zoology, Weldon's paper was a milestone in the quantitative approach to biology, the new field that would soon be dubbed "biometry." To show his gratitude, Weldon had taken the unusual step of including a testimonial to Galton in the text of the paper, writing, "My ignorance of statistical methods was so great that, without Mr. Galton's constant help, given by letter at the expenditure of a very great amount of time, this paper would never have been written." Not only had Galton tutored Weldon in the lore of the normal curve, but he had a large hand in shaping Weldon's conclusions to conform to his predictions as they had been set out in *Natural Inheritance*: in particular, the deviations from the average were shown to "obey the law of error in all cases, whether the animals observed were under the action of natural selection or not," thus supporting the idea that natural selection was no match for the power of regression.[2]

In December of that year Weldon left his lectureship at Cambridge to become a professor at University College London, where he struck up a friendship with Karl Pearson, who had been recruited six years earlier as a professor of applied mathematics. Brooding and solitary by nature, Pearson made a rare connection with Weldon, with whom the fit was nearly ideal. To Pearson, Weldon brought a vast knowledge of zoology, while Pearson provided Weldon with the benefit of a keen mathematical mind against which to sharpen his biological intuitions. Over lunches, as Pearson later recalled in his heartrending obituary of his friend, the two men talked shop, sketching out problems on napkins and using bread crumbs to work out questions of probability.

Pearson, a gifted polymath who was as versed in German literature and history as in mathematical physics, had already begun his immersion in the still largely unexplored subject of statistics before Weldon's arrival at University College. In a series of lectures given in 1891–92 he reviewed what he termed the "geometry of statistics," covering the various ways to represent data visually, and the laws that governed coin tosses and the roulette wheel.[3] In the course of his lunches with Weldon, Pearson began to see that statistics, and in particular the development of new techniques of graphing, might be the ideal tool with which to study evolution.[4]

In the spring of 1892, Weldon and his wife visited Naples, and Weldon took up the measurement of shore crabs in place of common shrimp. As had been the case with shrimp, the various organs of the crabs were for the most part normally distributed, but there was one notable exception. The measurements of the frontal breadth, the region of the shell that contained the crab's breathing apparatus, could not be neatly fit to a normal curve. That November Weldon wrote Pearson that he had succeeded in breaking up the frontal breadth measurements into a two-humped curve, and in June 1893 he wrote again, convinced that the departure of the crab data from simple normality constituted evidence that selection was at work.

Pearson was greatly excited by Weldon's result, in which he saw the perfect application for his newfound interest in curve fitting and the graphical display of data. In six months' time, Pearson worked out an entirely new theory of curve fitting, showing how skewed and other shapes might be produced by summing normal curves of different types. His "Contributions to the Mathematical Theory of Evolution," which appeared in 1894, marked the beginning of a series of papers in which he introduced a number of the seminal concepts of modern statistics.[5] Although he was often very little concerned with the actual biological objects behind the data, Pearson did not fail to note that the resolution of an asymmetric curve into *two* normal curves might be of particular interest because "a family probably breaks up into two species." Weldon's hunch that the crabs were a mixture of two normally distributed sub-races was thrillingly confirmed by Pearson's curve fitting.

For several years Weldon was riding the crest of excitement over his new statistical work. He had found a brilliant and energetic collaborator in Pearson and a mentor and friend in Galton, who himself was finally basking in the fame and acceptance that had so painfully eluded him in his first fifty years. As fate would have it, it was a person most dear to Weldon's heart, William Bateson, whose increasingly heated opposition would shatter his equanimity.

Bateson and Weldon had been friends since undergraduate days at St. John's College, Cambridge, where a fellow student described Bateson as a large and untidy classmate whose unconventional and iconoclastic attitude

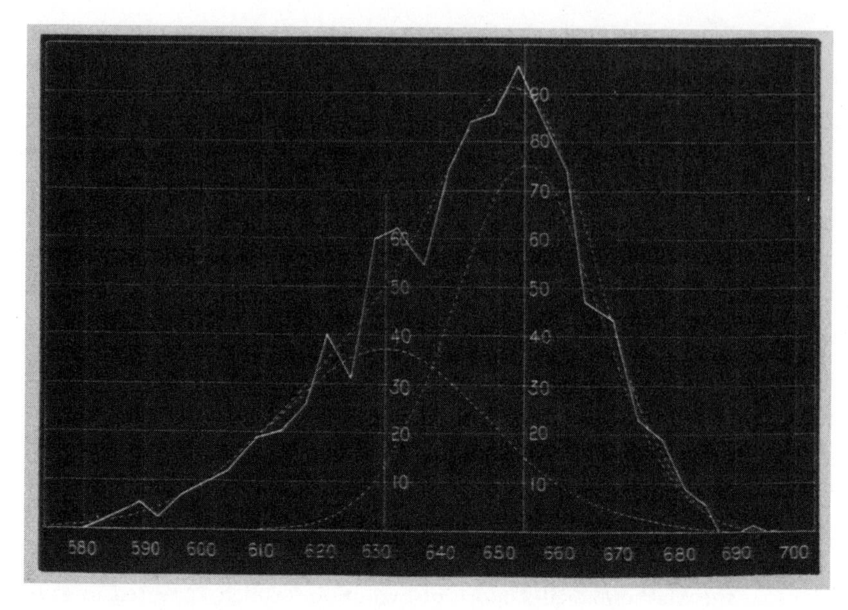


**Pearson's graph showing how two normal curves sum to make a skew curve.**

From Walter Frank Raphael Weldon, "On Certain Correlated Variations in Carcinus Maenas," *Proceedings of the Royal Society of London* 54 (1893): 324.

was apparent the moment you set eyes on him.[6] Weldon would come to rue the fact that he had ever worked to win Bateson over from the pursuit of Art and Truth, which included an intense devotion to the study of Greek and Latin, to the study of zoology. Both were dominant personalities, who were capable of sustained hard work and inspiring it in others. Each was also capable of summoning mountains of indignation at a perceived injustice, and each was fierce in defense of his own beliefs.

— LIKE WELDON'S, BATESON'S TRAINING as a zoologist had been in morphology with an evolutionary slant. As a graduate student in the early 1880s, he traveled to America to work with the great philosopher-biologist, William Brooks, to study a strange wormlike hemichordate called *Balanoglossus* in search of the origin of the chordates, the first vertebrates. The research was a failure, and although he retained a lifelong respect and affection for Brooks, by the time he left America, Bateson was deeply disillusioned with traditional zoology and desperately in search of a new direction.

Rather than study the evolution of the vertebrate in the manner that had become standard in the post-Darwin age, Bateson turned to study the origin of variation itself, and in particular to the kind of variation in the number of matching parts that was frequently seen in complex and repetitive organs like the vertebrate spinal cord. The fundamental challenge of biology, he began to believe, was to characterize the nature and mode of variation. By the time of the publication of his *Variation of Animals and Plants under Domestication,* Darwin himself had fully embraced the Lamarckian view that the environment was the cause of all variation in organisms,[7] and Bateson's first step was to test this idea by studying the marine life in the gradually drying up lake basins of western central Asia, which were known to vary widely in their degree of salinity. As he explained to a czarist authority, "Upon the drying up of this sea it is to be expected that the marine animals inhabiting it would be isolated in the various basins which would then be separated . . . a study of these variations will prove of extreme importance in determining upon what animal variation in general depends."[8] The results of his two-year excursion to the steppe proved to be highly ambiguous and unsatisfactory, representing Bateson's second major professional setback.

Upon his return to England in 1889 the ever-resilient Bateson took a new approach to the study of variation. Failing to find the origin of variation in varying environments, he now decided to approach the problem of variation from another angle. The idea that variation might not be nearly as continuous as Darwin thought had been impressed upon him by Brooks, who was a firm believer in discontinuous evolution. In his 1883 textbook, *The Law of Heredity,* Brooks had asserted that "very considerable variations may suddenly appear in cultivated plants and domesticated animals, and these sudden modifications may be strongly inherited, and may thus give rise to new races by sudden jumps."[9]

Bateson's own research had convinced him that the variations of repeated parts like vertebra, fingers, and teeth were subject to discontinuous variation. In an application for a professorship at Oxford, he had proposed studying domestic plants and animals, as had Darwin, and using them to gain insight into the "modes of occurrence of [discontinuous] variations and, if possible, of the laws which limit them."[10] In 1891 he published a paper on floral symmetry, in which for the first time he asserted his growing conviction that Darwinian evolution by selection of small variations—"Variations in degree which at any given moment are capable of being arranged in a curve of Error,"[11] as he expressed it—could *not* account for evolution, concluding, "It is difficult to suppose *both* that the process of Variation has been a continuous one, and also that Natural Selection has been the chief agent in building up things."[12]

As he had anticipated, Bateson was passed over for the Oxford professorship, but nevertheless he set out in earnest to document the existence of discontinuities in evolution. As his wife described it, "He ransacked museums, libraries, and private collections; he attended every sort of 'show,' mixing freely with gardeners, shepherds and drovers, learning all they had to teach him."[13] The result was his massive study, *Materials for the Study of Variation: Treated with Especial Regard to Discontinuity in the Origin of Species,* published in 1894. In it, Bateson gave more than 886 examples of what he believed to be discontinuous evolution.

Always eager for his friend's opinion, Bateson sent Weldon a copy of the unpublished manuscript in the spring of 1894. For years the two men had been attempting to crush one another in friendly battle over scientific principles.[14] Having read only the introduction, Weldon wrote to warn Bateson that he was ill disposed to the whole idea of discontinuity. "At present I do not quite grasp what you mean by discontinuity, by regression, and by oscillation about a position of organic stability," implying that Bateson did not properly understand the subtlety and care with which one must apply these statistical ideas. For more than a dozen pages Weldon preached to Bateson with the ardor of a new convert about the nature of statistics and the dangers of misinterpreting data on the basis of a too small sample. He described in detail his own computations with dice and provided numerical data in order to show Bateson the "importance of examining large numbers of individuals." In order to determine whether an extreme variation represented a sudden discontinuity or simply an extreme deviation from the norm in a continuously varying system, one had to examine a large number of cases, and furthermore it was very important to know just how many would suffice. Isolated examples of marked deviations, even in the absence of intermediates leading up to them, did not count for much. Until he had very good reason to believe otherwise, Weldon would continue to believe that an extreme deviation was determined by *chance*—that is, by the same law as that determining the stature of a *normal* population—not by a sudden leap. Daring to challenge the master himself, Weldon declared that he could not conceive why sports ought to prove an exception to the law of regression, and that "Galton was himself a good deal mixed, at least in his exposition, when he wrote *Natural Inheritance*."[15]

Two weeks later, Weldon's position had hardened: "I am still completely unable to understand your case for discontinuity," he wrote. He reiterated his feeling that the failure to find continuous variation of a structure did not itself constitute compelling evidence for discontinuity. In particular, Bateson's finding that certain types of teeth, such as wisdom teeth, seemed to appear or not in an on/off fashion was not compelling. In the time that had lapsed since his previous letter, he had turned up cases "much more frequent than those which you cite, in which a tooth is represented nearly by the smallest conceivable rudiment." He found numerous examples of well-documented continuous variation in human wisdom teeth, and other classes of teeth in humans, as well as in the teeth of dogs, cats, hedgehogs, marsupials, and even whales.[16]

Three days later, Bateson wrote back, confessing that Weldon's letter had, initially, given him "considerable alarm" and that he feared he had committed an "egregious blunder." However, when he had found time to consult his actual words, he was relieved: "I own that the point is stupidly put, is easily open to misinterpretation, and is overstated, but it is not the abject nonsense that it appears from your letter."[17] He had never intended to deny the reality of continuous evolution in certain classes of teeth of particular animals, Bateson explained. For example, the size of the human molar m3 is an "obvious and typical example of Variation that most probably is continuous or nearly so." What he meant to argue was that "in the case of *particular* teeth evidence of discontinuous variation has been given." The problem has been further compounded, he now realized, by his failure to give a more precise definition of discontinuity, Bateson feared, so that "each hostile reader remains free to attribute to it the one particular meaning which strikes him as especially foolish and that he can most easily show to be untrue."[18]

Two days later, Weldon hastened to assure Bateson that he had not approached Bateson's book with "any unduly hostile spirit," and yet, he added, "it does not even yet seem to me that you make a clear case for discontinuity in any single instance." Weldon proceeded to eviscerate Bateson's argument with elegant prose, much of which he included nearly verbatim in a *Nature* review that appeared in May. While acknowledging the vast effort that Bateson had put into collecting evidence for his book, Weldon argued that Bateson had failed to make a compelling case for discontinuous evolution. The concluding paragraph of Weldon's review, which was a masterpiece of understatement, cast serious doubts about the significance of Bateson's seven-year[19] effort to document the existence of discontinuous variation:

> The only way in which the question can be settled for a given variation seems to be by taking large numbers of animals, in which the variation is known to occur, at random, and making a careful examination and record of each. Mr. Bateson's chapter on teeth, like all his chapters, is of great interest, and will doubtless serve to throw important light on many things. But a careful histological account of 500 dogs would have done more to show the least possible size of a tooth in dogs than all the information so painfully collected. And so in many other cases.[20]

Weldon's negative opinion of Bateson's book was widely shared by the scientific community, which was still in the thrall of the Darwinian orthodoxy of evolution by incremental steps. Professors did not recommend the book to be read by their students, and it was soon remaindered. In the darkness of the dismal reception, there was one ray of light: Francis Galton loved the book. "It was," Galton wrote in the July issue of the widely read popular journal *Mind,* "with the utmost pleasure that I read Mr. Bateson's work bearing the happy phrase in its title of 'discontinuous variation,' rich with many original remarks and not a few trenchant expressions."[21] Galton's only reservation was that Bateson had not extended his inquiry into discontinuous evolution far enough, taking on the case of humans, whom he believed provided the richest store of examples of extraordinary leaps, upward and downward, from mediocrity.

Remarkably, Galton's strong support of the idea of evolution by sports, which was directly at odds with Weldon's views, did not come between them. Beginning in January 1894, Galton and Weldon served as president and secretary of the newly formed committee of the Royal Society, the Committee for the Measurement of Plants and Animals, devoted to the statistical study of evolution. In the summer of 1894 Weldon continued his study of the dimensions of shore crabs, now under the auspices of the Measurements Committee, focusing this time on two dimensions: the frontal breadth, which his earlier study of Naples crabs had suggested was subject to the forces of natural selection, and another dimension of the crab shell that he had reason to believe was not subject to the forces of selection.

Although he had yet to prove it, Weldon was convinced that it was natural selection acting on the slight variations in the size of frontal breadth of Naples crabs that had resulted in the asymmetric distribution of the measurements. Acting in response to a new environmental condition, natural selection had selectively killed off those crabs that were less fit. Likewise it was natural selection acting on small variations that was responsible for preserving the normal distribution of Plymouth crabs. Unlike their Naples counterparts, Weldon conjectured, the population of Plymouth crabs was at equilibrium, and the mean value of the frontal breadth was optimal.

By the summer of 1894 Weldon had conceived of an experiment designed to test his theory: If it was true that natural selection was in fact weeding out the less fit members of the normally distributed population of Plymouth crabs, one would expect the adult members of the population to show less variation than the juveniles, due to the selective destruction of aberrant individuals. Furthermore, the farther the deviation from the norm, the more pronounced the destruction. On the other hand, one would not expect to see any decrease in variation in the dimension of a character that played a less crucial role in the crab's survival. As Weldon reported in February 1895, his measurements on Plymouth crabs provided striking confirmation of his hypothesis. In particular, he found a significant decrease in frontal breadth variation and no decrease at all in the other character, which was presumably less important to the crab's survival. While the result did not rule out the possible effect of occasional sports, it seemed to show that small variations played a central role in evolution.

In the March 14, 1895, issue of *Nature,* William Turner Thiselton-Dyer, who was co-director of Kew Gardens and son-in-law of Joseph Hooker, the famous Darwinist, gave a ringing endorsement of Weldon's result, which he said "deserves to rank amongst the most remarkable achievements in connection with the theory of evolution."[22] With the authority of his position as a director of Kew Gardens, Dyer rejected out of hand the possibility that sports, which he asserted were rare in the vegetable kingdom, played any significant role in evolution. In support of his view, he referred to a recent meeting of the Royal Society where he exhibited the enormous differences between the flowers in the wild and cultivated forms of the tropical plant Cineraria. These differences, he asserted without evidence, were brought about by the systematic human selection of small variations.

The following month Bateson wrote a long letter to the editor of *Nature* contesting Dyer's account of the cultivated Cineraria and in its place, giving a detailed history of the origin of the modern plant. Contrary to Dyer's assertion, Bateson's exhaustive research using country records, horticultural treatises, and gardening magazines and manuals, some of which dated back to the end of the eighteenth century, made it clear that cultivated Cinerarias arose as hybrids of several very distinct species. Furthermore, sports of an extreme kind appeared after hybridization in the early years of "improvement" of the plants. More generally, Bateson contended, modern flowers arose first very rapidly due to sporting, and the new forms were subsequently perfected by the long, slow process of selection of small variations.

The indignant Dyer rushed his reply into the next issue of *Nature.* With the assistance of members of his large staff, Dyer had examined each of the species that Bateson claimed played a part in the parentage of cultivated Cineraria and found in each case reason to doubt Bateson's account. With a high-handedness that Bateson found infuriating, Dyer proceeded to predict that Bateson would regret that "he committed himself to print on a subject about which he evidently possesses little objective knowledge."[23]

The following week, Bateson wrote again, pointing out that in his second letter Dyer had failed to repeat the claim made in the first one that cultivated Cineraria was produced by the accumulation of small variations. With thinly veiled contempt for "authorities" who made *ex cathedra* pronouncements, Bateson challenged Dyer and his assistants to produce the missing evidence: "Of course these authorities may be right, and the rest who have written on the matter may be wrong, but I ask for proof of this, and the request can hardly be thought unreasonable."[24]

Adding greatly to his already much-strained relationship with Bateson, Weldon entered the fray over Cineraria with his own letter to the editor, which appeared in the May 16 *Nature.* Unlike Dyer, Weldon challenged Bateson's history of the origin of Cineraria by use of primary sources. By a careful reading of the articles, Weldon found that at least two of the alleged hybrids that had given rise to the modern form were actually descended from *Cineraria cruenta,* the wild form. In addition Weldon questioned the existence of Bateson's ancestral sports, showing that at least one such alleged sport might just as easily have been the final stage in a series of intermediates. With an uncharacteristic lack of tact, Weldon concluded that Bateson's "emphatic statements are simply evidence of want of care in consulting and quoting the authorities referred to."[25]

Disturbed by Weldon's letter, with its implication that he had intentionally misrepresented the evidence, Bateson arranged to talk to Weldon in person. During their meeting at University College London, Weldon expressed regret that his letter could be taken to imply that Bateson had intentionally misrepresented the evidence, which he had not meant to suggest, but he refused Bateson's request to withdraw it. Bateson was astonished to learn from Weldon that Dyer had in the past expressed the belief the Cineraria was a hybrid, and that Weldon believed him to be "bluffing." In his journal after the meeting, Bateson wrote, "Weldon's position is that of the accomplice who creates a diversion to help a charlatan." In a letter to Weldon the next day, he was only slightly more circumspect, writing, "It is through things like this that I have come to doubt whether the world of science is a school of truth or of deceiving."[26]

The following year, in June 1896, Bateson married, and Weldon, who had once been his closest friend, was not invited to the reception. That October, Bateson discovered what appeared to be a fatal flaw in Weldon's highly regarded experiments on selective death in Plymouth crabs. As he explained it in a long letter addressed to Francis Galton in his capacity as head of the Measurements Committee, which had sponsored Weldon's work, Weldon seemed to have overlooked an obvious confounding error. In order to study the variability of shell dimensions as a function of age, Weldon had grouped the crabs into seven groups according to size of shell. However, it appeared that Weldon had failed to take into account the fact that crabs molted in the course of development, sometimes shedding an old shell and forming a new one as many as a dozen times before reaching full adult size, and that individuals of the same size were not necessarily in the same molt. Therefore, differences that Weldon attributed to differences in age might in fact be due to differences in molting state. "It is so manifest a difficulty that I cannot but think that it has already been considered though no allusion is made to it in the Report," Bateson wrote, with what appeared to be genuine perplexity.[27]

Galton's first instinct was to brush aside Bateson's objections, writing back that he was "disposed towards a lenient view of special disturbing causes in large collections of statistics, and on the other hand indisposed toward too delicate handling of any statistics."[28] Nonetheless, he agreed to circulate Bateson's letter among the other members of the committee. Bateson wrote back the next day to say that it was evident from Galton's response that he had failed to make his point clear and asked if he might revise his letter before it was circulated to the committee. "The difficulty," Bateson wrote, "is not overcome by reason of the extent of the collection of statistics."[29]

On October 20, Bateson sent his revised critique to Galton, who sent it on to Weldon, and two days later Galton received Weldon's reply. "I quite see the importance of Mr. Bateson's letter," Weldon began.[30] It must have come as great shock to Galton to read Weldon's frank confession that the differences in variability among different size groups could be explained by heterogeneity with regard to molt, rather than by the action of natural selection. "Perhaps it would have been well if I had pointed out the need for paying attention to the question of molts in the first report," Weldon sheepishly admitted. However, the situation was salvageable, Weldon assured Galton, and he proceeded to give Galton a detailed description of some experiments that could be done to resolve the issue one way or another.

Hoping that the two men might be able to resolve their differences, Galton sent Weldon's response to Bateson, who at once took it upon himself to write another detailed critique, his third, and to mail it back to Galton on October 28. On the evening of October 29, Galton delivered Bateson's letter to Weldon in person. At their meeting, Weldon proposed to Galton that he write up an entirely new account of the work, taking into account all of Bateson's criticisms, by early the following year.

Anxious to rid himself of the increasingly time-consuming and thorny business once and for all, Galton wrote Bateson the next morning to inform him of the new plan and to direct him to send all further inquiries to Weldon himself. With an air of finality, he added, "Discussion by letter is so onerous. It is better that he should say his say thoroughly, once and in a printed form."[31]

On November 2, Bateson wrote Galton, promising to abide by his judgment and to look to Weldon's second publication for answer to his inquiries. He took the opportunity, though, to remind Galton that the reason he had not gone public with his criticism was partly that he simply had not believed that the Royal Society could have permitted the publication of a work that contained such an obvious potential problem.[32]

Any hope that Galton might have had of extricating himself from the conflict was dashed on November 7 with the receipt of an eight-page letter from Weldon written in response to Bateson's most recent critique. In what was now becoming a pattern, Galton sent Weldon's letter to Bateson, and Bateson sent back his response to Galton. Bateson spent just under a week preparing his final and most extensive critique. In a cover letter to Galton, Bateson now offered to issue printed copies of all four of his letters for distribution to the committee, which taken together made clear that Weldon's experiments were, at best, seriously flawed. With growing confidence, Bateson added, "As the committee is responsible for the Report they should have an opportunity of considering objections, of which the seriousness has not I think been diminished."[33]

At this point Galton must have felt deeply divided. On the one hand, he had almost single-handedly introduced the statistical methods that were used in Weldon's analysis, and Weldon in his collaboration with Pearson was in the throes of organizing an entirely new discipline around these ideas. On the other hand, with all the passion of his eugenical soul Galton believed that discontinuous evolution held out the only hope for the salvation of mankind. Without the appearance of sports, he had written in his 1894 review of Bateson's book, it was clear "that humanity must linger for an extremely long time at or about its present unsatisfactory level."[34] Furthermore, Galton could not have failed to notice that Bateson was proving himself to be a formidable opponent to Weldon, and that he represented the best hope for making the case for discontinuous evolution. The solution, when it finally occurred to him, must have seemed obvious: he would make Bateson a member of the Measurements Committee. The only difficulty was that this would be certain to upset Weldon.

The day after receiving Bateson's fourth and final critique, Galton broached his latest idea with Weldon. "It would in many ways be helpful if Bateson were made a member of our Committee," Galton wrote, "but I know you feel that in other ways it might not be advisable."[35] In a sign that he was serious about shaking things up, Galton wrote Bateson to inform him of his intention to introduce a resolution at the next meeting of the Measurements Committee to the effect that it would no longer take responsibility for the statements in papers that members presented for publication.

At the same time, Darwin's youngest son, Francis, did his part to smooth over relations between Bateson and Weldon, who had told Darwin that he believed Bateson had intentionally avoided him at a Royal Society meeting earlier that year. After Darwin spoke to him, Bateson wrote Weldon, "I am sorry that I should have given off this impression. It was not my intention to do so."[36] Bateson's words seemed to touch a chord in Weldon, who expressed remorse for the loss of their friendship: "The whole thing has been very painful to me: and I am glad to hope that we may now meet on better terms."[37] The ground was now cleared, and on New Year's Day 1897 Galton offered Bateson a place on the Measurements Committee: "Both Weldon and myself are desirous that you should join us. Would it be agreeable to you that we should propose your name?"[38]

Bateson rejected Galton's offer with the next day's post. Determined to finally say the truth as he saw it, he wrote, "I am not convinced that the present lines of inquiry of the committee are fruitful and I do not think it is likely that the results will be at all proportionate to the labour expended."[39] To Galton, who had eagerly sought acceptance his entire life, Bateson's response must have come as a surprise, but it seemed only to have increased his resolve to get him to join, and within the week he succeeded in persuading Bateson. On January 14, Bateson, along with several others, was appointed at the committee meeting. At the same meeting, a resolution was passed banning committee members from publishing their own work under the auspices of the committee, and in private Galton went further, making it clear to Weldon that he wanted to end once and for all the discussion of Weldon's crab work.[40]

Although the loss of the support of Galton and the others on the committee was a crushing blow, Weldon took it stoically. At the first meeting of the newly expanded Measurements Committee, Weldon gave Bateson a rough draft of his new paper, which Bateson returned by post the following day with a few minor corrections. Unable to leave well enough alone, Weldon pressed Bateson to level with him, inviting Bateson to "please say whatever you wish to say subject to the knowledge that I intend to write the whole thing again, to add to it, and to publish it."[41]

While Bateson did not relish the prospect of confronting Weldon with the full force of his opprobrium, he detested nothing more than sloppy or loose thinking. Refusing to take the easy way out, Bateson wrote, "As to this new work as a whole, I scarcely know how to say what I feel yet if I say nothing I may hereafter be reproached for my silence." He then proceeded to remind Weldon of the responsibility he would bear in using his authority to publish conclusions that most people would assume were true without attempting to judge the matter for themselves or in many cases even reading the paper. To him the evidence was "so wholly inadequate and superficial that I do not understand how a responsible person can entertain the question of accepting it," he wrote. "I very truly regret that you would give your countenance to such a production."[42] Weldon took his punishment like a gentleman, thanking Bateson for his "very plain speaking," and adding with great restraint that, of course, he did not agree.

At the first full meeting of the reconstituted committee on February 11, the Measurements Committee was rechristened "The Evolution Committee." The following day, Pearson sent an anguished letter to Galton, complaining that he felt "sadly out of place in such a gathering of biologists."[43] Because he had a gift for "creating hostility without getting others to see [his] views," he wrote, he had remained silent throughout the meeting, but he now felt compelled, even at the risk of "vexing" Galton, to share his true feelings. In a foreshadowing of the troubles ahead, Pearson insisted that nearly all the problems that had been discussed at the previous day's meeting were capable of solution "in one way only, by statistical methods and calculation of a more or less delicate mathematical kind."

In the meantime, Galton had written to Bateson to elicit his suggestion for a long planned experimental farm, which he hoped would take up work along the lines suggested by Darwin in *The Variation of Animals and Plants under Domestication.* In particular he drew Bateson's attention to Darwin's discussion of a cross between two closely related species of snapdragon (*Antirrhinum majus*). When the normal variety was crossed to the so-called Wonder, a variety with unusual five-petaled flowers, the hybrid progeny "perfectly resembled the common snapdragon." The self-fertilized progeny then gave rise to 88 normals, 37 Wonders, and 2 intermediates in the next generation. Although he was tantalizingly close to Mendel's discovery, Darwin had thrown up his arms in confusion, writing that the subject of "prepotency"—the tendency of the characteristic of one parent to overmaster the other—was "extremely intricate" and that it was "therefore not surprising that no one has hitherto succeeded in drawing up general rules on the subject."[44]

At the opening meeting of the newly named Evolution Committee, Galton tried to interest the other members in establishing the experimental farm. When the experimental farm seemed destined to die in committee, Galton took it upon himself to arrange funding for Bateson's experimental work. Bateson's goal, as he had explained it to Galton in his grant application, was to cross closely related varieties that differed in particular traits and to determine the degree to which the parental traits were blended or remained distinct in the offspring.

In addition to arranging for further experiments on heredity, Galton returned to work on the "law of ancestral heredity," a rule that could be used to predict the probability of transmission of a trait in a sibling group based on the frequency of the appearance of the trait in previous generations. In its first introduction in 1885, the law was applied to the continuously varying case of human stature in an attempt to precisely quantify the intuitively obvious idea that a person was more likely to be tall if his ancestors were also tall. Galton's law stated that the two parents each contributed one-half of their deviation from the norm to the deviation of their offspring, the four grandparents a fourth of their deviation, the eight great-grandparents an eighth, and the remoter ancestors in geometrically diminishing increments.[45]

When in 1896 Sir Everett Millais published *The Basset Hound Club Rules and Studbook,* which contained the pedigrees of nearly a thousand hounds, Galton was quick to see that the coat data provided the perfect opportunity to test his ancestral law as it applied to hound coat color, either *tricolor* or *lemon and white,* although the law would now be applied to the transmission of a discrete rather than continuous trait. Instead of relating the contribution of the average deviation from the norm of the various generations to the deviation of the offspring, the law related the fraction of tricolors in each generation to the fraction of tricolors in the sibling group. More exactly, it stated that the average parental coat color (which is either 0 percent, 50 percent, or 100 percent tricolor, depending on whether neither, one, or both parents, respectively, are tricolor) would determine the coat

**Galton's diagram illustrating the relative contribution of each ancestor to total heredity (or a particular trait). For example, squares 2 and 3 represent the contribution of each of the two parents, and squares 4, 5, 6, and 7 give contributions of each of the four grandparents.** From Francis Galton, "A Diagram of Heredity," *Nature* 77 (1898): 293.

color of one-half of the offspring, the average coat color of the grandparents a quarter of the offspring, the great-grandparents an eighth, and more distant ancestors would contribute with geometrically diminishing influence.

It was a relatively straightforward matter to test how closely the numbers gleaned from the studbook corresponded to the predictions based on the ancestral law. More than half of Millais's pedigrees contained the coat color of a hound, its parents, and all four grandparents, and 181 pedigrees gave the coat color for all eight great-grandparents as well. Galton's published results, which appeared in the *Proceedings of the Royal Society* in 1897,[46] showed an agreement between the expectation and the facts that was so striking that even the most hardened skeptic had to admit that there might be something to the theory (compare last two columns in Galton's pedigree chart).[47]

Shortly after the appearance of Galton's paper, Pearson set to work on his own revised ancestral law, and on New Year's Day 1898 he presented Galton with a preprint of his paper entitled "On the Law of Ancestral Heredity." It was a testimony to the ardor of his feeling for Galton that he began the paper with a public declaration of his affection: "A New Year's Greeting to Francis Galton, January 1st, 1898." Although Pearson retained the principle of geometrically diminishing contribution of more distant relatives, his law differed fundamentally from Galton's. While Galton had used a general mean that reflected the average value of a trait across all generations, in keeping with the idea that there was an "ideal form" toward which all individuals regressed, Pearson had measured his deviations from the mean of each corresponding generation. The result was that Pearson's law was compatible with evolutionary change while Galton's was not. In his paper Pearson glossed over the difference, announcing that he would in the future term his law "Galton's Law of Ancestral Heredity."

With his characteristic tendency toward hyperbole, particularly in all things concerning Galton, Pearson ended his paper with the prophecy that "the law of ancestral heredity is likely to prove one of the most brilliant of Mr. Galton's discoveries." Unable to contain himself, he reached even further, predicting that the law, suitably amended, of course, to allow for the action of natural selection of small variations, "must prove almost as epoch-making to the biologist as the law of gravitation to the astronomer."[48]

Despite the fact that Pearson's reworking of "Galton's Law" no longer reflected Galton's belief in the importance of discontinuous evolution and instead was now entirely compatible with traditional Darwinian evolution, Galton could not resist Pearson's intense flattery. On January 4 he wrote, "You have indeed sent me a most cherished New Year greeting . . . it de-

| No. of tricolours | | | Number of tricolours in great-grandparents. | | | | | Total tricolour offspring. | |
|---|---|---|---|---|---|---|---|---|---|
| in parents. | in grand-parents. | | 8 | 7 | 6 | 5 | 4 | Calcu-lated. | Observed |
| 2 | 4 | All cases ...... | 2 | 25 | 14 | 16 | | | |
| | | Coefficient...... | 0·96 | 0·94 | 0·92 | 0·90 | | | |
| | | Tricolours calc'd. | 2 | 24 | 13 | 14 | .. | 53 | |
| | | " observed | 2 | 25 | 14 | 15 | .. | | 56 |
| | 3 | All cases ...... | .. | 18 | 21 | 16 | 6 | | |
| | | Coefficient...... | .. | 0·87 | 0·85 | 0·83 | 0·81 | | |
| | | Tricolours calc'd. | .. | 16 | 18 | 13 | 5 | 52 | |
| | | " observed | .. | 17 | 19 | 14 | 6 | | 56 |
| | 2 | All cases ...... | .. | 3 | 2 | 3 | 3 | | |
| | | Coefficient...... | .. | 0·81 | 0·79 | 0·77 | 0·75 | | |
| | | Tricolours calc'd. | .. | 2 | 3 | 2 | 2 | 9 | |
| | | " observed | .. | 2 | 2 | 3 | 2 | | 9 |
| 1 | 4 | All cases ...... | .. | 2 | 1 | 9 | | | |
| | | Coefficient...... | .. | 0·69 | 0·67 | 0·65 | | | |
| | | Tricolours calc'd. | .. | 1 | 1 | 6 | .. | 8 | |
| | | " observed | .. | 1 | .. | 5 | .. | | 6 |
| | 3 | All cases ...... | 1 | 28 | 14 | 31 | 9 | | |
| | | Coefficient...... | 0·64 | 0·62 | 0·60 | 0·58 | 0·56 | | |
| | | Tricolours calc'd. | 1 | 17 | 8 | 18 | 5 | 49 | |
| | | " observed. | 1 | 16 | 12 | 8 | 9 | | 46 |
| | 2 | All cases ...... | .. | .. | 4 | 13 | | | |
| | | Coefficient...... | .. | .. | 0·54 | 0·52 | | | |
| | | Tricolours calc'd. | .. | .. | 2 | 7 | .. | 9 | |
| | | " observed | .. | .. | 1 | 7 | .. | | 8 |
| | | | | | | Grand totals .. | | 180 | 181 |

**Pedigree chart for tricolors. Compare "observed" and "calculated" numbers in the last two columns.** From Francis Galton, "The Average Contribution of Each Several Ancestor to the Total Heritage of the Offspring," *Proceedings of the Royal Society of London* 61 (1897): 412.

lights me beyond measure to find that you are harmonizing what seemed disjointed and cutting out and replacing the rotten planks of my propositions."[49]

— IN JULY 1899, BATESON and his collaborator Edith Saunders attended the international conference held by the Royal Horticultural Society in and about London. Miss Saunders's plant crosses formed the centerpiece of Bateson's address. Based on her success, he now saw how hybridization experiments could be used to prove his doctrine of discontinuous variation:

> It must be recorded how many of the offspring resembled each parent and how many showed characters intermediate between those of the parents. If the parents differ in several characters, the offspring must be examined statistically, and marshaled, as it is called, in respect of each of those characters separately. Even very rough statistics may be of value. If it can only be noticed that the offspring came, say, half like one parent, and half like the other, or that the whole showed a mixture of parental characters, a few brief notes of these kind may be a most useful guide to the student of evolution.[50]

Writing home to Beatrice, Bateson reported that Miss Saunders "chattered and talked as I never saw her do before,"[51] but the high point of the meeting, it was clear, had been meeting Hugo de Vries. After their first encounter Bateson wrote, "He is a really nice person, very simple and rough in style." Accustomed to skepticism and outright hostility, Bateson found it "delightful" to find someone who shared his views, and gleefully reported, "He is an enthusiastic discontinuitarian and holds the new mathematical school in contempt." It was not only a compatibility on abstract issues, but De Vries also shared Bateson's interest in practical gardening and breeding, and he had edited a gardening journal as well as served on horticultural juries in order to learn the tricks of horticulturists.

After their initial meeting, the two spent an entire day talking, and Bateson was entirely won over, now writing Beatrice that he was "so sorry that you should not have met him. He would have pleased you." It was certainly not a shared sense of culture and the finer points of life that drew the ultrasophisticate Bateson to De Vries. On the contrary, after showing him around London, Bateson, who was a passionate collector of Old Master drawings, concluded that "Art doesn't reach his soul" and thought him a "true farmer" for insisting on shutting the drawing room window on the hottest day of the year—"thank you yes—it is better so—I zink." He was also quite appalled that De Vries had declined a bath before dinner and did not seem to travel with a sponge. "His linen is foul," he noted, "I daresay he puts on a clean shirt once a week 'whether he requires it or not.'"[52] It was something deeper that drew Bateson toward this somewhat crude and parochial man, a raw vitality and scientific creativity that commanded his respect.

**Hugo de Vries.**

From Hugo De Vries, *Species and Varieties: Their Origin by Mutation,* ed. Daniel Trembly MacDougal (Chicago: Open Court Publishing Co., 1906).

CHAPTER 4

# Pangenes

— AS A YOUNG MAN, the Dutch botanist Hugo de Vries did not seem likely to be the long-sought protégé whom Darwin hoped would rescue his theory of pangenesis from oblivion. De Vries's doctoral work was in the relatively quiet backwater of plant physiology, far removed from the speculative world of pangenes and the debate over the inheritance of acquired traits. However, De Vries, whose father was the minister of justice under the king of the Netherlands and whose uncle was a professor of literature at the University of Leiden, was determined to make his mark in some way.

Born in 1848, De Vries grew up in the botanical Eden of Haarlem, where he acquired his lifelong habit of wandering the countryside looking for plant specimens. By the time he finished high school he had gathered a nearly complete collection of all the seed plants in the Netherlands. As a student at the University of Leiden, he had been greatly impressed with Darwin's *Origin,* which he read in its German translation. After graduating, he took a position as a high school teacher in Amsterdam, and spent summers at the University of Würzburg, where he conducted a series of elegant experiments on plant movement under the supervision of the great German plant physiologist Julius Sachs, a diehard reductionist and one of the principal promoters of the modern mechanistic approach to plant physiology. In Würzburg, home to Virchow and Haeckel and the richest vein of German cell biology, De Vries's sense of the world of biology and his possible role in it began to take shape. Over a period of three summers with Sachs, De Vries wrote several beautiful papers.

It was De Vries's good fortune that Charles Darwin shared his somewhat obscure interest in the nature of plant movement. In 1875 Darwin sent De Vries a complimentary copy of his new book, *On the Movements and Habits of Climbing Plants,* which contained no fewer than five glowing references to De Vries and his excellent studies! Exceedingly polite, De Vries wrote back that his papers "should not be entitled to such honor as you bestow on them."[1] This led to another exchange of letters in which Darwin, with his characteristic blend of scientific generosity and self-interest, gently tried to steer De Vries in the direction that Darwin himself found most interesting. Inspired by Darwin's high praise and the generally positive reception of his work, De Vries accepted a full-time position in a Prussian university, which he hoped would eventually lead him to a full-fledged academic career. After enduring two years in the intensely hierarchical German university system, De Vries leaped at the opportunity to return to Amsterdam as an instructor at the newly formed Athenaeum University.[2] A year later, in the fall of 1878, at the extraordinarily young age of 29, he was made a professor.

During the summer before he assumed his professorship, De Vries paid a final visit to Würzburg, where he was introduced to Darwin's youngest son, Francis. Through Francis, De Vries saw the means to renew his acquaintance with the old man himself and in so doing burnish his already brilliant debut in academic science. Several weeks later, with invitations to visit the cream of British botanists in hand, De Vries left for England. Once there, he quickly managed to get himself invited to dinner with the influential William Turner Thiselton-Dyer. Dyer, in turn, introduced De Vries to his father-in-law and Darwin's close friend, Joseph Dalton Hooker, who had been anxious to meet the Dutch botanist, mistaking him for the son of an old friend, Professor de Vriese, to whom De Vries was not related. Afterward, De Vries wrote his grandmother that he found Hooker and Dyer to be "cold"; but undeterred, he immediately turned his attention to the true object of his desire.[3] When he found that Darwin was away on holiday at his brother-in-law's country estate, De Vries promptly wrote to see if he might visit there. De Vries's persistence paid off, and Darwin, who was curious to hear about De Vries's latest experimental findings, agreed to see the young plant physiologist.

The brief visit, which would be forever seared into De Vries's memory, was destined to have a profound influence on the development of modern genetics. All told, De Vries spent two hours—from two to four in the afternoon—at Darwin's brother-in-law's estate in Surrey. After some initial discussion of their mutual interest in the habits of climbing plants, Darwin gave De Vries a tour of the gardens, including a hothouse with grapes and peaches. In the course of describing the plants, Darwin picked a peach and gave it to De Vries, a gesture that the impressionable young man endowed with great significance. At three o'clock, Darwin turned De Vries over to his nephew and retired to his bedroom for a half hour to recover his strength, as was his habit when people visited. When Darwin returned, De Vries had coffee with the extended Darwin family for half an hour before being driven by carriage back to the train station. On the trip to the railway station, Darwin's brother-in-law told De Vries that Darwin had been "exceptionally well and happy" that afternoon.

That evening, an awestruck De Vries wrote his grandmother: "I have talked with Darwin. I have visited him and was received so kindly and cordially as I never had dared hope for." He noted that Darwin looked as he expected, only taller. To his fiancée, he wrote, "It is such a pleasure to find that somebody is really interested in you and that he cares about what you have discovered."[4] As brief as the contact had been, Darwin's attentive kindness would prove to be one of the high points in De Vries's life.

In the following years De Vries and Darwin exchanged half a dozen letters, in which the two men discussed their mutual interest in plant physiology. In 1880 Darwin sent De Vries a complimentary copy of his second book on plant movement, *The Power of Movement in Plants,* in which he had adopted many of the younger man's ideas. De Vries acknowledged the honor that Darwin had paid him, but by then De Vries was actively experimenting on plant hybridization and inheritance and had mostly left behind his studies on plant movement. Darwin's kind judgment of his work would serve as "a stimulus to me in endeavouring to contribute my part to the advancement of science," De Vries would later write.[5]

The following year De Vries wrote Darwin for the final time, thanking him for the gift of Darwin's latest book, *On the Formation of Vegetable Mould through the Action of Worms.* De Vries also mentioned that he had taken up the subject of heredity and variation and expressed the hope that Darwin would one day complete the long-anticipated sequel to his 1868 *Variation in Animals and Plants under Domestication* about plants in the wild. Paving the way for another collaboration, this time over the facts of heredity and variation, De Vries wrote: "I have always been especially interested in your hypothesis of Pangenesis and have collected a series of facts in favour of it, but I am sure that your promised publication will contain much more evidence on all such points, as I would for many years be able to collect."[6]

Although Darwin must have been deeply gratified to learn that someone was still interested in Pangenesis, De Vries had arrived too late, and Darwin, who had long since abandoned his work on pangenesis and given up his plan to complete a sequel to *Variations,* died six months later in April 1882. Nonetheless, the memory of Darwin's previous encouragement seemed to sustain De Vries. Thirty-five years after Darwin's death, De Vries told an interviewer: "I was led to my study of heredity by my love for Darwin."[7]

— IN 1889, THE 41-YEAR-OLD De Vries published his own startlingly original theory of heredity in a slim monograph entitled *Intracellular Pangenesis.* Although the title paid homage to Darwin, and De Vries claimed to have preserved the main ideas, when all was said and done the revised pangenesis had little in common with Darwin's original theory. Not only had

De Vries profoundly altered the nature of Darwin's gemmules, but he had thrown out the idea that the gemmules were transported from all the cells of the body to the reproductive organs, which had been the heart of Darwin's pangenesis.

De Vries had, however, latched on to the one great virtue of Darwin's theory, the idea that there were discrete hereditary particles coding for individual cells or perhaps, as Darwin had tentatively suggested, for smaller entities inside cells. This was the same idea that had so excited Galton twenty years earlier when he first read Darwin's theory of pangenesis, but it was De Vries, in the reductionist spirit of his mentor Sachs, who pushed the idea to its logical limit. While Darwin's gemmules were of indeterminate complexity—almost, but not quite, as complex as full-fledged organisms—De Vries's pangenes (the term he coined to distinguish them from *gemmules* while at the same time honoring and benefiting from his association with Darwin) corresponded to enzymes and the most basic components of the cell. De Vries justified his radical approach as much on philosophic grounds as on scientific ones, writing:

> The whole organic world is the result of innumerable different combinations and permutations of relatively few factors . . . These factors are the units which the science of heredity has to investigate. Just as physics and chemistry go back to molecules and atoms, the biological sciences have to penetrate to these units in order to explain, by means of their combinations, the phenomena of the living world.[8]

In addition to using Sachs and Darwin as constant points of reference, *Intracellular Pangenesis* was also a response to the great German theoretical biologist August Weismann, whose penetrating critique of Lamarckian inheritance in his 1883 essay "On Heredity" had profoundly impressed De Vries.[9] By discrediting the doctrine of the inheritance of acquired traits, Weismann's critique had eliminated the rationale for Darwin's traveling gemmules: it was no longer necessary that information travel from distant organs back to the germ cells.

Weismann did more, though, than discredit Lamarck and by extension Darwin's theory of pangenesis. He proposed a grand hereditary scheme of his own. In the place of Darwin's vast numbers of gemmules, Weismann proposed the existence of "germ-plasm," a single indivisible unit that was thought to be capable of regenerating a complete individual. Unlike Darwin's changeable gemmules, the germ-plasm, located in specialized *germ cells,* was passed down through the generations, "uninfluenced by that which happens during the life of the individual which bears it."[10] In Weismann's developmental model, it was only the germ-plasm in the germ cells that remained inviolate, while the germ-plasm in the other cells of the body underwent a winnowing process, shedding hereditary material during successive divisions and retaining only the portion of the germ-plasm necessary to perform each cell's specialized function.

As De Vries hastened to point out, Weismann's theory flew in the face of one of the most prosaic of botanical facts, namely, that a whole plant could be regenerated from any of a variety of specialized cell types. To get around this obvious objection, De Vries proposed that the nuclei of all cells contained an identical collection of pangenes that were the repository of the totality of the hereditary information. The distinctive nature of different cell types was not a matter of the hereditary endowment of the cell, but was instead determined by the types and numbers of the particular pangenes that were transported from the nucleus to the cytoplasm. As he explained it in a popular magazine of the day, "We can say: the instructions lie in the nucleus, the realization takes place in the protoplasm."[11] This idea would turn out to be the crux of the modern understanding of the process of cell differentiation.

*Intracellular Pangenesis,* published in 1889, was the culmination of more than a decade of experimental and theoretical work that began in the late 1870s when De Vries had combed the countryside near Amsterdam looking for strange irregularities in plant shapes and forms—so-called monstrosities—that might illuminate the nature of the pangenes. His goal, as he later explained, was to create a "herbarium of monstrosities." Following the tried-and-true method of professional plant breeders, who were always on the lookout for promising new shapes and colors from which to create new varieties, De Vries self-fertilized his most promising specimens and selected the best examples from among the progeny. In one of his first successes, he isolated plants with strangely twisted stems and, by selecting the most twisted, showed that it was possible to increase the percentage of monstrosities. This strongly suggested that the monstrosity could be inherited and was not due solely to the effect of environmental influences.

Attacking the problem from another angle, he made hybrids from different varieties of beans and flowers, hoping to show that he could transfer particular qualities from one variety to another. The very fact that it was possible to transfer individual traits from one variety to another showed that the hereditary material was divisible into independent characters, a point very much contested by Weismann. In 1886 De Vries produced a hybrid seed by crossing a red-flowered male foxglove with a white-flowered female. The hybrid had pale red flowers, halfway between the colors of the two parents, and the offspring of the hybrid were of three types: pure red and pure white like the grandparents and pale reds like the hybrid parent. At this point, De Vries did not concern himself with the relative frequency of the different types, which would turn out to be the key to the whole puzzle. Instead he simply described the different classes and noted that the segregation of types among the grandchildren supported the idea that the factors coding for colors were inherited independently.

In addition to casting light on the transfer of individual qualities, De Vries saw that the hybridization experiments could be used to demonstrate the "intracellular" transport of pangenes from nucleus to protoplasm. His reasoning relied on the microscopical work of the German botanist Edouard Strasburger, who after years of trying had finally succeeded in observing the nucleus of a male pollen cell cross the pollen tube, enter the embryo sac, and fuse with the nucleus of an egg cell, proving that the nucleus was the only significant part of the male pollen cell that the red-flowered grandfather contributed to the hybrid seed.[12] Using Strasburger's observation, De Vries argued that the information for the red dye that was produced by the cells of the pink flowers of the hybrid must have been transferred from the nucleus of the grandfather's pollen to the nucleus of the egg cell. At some later stage in development, the information was transported from the nucleus of the flower cell to the cell machinery outside the nucleus where the dye was actually produced.

— DURING THE SUMMER OF 1886, while wandering near the small village of Hilversum, De Vries happened upon the flowering shrub *Oenothera lamarckiana,* popularly known as evening primrose for the way it bloomed in the early evening. Ironically, *O. lamarckiana,* which would mesmerize De Vries for nearly half a century and ultimately lead him astray, carried the name of a biologist who was immortalized for advocating a highly seductive but erroneous theory of evolution.

De Vries was filled with anticipation by the sight of the wildly proliferating *Oenothera,* which appeared to have spread from a neighboring park and halfway across an unused potato field to form a jungle of foliage. It was just such vigorously growing plants that he sought, following the conventional wisdom that vigorously growing plants were more likely to contain interesting variants and throw off sports. Upon closer examination he found that the plants were even stranger than he had dared to hope, showing an unusual variability in almost every organ and character. The plants held such a fascination for him that De Vries returned for repeated observations nearly every week that summer, and sometimes every day, for several hours at a stretch. That fall, he transplanted nine specimens of *O. lamarckiana* to his experimental garden in Amsterdam, and for the following two summers he continued his close observation of the flowers in the field. During his second summer of observation, two radically new types appeared, each type differing from the *O. lamarckiana* parent to such a degree that De Vries was certain that he had been witness to the birth of two entirely new species. In each case the new types appeared in a small cluster of similar plants, suggesting that a single new variant had given rise to a small colony of descendants. Meanwhile, the nine plants he transplanted to his experimental garden in the autumn of 1886 flowered and produced seeds in 1887 and 1888. Among the 10,000 plants grown up from these seeds, De Vries found what he believed to be 3 mutant plants, which differed so markedly from their *O. lamarckiana* parent that De Vries added them to his list of examples of the births of new species. Remarkably, all of the new forms remained true to form when interbred.

Over the next decade he cultivated over 50,000 *Oenothera* plants and found over 800 mutated ones. Among these, he singled out 7 that he classified as new species. Emphasizing his belief that each new form represented a new species, he christened each of them with the traditional Linnaean binary names. He presented the results of these experiments in his 1901–1903 *Die Mutationstheorie,* a monumental two-volume treatise that set out to unify the facts of evolution, inheritance, and cell biology. Beautiful color lithographs accompanied long descriptive passages in the *Mutationstheorie* devoted to *Oenothera gigas,* a giant form, *Oenothera albida,* a variety with pale-green delicate, narrow leaves, *Oenothera rubinervis,* with red veined leaves, *Oenothera oblonga* with narrow leaves on long stalks, and the dwarf variety, *Oenothera nanella.*

From the beginning, De Vries was convinced of the monumental importance of the *Oenothera* work. Indeed, he was convinced that the discovery of the *O. lamarckiana* and its new forms was the most significant discovery of the modern era. By delaying before publishing on *Oenothera* he would give himself time to build the case slowly and methodically, time to collect more data and sharpen his arguments. Then, when the time was ripe, he would dazzle the world with his new theory. Throughout *Intracellular Pangenesis,* which is otherwise rich in specific examples, he never referred to *Oenothera,* and only once, buried in the middle of *Intracellular Pangenesis,* did a flash of the grand importance he attached to his species-forming mutations slip out. "On the species-forming variations," he wrote, "depends the gradually increasing differentiation of the entire animal and vegetable world of variation." Hamlet-like, he equivocated, withdrawing from his grand ambition in the very next line: "But the farther we get away from the facts the more likely we are to get lost in false speculations. My object was only to place the fundamental idea of Darwin's pangenesis in the right light. I hope I have succeeded in this."[13]

Although De Vries did not dare to reveal his grand intentions, *Intracellular Pangenesis* laid out the theoretical framework on which all his subsequent work would build. There were two types of variation, he hypothesized: The first he dubbed "fluctuating variation," which was determined by the number of pangenes that were found in the cytoplasm of the cells. The greater the number of a particular pangene, the greater the intensity of the character it embodied. This was the kind of continuous variation that Galton had plotted in bell curves. Like Galton, De Vries believed that such continuously varying differences could be selected for, and that when selection was relaxed, characters regressed to the mean. The second type of variation corresponded to what were commonly known as "sports." Sports, De Vries hypothesized, were the result of a qualitative rather than a quantitative change in a pangene, and they resulted in discontinuous variation. The new species of *Oenothera* were the perfect embodiment of this discontinuous change that he believed to be caused by the introduction of a new pangene.

In a letter to the editor of the prestigious *Botanical Gazette,* De Vries's close friend and colleague Jan Willem Moll offered to contribute a review of *Intracellular Pangenesis,* writing, "I have no doubt that for this hypothesis a new era begins with the publication of this book." He went on to predict that the book would create something of a sensation and that much would be spoken and written about it. De Vries, who never appeared to suffer any doubt about the significance of his own work, shared Moll's high hopes. To both men's surprise, Moll's review in *Botanical Gazette* would turn out to be the only review of the book to appear in an English-language journal. The reception of the book was disappointing at best, and from 1889 on, De Vries published a large number of papers related to the theory of intracellular pangenesis but he rarely mentioned the theory explicitly.[14] By the time De Vries was ready to begin work on his next book, Moll urged his friend to drop the pangenesis theory altogether for fear that it might prejudice his readers against the new effort.[15]

Although he grudgingly agreed with Moll that it might be a good idea to suppress any mention of his pangenesis theory, De Vries never lost confidence in the principles of pangenesis. From 1889 on, De Vries single-mindedly devoted himself to the development of the pangene model and to strengthening the experimental evidence for it. His highest priority was to look for evidence of qualitative changes in pangenes, and he believed that the best way to do this was by studying monstrosities. If an altered pangene was in fact responsible for the appearance of a monstrosity, he reasoned, then it ought to be possible to transfer the monstrosity-causing pangene from one variety to another. For his model system, he chose the exotic poppy *Papaver somniferum polycephalum,* valued for the strange beauty of its large crown of extra fruits surrounding the central capsule, which were caused by the conversion of stamens to pistils (male parts to female). In 1893 he formed a hybrid between a polycephalous black-flowered poppy (the variety known as mephisto) and a normal white-flowered variety (known as Danebrog). If the polycephalous monstrosity was caused by a pangene, De Vries reasoned, it ought to be possible to transfer the polycephalous-causing pangene of the black-flowered variety to the white-flowered one by crossing.

As it turned out the experiment was a success, clearly demonstrating that the monstrosity was heritable and could be passed from one variety to another.[16] De Vries noted the relative numbers of white- and black-flowered poppies among the progeny of the hybrid in a brief aside. In later years, he would claim that it was the data on flower color from the poppy crosses that had led him to independently discover the laws of Mendelian segregation, and would point to the nearly perfect 3:1 ratio of black- to white-colored flowers among the progeny of the hybrid as evidence of his prescience, but the real progression of his ideas was far less direct and considerably more interesting.

In fact, it was De Vries's encounter with statistics and probability that triggered the line of inquiry that would lead him to the brink of Mendelism. Like Galton, De Vries had been won over to the statistical approach by Adolphe Quetelet, whose work he learned about at a conference in 1887, where a Ghent botanist, named Julius MacLeod, whose father had corresponded with Quetelet, read his paper on the statistical analysis of flower fertilization.[17] MacLeod led De Vries to read Quetelet's *Anthropométrie ou mesure des différentes facultés de l'homme,* which emphasized the appearance of the normal distribution in human traits, as well as in the traits of plants and animals.

In a paper of 1890, De Vries showed that the number of rows in ears of corn followed a normal distribution, and the following year he did a statistical study of the length of seeds of sunflowers and primrose seeds isolated from *O. lamarckiana*. He was so pleased with the *O. lamarckiana* results that he made a lecture plate on which he plotted the curve generated from his *O. lamarckiana* data next to the theoretical bell curve.[18] Three years later, in the fall of 1893, De Vries read Weldon's 1890 paper in which Galton's statistical methods were applied to the analysis of claw length in different races of *Crustacea*, and found the approach so exciting that he immediately undertook another round of data collection from *O. lamarckiana*. From these data he produced a beautiful symmetric curve of the size of *O. lamarckiana* fruit in the Galton style, but he did not have enough information to make his data fit the theoretical curve that Weldon had presented, and he put the experiment aside. Nonetheless he had been sufficiently inspired by Weldon's paper that he wrote to his friend Moll in November to inquire if Moll had any of Galton's books, and early in 1894 he had managed to procure copies of Galton's *Hereditary Genius* and *Natural Inheritance*.

While *Hereditary Genius* had little of the nuts and bolts of the statistical methods that De Vries was seeking, De Vries found more of what he was looking for in *Natural Inheritance*, including a description of Galton's quincunx, with its vivid real-time demonstration of the principles of the binomial distribution. As De Vries had recognized, Quetelet had illustrated the same principle in his *Anthropométrie* by considering the now-classic example of a bowl filled with a large number of black and white marbles. Instead of summing up the random right and left motions of a steel ball as it fell through a fixed number of rows of pins, Quetelet tallied the number of white (or black) marbles resulting from a fixed number of blind draws from the bowl. As was the case with the quincunx, the most common value was the central value (when half the marbles were white and half were black) and the more extreme values were less frequent. The greater the number of draws in each run, the better the frequencies of the various outcomes approximated the idealized bell curve.

For De Vries, the advantage of Quetelet's approach over Galton's was that it suggested a simple way to compute the probabilities of the expected outcomes. It was relatively easy to see, for example, that the chances of getting 0, 1, or 2 white marbles in two blind draws from a bowl containing an equal number of blacks and whites was as 1:2:1.[19] In his book Quetelet gave a table with the probabilities for all combinations of blacks and whites for samples of size up to 20, and De Vries studied these tables in detail.[20] In 1895 he included a detailed discussion of combinatorics and their use in approximating the normal curve in the third edition of his *Textbook of Plant Physiology*.[21]

By the summer of 1895, steeped in the theory of probability, he began a new experiment designed to find a "law of pangenes" that could explain the seeming regularity of the distribution of a trait among the progeny of a hybrid plant. Rather than lose time by breeding his own hybrid, he used a hybrid that had been given to him several years before by his friend Moll, a cross between a white-flowered and a blue-flowered variety of the perennial *Veronica longifolia* (commonly known as speedwell). In the spring of 1895 he sowed several dozen seeds harvested from Moll's hybrid and collected seeds from each of the resulting progeny later that summer. In the following year he sowed these seeds, and by June 1896 he could make a tentative classification of the flower colors of the grandchildren of Moll's plant.

As a starting point, De Vries assumed that the hybrid parent contained a large pool of white and blue pangenes in roughly equal quantities and that in the process of fertilization each progeny inherited a fixed number of color pangenes from the parent. Furthermore, he assumed that the relative frequency of blues to whites in any fertilized egg was simply a matter of chance, and that the probabilities of a given number of blues in a sample of a fixed size was governed by the same simple laws of probability that Quetelet had used to compute the frequencies of the different outcomes in the series of random draws of black and white marbles.

The complicating factor, De Vries realized, was that the outward appearance of the progeny might not accurately reflect their inner "pangenic" structure. For example, the color of a flower whose nucleus contained three blue pangenes and one white might be indistinguishable from that of a flower whose nucleus contained four blue pangenes. To get around this problem, De Vries invented the concept of "hereditary number," which posited that the intensity of a trait, as measured by the percentage of grandchildren that expressed it, was a reflection of the inner structure of the parent.

That June De Vries attempted to divide the children of Moll's hybrid (based on the percentage of whites among their grandchildren) into five distinct types—ranging from those that produced the lowest percentage of white-flowering grandchildren plants to those that produced the most—in such a way that they most nearly approximated a ratio of 1:4:6:4:1—the relative frequencies of drawing 4,3,2,1 or no blues in a random sample of four pangenes. By August, when he was able to classify the grandchildren types more exactly than had been possible in June, it became clear that many of plants that had been classified as nearly pure white breeders were in fact blues and that, as a result, the data now fit better with a 1:2:1 law.[22]

Further confirmation of the 1:2:1 law was provided by an independent experiment, which was coming to fruition in the same period. As had been the case with many of De Vries's most significant discoveries, this one had begun with a walk in the countryside when, in the summer of 1895, he happened upon a rare white-flower variety of the normally purple *Aster tripolium.*[23] Following his common practice, he dug up the rare specimen and took it back to his experimental garden in Amsterdam, where the nascent seeds were allowed to ripen. The existence of the seeds indicated that the plant had already been pollinated, most likely by a combination of pollen from the surrounding blue plants and self-fertilization. He sowed the seeds the following June, and they yielded 72 plants. In a July note, he recorded that 10 plants had flowered and none of them showed white flowers. By August, one of 8 flowering plants now had white flowers and the remaining 7 were purple. On August 10, he paused to write a short paragraph exploring the meaning of the *Aster* mutant and his future plans for it: "The purple individuals, from a white mother, must have purple fathers and be central hybrids according to the law of the cross of pangenes." He went on to speculate that 95 percent of the time, white-flowering plants were pollinated by purple individuals. This was a hypothesis that could be tested experimentally, in the same way that he had tested the children of Moll's hybrid. Under the assumption that the white pangene was "latent," and would thus be overmastered by the "active" purple pangene, the 1:2:1 law predicted a 3:1 ratio of purples to whites in the grandchild generation. Reflecting his now-growing conviction that flower color at least followed a 1:2:1 law, he wrote: "If there are no white flowering individuals this year, and consequently all individuals are central hybrids, then the seed must yield 75% purples and 25% whites. This to be investigated."[24]

While De Vries was very much interested in deducing the laws governing the inheritance of pangenes, the transfer of monstrosities, which he believed held the key to evolution, burned brighter in his imagination, and it was to this theme that he next returned. In a 1897 paper, "Erfelijke Monstrositeiten" (Hereditary Monstrosities), De Vries reported on the transfer of a character between two closely related varieties of evening campion, *Lychnis verspertina.* Ordinarily campion stems were coated in fine white hairs, but De Vries had discovered a rare hairless variety near Hilversum in August 1888, and by 1892 he had isolated an essentially pure hairless "glabrous" type in his experimental garden. Pollen from this hairless variety, which he called *Lychnis vespertina glabra* was crossed to the normally hairy *Lychnis diurna.* The hybrid generation was hairy, as hairiness was dominant (in the soon-to-be rediscovered Mendelian terminology). When the hybrids were selfed, they gave rise to both hairy and hairless forms: "99 hairy and 54 hairless, in all 153 plants," as he reported in his 1897 paper, which he noted in passing gave a 2:1 ratio of "pubescent" to hairless forms. He made no mention of the 1:2:1 law of pangenes, which would have called for a 3:1 ratio of hairy to hairless forms. In the same paper he mentioned another ratio of progeny from a self-fertilized hybrid made from crossing two strains of *Linaria,* one with orangish flowers, the other with pale yellow. "About 80% had an orange labellum, about 20% had wholly yellow flowers," he wrote, rounding off the data without regard to a 1:2:1 or any other law of pangenes.

De Vries returned again to the *Lychnis* experiments in July 1899 when, along with Bateson, he was a featured speaker at the meeting of the International Hybrid Conference sponsored by the Royal Horticultural Society of London. Despite the fact that he had reported a 2:1 ratio among the progeny of the *Lychnis* hybrid in his 1897 paper and given the actual data of 99 hairy to 54 hairless, the published transcript of the lecture, which appeared in April 1900, now reported a 3:1 ratio of hairy to hairless progeny—the expected ratio under the 1:2:1 law of pangenes for traits determined by two pangenes.[25] Somehow, in the period between the lecture and the writing of the manuscript, De Vries appeared to have become certain that the 1:2:1 law was correct, and therefore reported the expected 3:1 ratio of hairy to hairless. What had precipitated the revision of the ratio?

Early in 1900, just before his father's death in March, De Vries was sent an obscure reprint entitled *Experiments on Plant Hybrids,* written by a Moravian monk in 1865. The paper had arrived from his old friend and colleague Martinus W. Beijerinck, who was one of the first to develop the modern idea of the virus, which he described as an *heritable infectious agent,* and whose 1882 paper on crossing had first piqued De Vries's interest in hybridization experiments.[26] In his paper, Beijerinck had noted that the offspring of hybrids sometimes showed the parental types and other times produced entirely new forms. In fact, De Vries's debt to Beijerinck was reflected in his Royal Horticultural Society address, which was first titled "Hybridization as a Means of Pangenetic Infection" and only later amended to "Hybridizing of Monstrosities."[27]

From many other sources, an obscure reprint, which was thirty-five years old and written by a complete unknown, might not have claimed De Vries's attention, but De Vries was predisposed to listen to Beijerinck. In his cover letter, Beijerinck wrote, "I know that you are studying hybrids, so perhaps the enclosed reprint of the year 1865 by a certain Mendel which I happen to possess, is still of some interest to you."[28] Several of the remarkable features of Mendel's paper would have jumped out at De Vries the moment he started reading. In fact, much of the experience of reading it must have been a kind of déjà vu, for Mendel had zeroed in on the key feature of pangenes.

Like De Vries, Mendel had assumed that an organism could be broken down into different traits that each corresponded to an individual hereditary factor, and that these factors were shared among many different organisms and could be transferred independently from one to another by the process of hybridization. Both men had suggested that the factors might come in alternative forms, and that one form could overpower the other in hybrids. Mendel had coined the terms *dominating* for those traits that were passed on to the hybrid unchanged, and *recessive* for those that receded or completely disappeared, while De Vries had called them *active* and *latent.* What was perhaps most striking of all, and must have given De Vries many sleepless nights, was that Mendel had demonstrated the 3-to-1 ratio of dominants to recessives among the progeny of hybrids exactly as De Vries had predicted in his 1:2:1 law of pangenes.

**The "handsome" portrait of Mendel.**

From *Gregor Johann Mendel: Leben, werk und wirkung* (Berlin: Julius Springer, 1924), after p. 7, Tagfel 7. With kind permission of Springer Science and Business Media.

CHAPTER 5

# Mendel

FRANCIS GALTON AND GREGOR MENDEL were born in the same year, 1822, but the circumstances of their births could hardly have been more different. In the marriage of Galton's father, Tertius, and Violetta Darwin, two of England's leading families were joined together. Gregor Mendel's father, Anton, who was born of peasant farmers in the small village of Heinzendorf in Austrian Silesia,[1] married Rosine Schwirtlich, the daughter of a gardener, also from Heinzendorf.

Though he was forced to work three and a half days a week for his landlord, the industrious Anton managed to eke out a living as a tenant farmer. When time permitted, Anton tended his fruit tree orchard, grafting his trees with finer French varieties provided by the local parish priest, Johann Schreiber, who was a cultivator of fruit trees. At the village school, Schreiber taught the younger Mendel the art of fruit tree improvement, and recognizing the boy's unusual intellectual ability, urged his parents to send him to a special school to prepare him to enter Gymnasium. The younger Mendel having proved himself at the preparatory school, the following year Rosine and Johann prevailed over the objections of Anton, who hoped his son would take over the family farm, and the 12-year-old Mendel was permitted to enroll at the Troppau Gymnasium.

For his first four years at Gymnasium, Mendel's parents paid his board and half rations, but in his fifth year at school his father suffered a bad farming accident that significantly diminished his earning capacity and, at the age of 16 with two years of Gymnasium left to complete, Mendel was forced to earn his own way.[2] Supporting himself on the meager income he earned from private tutoring proved to be difficult going. He had to return home in late May 1839 after suffering a nervous collapse, but he returned again to Troppau in September, finally graduating from Gymnasium in August 1840.[3]

When Mendel returned home, his father, who was now in poor health, renewed the pressure on his only son to take over the family farm, but Mendel steadfastly declined. Instead, he moved to Olmütz, which was the nearest place where he could complete the two years of post-Gymnasium coursework required to enter university. Without friends or references in the mostly Czech-speaking city, Mendel was unable to find tutoring work, and once again he suffered a nervous breakdown. "Distress at the prostration of his hopes," he wrote in a third-person autobiographical statement, "and the gloomy outlook upon the future, had so marked an effect upon him that he fell sick." This time he ended up spending a year at home.[4]

In August 1841, Mendel's mother and now-infirm father sold the family farm to Mendel's older sister and her husband. In the sales agreement it was stipulated that Mendel would be provided with a small annual income to support his studies, but the sum was not nearly enough to cover the expenses of his education, and once again Mendel despaired of ever realizing his hopes. But his younger sister, who had always been sympathetic to his plight, came to his rescue, turning over a portion of her dowry to support him, an act of generosity he would never forget.[5]

With the help of his sister's money, he returned to Olmütz in 1841 and over the next two years completed the training in the natural sciences that was required for admission to university.[6] But the years of struggle had taken their toll, and he was now convinced that the only way to pursue a life of the mind was to "enter a profession in which he would be spared perpetual anxiety about a means of livelihood."[7] In October 1843 at the age of 21 he took the name "Gregor" and was admitted to the Augustinian monastery of St. Thomas in Brünn as a novitiate.[8]

In one stroke, Mendel had transformed his life. The monastery in Brünn, the provincial capital of Moravia, was particularly well-to-do, and the facilities included a twenty-thousand-volume library and spacious, tastefully furnished, two-room suites for each of the monks. Under the leadership of Abbot Cyril Franz Napp, who was himself a well-known scholar, the monastery was one of the region's leading centers of intellectual life with a particular emphasis on the natural sciences. In his second year Mendel took the vows of obedience, chastity, and poverty in accordance with the rule of St. Augustine, and on July 22, 1847, his twenty-fifth birthday, Mendel was ordained subdeacon.[9]

While he had applied himself to the required ecclesiastical studies at Brünn Theological College, Mendel still had time to pursue his ever-growing interest in the natural sciences.[10] In his fourth year, he attended lectures on economics, fruit growing, and viticulture at the Philosophical Institute,[11] as well as learned the techniques of artificial pollination of plants, which would be the fundamental tool of his future research.[12]

Following the monastery's tradition, Mendel began to work as a parish priest after his graduation from Theological College in June 1848.[13] His duties included ministering to the sick at a nearby hospital. Seeing no future for the study of natural science in his job as a curate, and unable to cope with the possibility that his hopes might be disappointed, Mendel once again succumbed to his nervous disposition. In January 1849 he fell gravely ill and was confined to bed for more than a month.[14]

Due to the enlightened intervention of Abbot Napp and the fact that there was a shortage of teachers, Mendel was excused from his duties ministering to the ill in hospitals and allowed instead to become a high school teacher in October 1849. Based on his obvious aptitude for teaching, after a year Mendel was encouraged to take the exams that would qualify him as a high school teacher of natural history for all grades and physics for the lower grades. The physics examiner, Baron von Baumgartner, who had been a professor at Vienna University, was more than happy with Mendel's performance in the physics exam, but the natural history section was a disaster. Mendel's first essay, written in his characteristically lucid style, contained an entirely scientific account of the origin of life on earth without resort to any supernatural intervention, concluding with the observation that "the history of creation is not finished."[15] Whether it was the sentiments expressed, or the fact that they were expressed by a monk, the essay aroused the examiner's intense displeasure, which was compounded by the second half of the exam, in which Mendel demonstrated a nearly complete ignorance of the classification of mammals, the examiner's field. In the end, he was denied accreditation as a teacher, even of elementary school.

Had he passed his exam, Mendel would have undoubtedly remained a Gymnasium teacher, but once again Napp intervened. Based on the extremely positive reference provided by Baumgartner, Napp arranged for Mendel to study at a new institute for physics headed by Professor Christian Doppler (famous for the Doppler effect) at Vienna University. At University, Mendel was exposed for the first time to science presented by leading research physicists, mathematicians, and biologists. Doppler died during Mendel's two-year stay and his place was taken by A. Ettingshausen, a mathematician with a specialty in combinatorial analysis.[16]

Not only did Mendel receive top-notch training in physics and probability theory, but he was steeped in cell biology and cytology under the instruction of F. Unger. In Unger's course, Mendel also read the work of the great German botanist Carl Friedrich von Gaertner, whose *Experiments and Observations upon Hybridisaton in the Plant Kingdom* had just then appeared in German for the first time.[17] In his book, Gaertner reported on over 10,000 experiments on 700 species belonging to 80 different genera of plants, from which he obtained some 350 different hybrid plants,[18] and it is clear from Mendel's underlining that he studied the book in great detail.

As was characteristic of the hybridists of the era, Gaertner believed in the "inner nature" of the species, a nature that was a reflection of a Divine purpose and therefore could not be altered in the process of hybridization. In the same sense, the infertility of hybrids was also an expression of divine will, preventing man from tampering with God's creation. In his copy of

Gaertner's book, Mendel had underlined references to the English botanist Andrew Knight, who expressed the view that while hybrids formed between different species were infertile, hybrids formed between varieties were productive.[19] He had also underlined references to articles on pea-crossing experiments by John Goss and Alexander Seton, both of whom had also made hybrids from different varieties of pea plants.[20] Goss's article was particularly suggestive, and it is possible that these two articles played a part in the planning of Mendel's own experiments on hybrids made from garden peas that were to commence a few years later.[21]

As Mendel was well aware, Unger held views very much at odds with Gaertner's idea that species were possessed of an inner nature or that they were immutable. Instead, Unger maintained, species were composed of different elements, these elements were combined in new "law-combinations" to make new forms, and the diversity of life was explained by a process of transformation from less- to more-complex forms.[22] The idea that it might be possible to explain the inner workings of organisms and cells and by so doing to account for the diversity of life incited the wrath of the church, and in 1851, during Mendel's first year as his student, Unger was attacked for his "pagan views" by the editor of a Catholic journal and was nearly forced to resign his post as dean of science.

After the highly stimulating two-year stint in Vienna, Mendel returned to the monastery in July 1853, and the following year he filled an opening as a teacher of physics and natural history at the Realschule, a newly opened technical school in Brünn, where he would remain for fourteen years. At the end of his first year of teaching, which was praised by students and other faculty, Mendel once again applied for his certification to teach physics and natural science (this time physics at the upper levels and natural sciences at the lower ones), and once again, in May 1856, he failed to be accredited. This time the cause appears to have been a dispute with his natural history examiner, Professor Fenzl, who was a diehard "spermist," insisting that the egg made no contribution to the hereditary endowment of the embryo.[23] Upon his return from Vienna, Mendel was described by a fellow monk as suffering from an "ailing head" and "very much out of humor," but this time he managed to stay off a full-fledged nervous breakdown.

Rather, it appears, Mendel plunged into his own experiments on hybrids, following the injunction of his teacher Unger to "reduce the phenomena of life to known physical and chemical laws."[24] He'd already been crossing grey and white mice in his rooms, observing the coat color of the offspring, and at the same time he'd been producing new color varieties from crosses of different flowering plants. While he was not free to discuss the results of his rodent-breeding experiments, which would have been viewed as immoral by the church authorities, he was deeply impressed by "the remarkable regularity with which the same hybrid forms continually recurred when the fertilization took place between like species," as he put it in his later account.

By 1854 Mendel had settled on the common garden pea, *Pisum,* as the most suitable object for his studies of the behavior of hybrids. *Pisum* offered great advantages that had been recognized by several other investigators, including Darwin. Not only did crosses of *Pisum* varieties yield fertile hybrids, but the fact that the sexual parts of each flower were enclosed in a capsule (known as the keel) made *Pisum* a wonderfully convenient plant in which to make artificial crosses. Ordinarily the pollen-bearing anthers burst inside the closed capsule, but it was possible to cut open the keel of still-immature flower buds, cut out the anthers, and immediately dust the female parts with foreign pollen.

In the legendary paper describing the account of his research, Mendel created the impression that the conclusions of one experiment led him to the design of the next, but in this he was likely bowing to the conventional view that the only way to scientific truth was through experiment. As was first suggested by R. A. Fisher, Mendel's experiments were surely designed to prove a point and not "for his own enlightenment."[25] Unlike Gaertner and many of the other hybridists, who viewed species as indissoluble entities, and hybrids between species as a test of supremacy of one type over another, Mendel appeared to assume that a species was a mosaic of factors that were randomly shuffled in the process of sexual reproduction.

In the simplest case, Mendel considered a hybrid formed from parents that differed in only one major factor, which he denoted by capital *A* or lowercase *a*. In the process of reproduction, egg and pollen cells were produced containing one or the other of the factors with equal probability, and it was strictly a matter of chance which kind of pollen joined with each

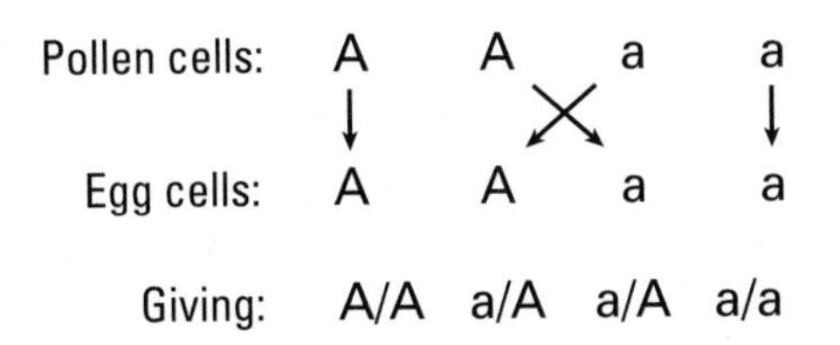


**Four kinds of fertilization.** Illustration by Tamara L. Clark.

egg.[26] Thus, four kinds of fertilizations took place with equal probability (see the diagram).

The result was four types of fertilized eggs—one true-breeding for *A* (*A*/*A*), two hybrids identical to the hybrid parent (*A*/*a*), and one pure-breeding for *a* (*a*/*a*)—in a ratio of 1:2:1. If *A* was *dominant* and *a recessive,* in the terminology that Mendel would soon introduce, then the ratio of dominant to recessive types would be expected to be 3:1.

Already in 1854, he had begun testing different varieties of *Pisum* in order to make sure that they bred true—that is, that their offspring were essentially identical to the parent and that they remained constant.[27] By 1856 he had selected 22 from a total of 34 closely related varieties and from these he ultimately chose seven pairs of plants, each of which differed in one essential trait, including the two most famous pairs of Mendelian traits, yellow and green peas and round and wrinkled ones.[28]

In the summer of 1857 he produced hybrids for seed shape and color, as well as for flower color and plant height. While the color of the hybrid flowers in plants raised from fertilized seeds would not appear until the following year, the hybrid seed forms began to develop immediately after the cross-fertilization was performed.[29] Already by the end of the summer of 1857 Mendel would have seen that the seeds resulting from crossing green- with yellow-seeded parents were all of the dominant yellow type, and likewise the seeds produced by crossing smooth and wrinkled parents were all smooth. The form of the hybrids would not have come as a surprise to Mendel, who had deliberately selected them that way. Already in his previous work on ornamental plants, he had seen that hybrids were rarely equal blends of the two parental traits, and that the hybrid often resembled one parent so closely that that the contribution of the other parent was totally invisible, and the same turned out to be true for *Pisum.*[30] But the experiment also confirmed the fundamental fact, which had been observed by other hybridizers, that the form of the hybrid was unaffected by which of the parental types was the seed or the pollen plant.

The next step in the experimental program was to look at the distribution of traits in the progeny of the hybrids.[31] In the summer of 1858, the second year of the experiment, the seeds from the smooth-wrinkled hybrid and as well as the green-yellow cross were grown up and self-fertilized. By the winter of 1858–59, Mendel had demonstrated the 3:1 ratio of dominant to recessive types among progeny of hybrids—the biggest moment in his scientific life—and he'd done it with an exactitude that may have defied even his dreams. Of 7,324 seeds collected from 253 round-by-wrinkled hybrid plants, he found that 5,474 were round and 1,850 were wrinkled, giving a ratio of 2.96 to 1. Among 8,023 seeds from 258 yellow-by-green hybrids, he found 6,022 yellow and 2,001 green, giving a ratio of 3.01 to 1.

The last step in the analysis of seed characters was the demonstration that the dominant classes of the offspring of hybrids were actually composed of two types, pure breeders and hybrids, in a ratio of 1:2. In the case of seed characters, it was a relatively straightforward matter to test. Pure-breeding yellow seeds, for example, would be expected to give plants bearing exclusively yellow seeds, while hybrid yellow seeds would once again yield plants with a 3:1 ratio of yellows to greens. Once again, the experiments conformed to Mendel's expectations with remarkable precision.[32]

— WITH THE DEMONSTRATION of the 1:2:1 ratio of types among the progeny of hybrids Mendel had verified half of his theory. The second half, later named "Mendel's second law,"[33] posited that pairs of differing characters behaved independently of all other pairs. The demonstration of the independent assortment of traits was the most time-consuming and laborious of Mendel's experiments. He began in 1860 by showing that inheritance of seed color and seed shape were independent.[34] In particular this meant that the progeny of yellow-round hybrids, which also contained the factors for green and wrinkled, would be expected to produce equal numbers of sex cells for yellow-round, yellow-wrinkled, green-round, and green-wrinkled.[35] Once again the numbers were close to expectation, as was the case in the still more technically demanding experiment demonstrating the independence of three factor pairs.

Seventy years after Mendel's paper, the renowned British statistician R. A. Fisher published a dramatic article in the *Annals of Science* in which he claimed that Mendel's data were simply too good a fit to the theoretical expectations to be real.[36] In applying modern statistical methods, many of which he had invented, Fisher showed that, taken as a whole, data as good as Mendel's would arise in fewer than 1 in 14,000 trials. While such a discrepancy could be due to several causes that do not involve any deliberate attempt to deceive on Mendel's part, Fisher pointed to two particularly damning cases that seemed to suggest a more direct manipulation of data. In the first case, Mendel provided near-perfect data in support of a 2:1 ratio of heterozygotes to dominant homozygotes among the offspring of hybrids, but he failed to take into account a built-in bias in his experimental procedure that altered the expected ratio.[37] The chance of obtaining the "near perfect" data that Mendel reported in his paper was, Fisher computed, less than 1 in 444.[38] Secondly, Fisher pointed out that it was highly improbable that Mendel could have picked out seven independent characters without in the process coming across some pairs of traits that were in some degree associated. In fact, Fisher asserted, Mendel's own data did not definitively rule out the possibility of a loose association between the genes for pea shape and color, nor did he systematically study the linkage of all of the pairs.

Mendel's veracity has been fiercely disputed ever since Fisher published his paper in 1936. In 2001, the botanists Daniel J. Fairbanks and Bryce Rytting showed that Fisher had overlooked several subtleties of Mendel's experimental procedure that could only be appreciated by an experienced gardener and that, as a result, Mendel's 2:1 expectation may not have been as far off as Fisher believed.[39] More puzzling, however, is Mendel's demonstration that all seven of the traits he studied assorted independently. That is, an arbitrarily selected set of seven traits would have been expected to contain two or more that showed some degree of linkage.[40] A conceivable explanation is that Mendel singled out the particular set of seven traits from several dozen candidates precisely because they could serve to best illustrate the principle of independent inheritance of traits.[41]

A less sinister interpretation of the too-good-fit of the data to theoretical expectations is that Mendel simply stopped counting when his experimental ratios approached a desired outcome. Although this practice would be unacceptable by modern standards, in the era before the formalization of the concept of a random sample and probable errors, the decision to cut the experiment off at an arbitrary point might have seemed entirely justified. Alternatively, it is possible that Mendel presented only those experiments that best supported his theory and put the data from puzzling or nonconforming experiments aside to be interpreted at a later date. While this is clearly a risky approach, many great discoveries have sprung from ideas that are too compelling to yield to contradictory data. Ironically, the law of independent assortment of factor pairs, which was, along with the law of segregation of single traits, Mendel's most important contribution, would turn out to be an oversimplification. However, it was the great clarification offered by Mendel that paved the way to a more complex view.

— ON NEW YEAR'S EVE of 1866, Mendel wrote a letter to Karl von Naegeli, the renowned German plant physiologist. If there was anyone who was in a position to understand the significance of his work, it ought to have been Naegeli. Naegeli, who had given a remarkably accurate description of cell division and seed formation in flowering plants in 1841, was not only versed in physiology but a world authority on hybridization. Furthermore, he was a committed reductionist, as Mendel had undoubtedly learned from Unger's lectures during his two-year stint at Vienna University.

Along with a reprint of his *Pisum* article, published early that year in the *Transactions of the Brünn Natural Science Society,* Mendel enclosed a carefully worded letter, printed in his impeccable script.[42] Though he is unfailingly deferential and polite, addressing the great Naegeli, as "Highly Esteemed Sir," the pride and confidence in his achievement is unmistakable. After briefly summarizing his own results, Mendel tried to give a brief indication of where Gaertner had gone wrong. His mistake, Mendel asserted, had been to classify hybrids based on the degree to which a plant was like its father or mother rather than to track the inheritance of individual traits, and as a result his experiments were "too general, too vague, to furnish a basis for sound judgment." However, he hastened to reassure Naegeli, there was no inherent incompatibility between the vast collection of Gaertner's observations and Mendel's own results. On the contrary, it was possible that if his experiments were repeated and properly designed, Gaertner's results would provide confirmation of Mendel's theory.

A committed believer in the priesthood of science, who traveled only in the most illustrious scientific circles, Naegeli did not consider the work of a mere high school teacher worthy of his full attention, and Mendel waited more than three months for Naegeli's reply, which was dated February 27, 1867.[43] Although he addressed Mendel as "honored colleague," and showed sufficient interest in his work to request samples of the seeds in order to attempt to duplicate his results, his letter showed Naegeli's utter failure to grasp the significance of Mendel's work. Naegeli rejected the key result—that self-fertilized hybrids produced pure-breeding progeny—and ignored the mass of evidence amounting to more than 355 artificial fertilizations and 12,980 recorded observations that Mendel had conducted to support his theory. Even more distressing was his injunction to Mendel that he "should regard the numerical expressions as being only empirical, because they can not be proved rational," which implied that Naegeli had rejected Mendel's underlying "rational" model.[44]

With respect to Mendel's plan to test his theory in other species, Naegeli confidently predicted that "with these forms you will get notably different results (in respect of inherited characters)." In particular, Naegeli recommended *Hieracium,* commonly known as hawkweed, an unprepossessing flowering perennial bearing a resemblance to the dandelion, for Mendel's consideration. As the world authority on the hybridization of hawkweed, Naegeli was certain that Mendel would not find confirmation of his 3:1 law in *Hieracium* hybrids. In fact, Mendel was well aware of the existence of hybrids that defied explanation under his theories. Already in his *Pisum* monograph, he'd mentioned certain hybrids whose progeny were "exactly like the hybrid and propagate unchanged," and in his letter to Naegeli he'd speculated that some species of *Hieracium* might also produce "nonvariable progeny."

Although Naegeli's letter must have been a crushing blow, Mendel quickly regained his stride, and six weeks later he had completed a twelve-page, single-spaced reply.[45] He fully understood Naegeli's tone of "mistrustful caution," Mendel wrote, and he would undoubtedly have felt the same way had their positions been reversed. While acknowledging that his theories were not "easily compatible with our contemporary knowledge," he pressed his case: "Two points in your esteemed letter appear to be too important to be left unanswered." On the first, concerning the constancy of true-breeding progeny, Mendel appealed to the record of his experiments: "Permit me to state that, as an empirical worker, I must define constancy of type as the retention of a character during the period of observation." In the course of his six years of pea experimentation, he observed, the true-breeding progeny of hybrids stayed constant for four generations.[46]

Regarding Naegeli's second criticism—that Mendel had ventured beyond the empirically determined facts—Mendel showed a nearly saintly restraint. In the first place he pointed out that the 1:2:1 ratio of different types among the progeny of hybrids was established empirically, as was the law of independent segregation of traits for two, and also for three, traits. With the extension to an arbitrary number of traits, he granted, "I have indeed entered the rational domain." But such a leap, he explained, was justified by the fact that "the inheritance of one pair of differing characters (giving rise to a 3:1 ratio) proceeds independently of any other differences."[47] For Mendel, who had been trained in physics and mathematics at the University of Vienna, the logic of the argument was so transparent as to be almost self-evident.

Once again, Mendel stoically endured Naegeli's silence. During the summer of 1867, he grew up the hybrid *Hieracium* seeds he had produced the previous year, and found that the plants were intermediate in character between the two parental strains. Having waited for more than six months for a response to his second letter, Mendel wrote yet again on November 6, 1867. This time he made no mention of *Pisum* or of his theory of heredity. Striking a markedly lighter tone, he began by explaining the reason he had been unable to collect wild specimens: "Heaven has blessed me with an excess of avoirdupois which becomes very noticeable during long travels afoot, and, as a consequence of the law of general gravitation, especially when climbing mountains."[48]

Another three months passed, and Naegeli still hadn't replied, but the indomitable Mendel wrote again on February 9, 1868, this time with a request that he hoped Naegeli would be unable to refuse. If Naegeli would send him a collection of strains, Mendel proposed to undertake a systematic study of *Hieracium* and to send him any hybrids he happened to produce. Two months later, more than a year after his first communication, Naegeli finally wrote again promising to send Mendel some live plants seeds with which to continue his experiments with hawkweed.

On May 4 Mendel wrote thanking him for the seeds and sharing news of a major new development in his life: "Recently there has been a completely unexpected turn in my affairs. On March 30 my unimportant self was elected life-long head, by the chapter of the monastery to which I belong. From the very modest position of teacher of experimental physics I thus find myself moved into a sphere in which much appears strange to me." In the same letter, Mendel assured Naegeli that his new position would not get in the way of his continued research. Five weeks later, Mendel wrote to report on the entirely unwelcome and perplexing fact that all 112 progeny plants resulting from self-fertilization of his first *Hieracium* hybrid resembled the mother plant. By the summer of 1870, Mendel had been crossing *Hieracium* for five years and he wrote to Naegeli in a more revealing way, owning up to the sense of disappointment he had experienced in his *Hieracium* experiments. His *Pisum* experiments had led him to expect absolute uniformity in the hybrid form resulting from the cross of two pure-breeding parents, Mendel explained, and the enormous variety of the *Hieracium* hybrids was a shock: "I must admit to you, honored friend, how greatly I was deceived in this respect," he wrote. "No one would take them for siblings if he found them growing in the wild."[49]

Not only the hybrids had been disappointing; so had the results of the later generations. In all his crosses, he'd never seen a hybrid *Hieracium* give rise to progeny that were other than exactly like the mother plant. In a paper delivered to the Brünn Natural Science Society in 1869, he spelled out the surprising and disappointing facts:

> In *Pisum* the hybrids which are the immediate outcome of the crossings of two forms are in all instances of the same type, but their offspring are changeable, varying in accordance with a definite law. In *Hieracium*, however, the experiments hitherto made would seem to give results the precisely opposite of this.

In the same paper, Mendel also mentioned the work of the German botanist Max Ernst Wichura, who had seen the same constancy among progeny of hybrid willows that Mendel had observed in hawkweed. Mendel speculated that the constancy might in some way be connected with the process of hybridization, but the mystery remained unresolved. For the remainder of Mendel's life, the behavior of *Hieracium* cast doubt on the beautiful model in which pairs of hereditary factors were divided in the production of sex cells and recombined in all possible combinations during the process of sex conjugation.[50]

The choice of hawkweed as an experimental plant was twice cursed. Not only did it deny Mendel the confirmation of his theories that almost any other species would have provided, but it had led him to Naegeli in the first place. After Mendel had become involved with Naegeli, who fed Mendel's natural tendency to bear down on a single problem, any hope that Mendel might have turned back from the long dead end of the *Hieracium* experiments was extinguished. Although he never referred to Mendel in his published articles or in his personal letters, Naegeli's 1884 tour de force, *Mechanisch-physiologische Theorie der Abstammungslehre,* showed that in fact he had absorbed the key lessons of Mendelism. In this highly influential book, Naegeli laid out his theory of the "idioplasm," the bearer of the hereditary qualities. As in Mendel's factor theory, every kind of quality was represented as a rudiment in the idioplasm. Furthermore, like Mendel's germ cells containing every possible combination of parental traits, Naegeli posited the existence of as many kinds of idioplasm as there were combinations of qualities. Naegeli also noted the difference between visible characters and underlying hereditary factors, and gave as an example a recessive coat color reappearing among the progeny of hybrid cats.

In November 1873, Mendel wrote Naegeli for the final time, explaining that his administrative responsibilities had prevented him from doing any research for the previous two years. The following spring the reform-oriented Liberal Party passed a series of measures aimed at wresting control of home and family life from the Church. Among these was a new law, passed in 1874, that would require monasteries to increase their contributions to a religious fund used to support Church institutions. In accord with the provisions of the newly enacted law, on October 10, 1875, the monastery was assessed 7,336 guldens (10 percent of the full value of the property) for each of five years. Although the Abbot Mendel had been an outspoken supporter of the liberal government in Vienna, and had in the process alienated Church officials in the opposition Conservative Party, he was adamant in his opposition to the new tax, which he considered unjust and illegal. As a token of his goodwill, Mendel sent in a small "voluntary contribution" to the government that was promptly returned.[51]

In fact, even under the terms of the new law the assessment was excessive, because Mendel had not deducted the salaries paid to the abbot and the brothers from the monastery income, as was permitted under the new law, a fact happily acknowledged by the government. This would have significantly reduced the tax burden, and the deduction could have been legitimately increased by raising the compensation made to the monks. It was in fact, by such loopholes, that all the other monasteries managed to evade the bulk of the new tax. Although it was clear that the government was open to compromise, Mendel refused to relent as a matter of principle. The lord lieutenant, who was the local Brünn official in charge and up until this point had been on cordial relations with Mendel, now turned to the minister of culture and education in Vienna for assistance.

The problem for the authorities, who had shown themselves ready to negotiate behind the scenes, was that the government could not be seen to back down in an individual case on the enforcement of a law passed by Parliament and the Monarch, and the problem was compounded by the facts that Mendel played a prominent role in the life of Brünn and that the monastery was a major landholder. As might have been expected, the minister of culture and education reacted harshly, ordering the lord lieutenant, his local representative, to "enforce compliance" if Mendel refused to cooperate. In April 1876, police officers were dispatched to the monastery to seize the money from the monastery safe. When they arrived, Mendel refused to turn over the keys to the strongbox and informed them that they would have to take the key out of his pocket if they wanted it and use force to remove any property for sale.

Opting against the use of force, the exasperated lord lieutenant decided instead to place an embargo on the monastery revenues. By October 1877, the state had succeeded in collecting all the back taxes and was prepared to end the sequestration if Mendel agreed to pay the tax in the future, but Mendel refused. His fellow brethren, many of whom had egged him on in the early days of the struggle, were now ready to relent, but Mendel would not hear of it.[52]

For the next five years, the government continued to confiscate rents from the monastery's dairy farms and its sugar factory, and eventually sequestered one of its estates. In the meantime, Mendel continued to keep track of all sequestered funds, which he viewed as a loan to be repaid by the state with 5 percent interest. The authorities had now come to view Mendel as suffering from a "litigious mania," and the members of the monastery also began to spread rumors. Eventually Mendel withdrew and communicated only with one or two trusted younger priests. He turned increasingly for company to his two nephews, his beloved younger sister's sons, confiding to them that he was being persecuted and that there was a plan to send him to a lunatic asylum.[53]

On May 4, 1883, Mendel sent his last protest to the lord lieutenant, steadfastly sticking to his position. By July he had suffered a rapid deterioration in his health and was forced to turn over the administration of the monastery to a surrogate, Father Poje. For several months Poje deflected the demands for payment of taxes, explaining that Mendel was suffering from organic heart disease and "all emotional disturbance must be avoided." In September Poje sent a petition to the Ministry for Public Worship protesting the *amount* of the assessment, signaling the end of an era, and two months later, the ministry accepted the proposed reduction and in addition granted a retrospective reprieve.

Mendel died in his sleep on January 6, 1884. He was 61. His funeral was attended by a huge crowd of local peasants and the poor, who had always appreciated his humility and kindness, as well as a large number of local dignitaries, including the government officials who had darkened his final years. A fortnight later, the monastery acknowledged the "erroneous views" of Abbot Gregor Mendel and asked for a complete reprieve for the entire decade 1881–1890, which was partially granted two years later. The tax law was never repealed, but Moravian monasteries ended up contributing very little to the religious fund through various loopholes, including the deduction offered on salaries paid to monks.

Even as a schoolboy Mendel seemed to have a premonition of the fate that awaited him. In a poem describing Gutenberg and his invention, he wrote:

But unfading are the laurels of him
Who earnestly and zealously strives
To cultivate his mind
Who with the full light of his understanding
Seeks and finds the mysterious depths of knowledge,
Of him in whose development the germ
Of glorious discovery implants itself,
Nourishing him, and sending abundant blessing
To the thirsty crowd of mankind—
Yes, his laurels shall never fade,
Though time shall suck down by its vortex
Whole generations into the abyss,
Though naught but moss-grown fragments
Shall remain of the epoch
In which the genius appeared.

The poem proceeds to celebrate the immortality made possible by the invention of the printing press, and concludes with the following remarkably prescient wish:

May the might of destiny grant me
The supreme ecstasy of early joy,
The highest goal of earthly ecstasy,
That of seeing, when I arise from the tomb,
My art thriving peacefully
Among those who are to come after me.[54]

Although destiny would grant Mendel's wish, his final years on earth were plagued by misery that was largely of his own making. But there was something inevitable about the way the end played out. Mendel never wavered in his defense of what he considered to be "truth," whether it concerned a scientific principle or a civil right. In the case of his scientific theory, the misery that followed from steadfastly adhering to his own vision was the cost of making one of humankind's greatest discoveries.

**William Bateson.**

John Innes Archives care of the Trustees of the John Innes Foundation.

# CHAPTER 6

# Rediscovery

— DE VRIES'S FATHER DIED on March 4, 1900. It may have been the intense reminder of his own mortality, or the need to distract himself, or the premonition that others were about to publish and there was no time to lose, but whatever the cause, the death of his father marked the beginning of a period of frenetic activity. In the following two weeks he completed two long, detailed accounts of the basic principles that would come to be known as Mendelism, one in German and the other in French, and one short summary to be delivered to the French Academy of Sciences.[1] Because papers that were read before the French Academy were published quickly in the *Comptes Rendus,* the short summary was the first of De Vries's three accounts to appear, officially bringing to an end thirty-five years of neglect of Mendel's work.

Soon after it was published, De Vries sent a reprint of the *Comptes Rendus* paper to the German botanist Carl Correns. As it happened, Correns, who had no idea that anyone else was working on a similar problem, had been planning to publish his own paper on Mendel's laws and the progeny of hybrids, and it was a terrible jolt to find that De Vries had already published on the same topic. What made it even more difficult for Correns was that this was not the first time that De Vries had beaten him to the punch. In fact, the previous issue of *Comptes Rendus* contained an article by De Vries on a subject of great interest to both men, and De Vries had managed to publish first in that case as well. But this time was different, and the normally mild-mannered Correns later recalled the fury he felt when he discovered that De Vries's article did not contain a single reference to Mendel's paper, although it was absolutely clear that De Vries had read it.[2] This time, Correns was convinced, the attention-hungry De Vries had gone too far. Hoping to destroy the cheater once and for all, Correns set immediately to work. He received De Vries's reprint on Saturday morning, April 21, and mailed off his own finished article on the evening of the 22.

Even the title, "G. Mendel's Laws concerning the Behavior of Progeny of Varietal Hybrids," was meant to call attention to Mendel's priority and highlight De Vries's deception. "The same thing happened to me which now seems to be happening to de Vries," Correns wrote in the first paragraph.[3] "I thought that I had found something new. But then I convinced myself that the Abbot Gregor Mendel in Brünn, had, during the sixties, not only obtained the same result through extensive experiments with peas, which lasted many years, as did de Vries and I, but had also given exactly the same explanation, as far as that was possible in 1866."

Not content to set the record straight, Correns was determined to expose the magnitude of De Vries's duplicity, and he believed he had irrefutable proof of it. It was either the height of arrogance or stupidity, but De Vries had had the audacity to use Mendel's exact terminology for the alternative forms of a trait. When it came to describing the phenomenon of dominance in his own paper, Correns could not resist taking a dig: "This one may be called the dominating, the other one the recessive, anlage. Mendel named them in this way, and by a strange coincidence, de Vries now does likewise."

In a calmer state, Correns might have thought twice before committing himself in print to the implication that De Vries was a plagiarist. Even if De Vries, realizing that he had stumbled on the discovery of the century, had been willing to sacrifice all honor, he was not fool enough to believe it was possible to get away with the theft of Mendel's ideas. Despite the fact that Mendel's 1865 paper had received very little notice, the journal in which it had been published, the *Transactions of the Natural Science Society of Brünn,* had been sent to 120 libraries throughout the world, and in addition Mendel had himself sent out forty reprints, one of which De Vries himself had read. Focke's book, the standard reference on hybridization, had fifteen references to Mendel.[4] In fact, Correns himself had seen the reference in Focke's book. Furthermore, if De Vries had been hoping to get away with an outright theft, surely he would not have risked drawing attention to his crime by using Mendel's exact terminology.

On April 25, three days after Correns sent off his paper, De Vries's long German paper appeared in print. Although the short French paper appeared earlier, the longer German paper had been submitted on March 14, and thus it predated the *Comptes Rendus* paper. On the final page De Vries had written an acknowledgment of Mendel: "I draw the conclusion that the law of segregation of hybrids as discovered by Mendel for peas finds very general application in the plant kingdom." Correns, who must have felt somewhat chastened after reading De Vries's *Berichte* article, hastened to add a note in proof to his own. In the late added postscript, Correns grudgingly admitted, "He [De Vries] refers to Mendel's investigations, which were not even mentioned in the *Comptes Rendus.*" In De Vries's French article, which was submitted on March 19 and appeared after the German article and therefore had not been read by Correns, De Vries had been more generous in his praise of Mendel's paper, writing, "This memoir, very beautiful for its time, has been misunderstood and forgotten."[5]

While it was almost certainly unfair of Correns to imply that De Vries was attempting to steal Mendel's ideas without crediting him, he would have been perfectly justified in suspecting De Vries of exaggeration in his claim to be an independent discoverer of Mendelism. For it seems clear that De Vries hadn't discovered Mendel's laws independently, as he claimed, and then, after the fact, found that Mendel had preceded him by thirty-five years. Rather, De Vries had recognized that Mendel's binary model, in which each pangene, or "anlag" as Mendel called them, was present in exactly two copies, each of which was segregated into separate sex cells that were then reunited during the process of fertilization, represented a wonderful refinement of the models he'd been contemplating in 1896. In fact, he would have quickly understood that Mendel's model was a special case of the law of pangenes, the one he had referred to as the 1:2:1 law, and having seen how well the 1:2:1 law described traits in peas, he had immediately reinterpreted a slew of old data in the new light.

Twenty-five years later, Correns was asked to recall the circumstance surrounding his own independent rediscovery of Mendel's laws. In a letter, he described the sleepless night in the autumn of 1899 when he had the great insight of his life: "It came to me at once, like a flash, as I lay toward morning awake in bed, and let the results again run through my head. Even as little do I know now the date upon which I read Mendel's memoir for the first time; it was at all events a few weeks later."[6] As exciting as it had been for him personally, a philosophical Correns downplayed the significance of the exact circumstances of the rediscovery, observing, "It is an indifferent matter for science whether someone discovered the laws independently, or first read Mendel and then repeated the experiments." On the surface, it would seem that Correns was offering De Vries a form of absolution for his sins, and that the years had brought him a philosophical detachment.

For nearly a century, Correns's story has been viewed as an object lesson in what happens to honest players, but a reexamination of Correns's lab notebooks revealed that Correns too had been guilty of his own misrepresentation of the facts of his rediscovery.[7] Not only had he read Mendel's paper before 1900, but he had taken detailed notes on it, including an explanation of dominant and recessive traits and the 3:1 ratio among the progeny of hybrids. The page on which the notes were written was clearly dated April 16, 1896. In the end, then, it was not only De Vries who was in need of absolution.

— AS LEGEND HAS IT, William Bateson first read Mendel's paper on May 8, 1900, while taking the train from Cambridge to London, where he was to deliver a lecture that evening to the fellows of the Royal Horticultural Society. The story, as told in Beatrice Bateson's memoir of her husband, is meant to show the laser-like intensity with which Mendelism acted on Bateson's mind: "As a lecturer he was always cautious, suggesting rather than affirming his own convictions. So ready was he however for the simple Mendelian law that he at once incorporated it into his lecture."[8] According to his wife, Bateson had not only read and assimilated Mendel's paper, but he had rewritten his evening lecture to make it reflect the new Mendelian reality that would soon transform the world.

A short summary of Bateson's lecture that was written by the editor of the *Gardeners' Chronicle* and ran in the May 12 issue revealed that Bateson had not actually referred to Mendel in his lecture at all, but instead had mentioned "recent work of Professor de Vries." Furthermore, it is clear from the summary in the *Gardeners' Chronicle* that Bateson had only the faintest glimmer of the significance of what he took to be De Vries's theory of inheritance, and it was not a matter of a short train ride from Cambridge to London but nearly two years before Bateson came to fully appreciate the monumental significance of Mendel's achievement. Contrary to Beatrice Bateson's assertion, it was almost certainly De Vries's short *Comptes Rendus* paper that Bateson read on the train.[9]

On the May evening he gave his lecture, Bateson declared that the best insight into the laws of heredity was "almost wholly due to the work of Francis Galton." In particular he singled out Galton's law of ancestral inheritance based on the 1897 analysis of coat color in basset hounds as an example of the kind of hereditary law that one could hope for. Nonetheless, Galton's ancestral law was not universally true, Bateson insisted, and in particular he pointed out that it failed for traits that showed dominance, where, as he put it in the pre-Mendel language, one parental type "wholly predominates the first generation of crossing."[10] The significance of De Vries's new result, as he explained that night, was that it showed that Galton's law might be more generally applicable than had previously been supposed. In particular, he saw that the 3:1 ratio of parent types in the second generation might be consistent with a "modified" version of Galton's law.

On October 10, 1900, De Vries sent Bateson an early printing of the first installment of his new work, *Mutationstheorie,* writing, "I have now the pleasure of offering you my work on the origin of species, as discontinuous as you could hope it."[11] It would end up being the first of a two-volume, 1,500-page treatise, published in 1901 and 1903. In his letter, De Vries expressed his great disappointment in the poor reception that his theory of intracellular pangenesis had gotten in England, but he had high hopes for the new book. "I hope that you will see in Mutationstheorie proofs for the Pangenesis, and that I will succeed in persuading you," he implored, "that your discontinuity and Darwin's pangenesis are founded on exactly the same principle." Almost as if to reassure himself, he added, "I feel quite sure that Darwin, if he could have read my little book, he would have approved of my conception of his Pangenesis."

While it would not seem necessary to defend the role of discontinuous variation to Bateson, who was, after all, a passionate discontinuitarian, De Vries might well have been anxious about Bateson's growing infatuation with Mendelism. Although he didn't confront the issue straight on in this first letter, De Vries did try to prepare Bateson for the fact that he had grave doubts about the importance of Mendel's results. In what could be read as a casual aside, but was in fact a matter of crucial importance, De Vries made reference to the behavior of "false hybrids," first pointed out by the French hybridist Millardet. In contrast to Mendel's crossbred peas, Millardet had found that hybrids formed from different varieties of strawberries often remained stable for generations, expressing either the paternal or the maternal traits. This phenomenon, De Vries wrote Bateson, was "more common than could be deduced of [*sic*] Millardet's paper," and he had "found other cases which clearly show that Millardet is as right as Mendel is."

Two week later, De Vries wrote again, once again emphasizing the similarity of their views on discontinuous variation, insisting that "there must be no discontinuity between us, not even in the use of the word."[12] He hoped that Bateson and he could agree that discontinuous variations were identical to what he now called "mutations." In response to an inquiry of Bateson's concerning the behavior of hybrids, De Vries returned to the contrast between false hybrids and what he proposed to call "true hybrids" like those studied by Mendel. "Millardet's hybrids are not the same cases as those of Mendel," he declared. To avoid any possible confusion that may have resulted from his less than perfect command of English, he found three different ways of saying the same thing: "The Millardet hybrids never can be 'disjoints' (= split up?); all their seeds ever give the same type as they bear themselves. Mendel's hybrids of the first generation *always* split up in the next."

Though he had resisted spelling out his observations with *Oenothera* in his first letter, rightly fearing that they would arouse the suspicions of the highly skeptical Bateson, De Vries now came forward with his new results that were about to appear in print. "There are very nice Millardet hybrids in the genus *Oenothera,* which I have cultivated for some generations," he wrote, adding that "false hybrids is a bad name; they are as numerous as true or Mendel hybrids." Lastly, he insisted that the evidence from his crosses was "absolutely opposed to Mendel's law."

Although he had been aware of the unusual behavior of his *Oenothera* crosses for many years, De Vries published his results only in November 1900 in the same German journal in which he'd published his first full treatment of Mendelism seven months earlier.[13] As he'd warned Bateson, De Vries reported on a number of anti-Mendelian behaviors of his *Oenothera* crosses, beginning with Millardet-like false crosses. Already in 1894 he crossed *Oenothera lamarckiana* with the pollen from the closely related variety *Oenothera biennis,* and found that the resulting progeny resembled the male parent and bred true when self-fertilized. And the same pattern was repeated when *O. lamarckiana* was crossed with *O. muriata,* another closely related variety, and again when *O. muriata* was crossed with *O. biennis.*

But the behavior of the *Oenothera* was even stranger than De Vries admitted to Bateson. In the summer of 1898 he found that the dwarf variety *O. nanella* when pollinated by *O. lamarckiana* gave two types of progeny, some like the mother and some like the father, and the resulting progeny had each bred true, and the same was true for *O. lata* crossed with *O. lamarckiana* and for *O. rubrinervis* crossed with *O. nanella.*[14] In fact, so long as the *O. lamarckiana,* or one of its mutant forms (*nanella*), was the pollen parent, hybrids made from a large number of different females split into two types, one closely resembling the *O. lamarckiana* male parent and the other the mother. Later, in 1907, he would call the result of such crosses "twin hybrids" and he would even give them each names, "laeta" for the smooth-leaved, *O. lamarckiana*-like class, and "velutina" for the other.[15]

As De Vries pointed out to Bateson, Mendel had seen the very same phenomenon in *Hieracium,* in which the immediate cross of two pure-breeding forms gave not one but two hybrid types and each of the types bred true when self-fertilized. But Mendel had not pursued the *Hieracium* research far enough, De Vries insisted, and as a result he had not determined whether peas or *Hieracium* illustrated the more general principle. "It was for this reason," De Vries now argued, "that his [*Hieracium*] results were regarded, until recently, as isolated cases without an important principle being involved therein, and they consequently lapsed into oblivion."[16]

That winter a much-revised version of Bateson's May 8 lecture to the Royal Horticulture Society appeared in print. By then it was clear that he had read De Vries's first German paper, which had led him to Mendel's paper, as well as well as the paper by Correns. Six months after his first exposure to Mendel, Bateson was now convinced that Mendel's hereditary theory was an independent contribution of significance comparable to, or perhaps even greater than, Galton's. "That we are in the presence of a new principle of the highest importance is, I think, manifest," he wrote.[17] Furthermore he asserted that Mendel's results would "certainly play a conspicuous part in all future discussion of evolutionary problems" and that he no longer believed that Mendel's result might be a special case of Galton's law of ancestral inheritance.

Bateson now believed that the heart of Mendel's theory was the notion that there was a "perfect purity in the reproductive cells," as he referred to Mendel's assumption that the egg and sperm cells produced by hybrids each contained only one of the alternative characters. But he still insisted that the purity of the gametes was only a working hypothesis, and that further experiment was required before it could be elevated to the level of established fact. The most troubling piece of contrary evidence, as he saw it, was that the characters found in hybrids produced by the combination of allegedly pure sex cells showed a wide range of variability. "Be this, however, as it may," he wrote, "there is no doubt we are beginning to get new lights of a most valuable kind on the nature of heredity and the laws which it obeys."[18]

Meanwhile Bateson's lab notebooks from 1901 show that he conducted three times as many crosses as in the previous year, and they also reflect the fact that he had changed his approach, now routinely examining the grandchild generation instead of simply looking at plants from the first generation as had been his practice prior to reading Mendel.[19] As the months passed, Bateson grew ever more certain that Mendel was right, and ever more doubtful about De Vries's work. In a letter to De Vries, Bateson raised serious doubts about the interpretation of his *Oenothera* crosses, suggesting that *O. lamarckiana* might not be a pure-breeding strain. Furthermore, while he granted that there was little doubt that new characters arose just as De Vries had suggested by mutations of individual pangenes, Bateson raised the possibility that De Vries's alleged "mutations" of *Oenothera* might in fact be the result of unintended crossing.[20]

De Vries, who had a remarkable ability to tune out criticism, was hardly fazed by Bateson's comments and had somehow managed to persuade himself that Bateson would soon come to understand things his way. He was now firmly convinced that mutations that changed a pangene from an active to a latent state (dominant to recessive, in Mendel's terminology) were merely "derivative cases," and that the same was true for mutations that changed inactive pangenes to active pangenes. For crosses involving these special kinds of mutations De Vries granted that Mendel's laws held; however, these kind of changes explained only the small and relatively insignificant differences between closely related varieties[21] and were therefore of limited interest in the grand scheme of things. The key to the whole puzzle was provided by the behavior of his *O. lamarckiana* mutations. In the process of composing the second half of his book, De Vries explained to Bateson, "it becomes more and more clear to me that Mendelism is an exception to the general rules of crossing." The more general rule of crossing, he was convinced, was exemplified by his new *Oenothera* hybrids that remained constant. "My Oenothera, and other Oenothera too, don't Mendelize at all," he wrote.[22]

That winter, in December 1901, Bateson delivered the first of five reports to the Royal Society's Evolution Committee. With Mendel's law he now boldly declared the "whole problem of heredity has undergone a complete revolution." He had now decided that the claim that the sex cells produced by hybrids were pure with respect to each of the parental characters was no longer a hypothesis to be tested, but rather "the central fact proved by Mendel's work."[23] In this paper he introduced the now-standard terminology for the new field that he would soon dub *genetics.* In place of the traditional notion of a *unit character,* he proposed to call each member of a pair of antagonistic characters an *allelomorph* for "like shape."[24] He further adapted the existing term *zygote,* which had been first used in 1891 to describe the cell formed from the union of two germ cells, to reflect the new Mendelian insight, proposing to call zygotes that were formed from germ cells containing identical allelomorphs "homozygotes," and those from germ cells containing contrasting characters "heterozygotes."[25]

Written in the first flush of the rediscovery, Bateson's 1900 first report to the Royal Horticultural Society had given only very brief consideration to the principle of independence of traits, but in his 1901 lecture, he reported on an exception to the independence rule found by Miss Saunders in her crosses of *Matthiola:* in plants that were dihybrids for seed color and character of the leaves (either smooth or hairy) there was sometimes a close correlation between the green seeds and hairy leaves, on the one hand, and brown seeds and smooth leaves, on the other.[26] The importance of this observation did not escape Bateson, who proposed a theory of coupling (or repulsion) between certain alleles to explain the nonrandom segregation of factors for different characters.[27]

— IN FEBRUARY 1902, just as he had begun to see that the spreading of the gospel of Mendelism would be his lifework, Bateson received a copy of the second issue of *Biometrika,* the new journal published by Karl Pearson and Bateson's former friend turned bitter antagonist, W. F. R. Weldon. The journal, which had been sent by Pearson, contained an article by Weldon. With a self-confidence bordering on arrogance and a dismissiveness that was all too familiar to Bateson, Weldon picked apart Mendel's arguments, attempting to undermine the notion that Mendel's laws had any claim to be a universal mechanism of heredity or that they even held for the characters of peas.

In his article, Weldon pointed to the "grave discrepancy" between Mendel's results and the work of others "equally competent and trustworthy." Indeed, it was nearly inconceivable, Weldon argued, that Mendel could have discovered a universal law of heredity that had escaped the notice of Knight and Laxton, "the great founders of our modern races of Peas," as Weldon described them. However, it was not simply sloppiness, lack of rigor in his experimental technique, or possible dishonesty that had prevented Mendel from arriving at the truth, but his failure to take into account Galton and Pearson's law of ancestral inheritance. As Weldon explained, "the condition of an animal does not as a rule depend upon the condition of any one pair of ancestors alone, but in varying degrees upon the condition of all its ancestors in every past generation, the condition in each of the half-dozen nearest generations having a quite sensible effect." Mendel was simply wrong, Weldon insisted, to see the "whole effect upon offspring, produced by a parent, as due to the existence in the parent of particular structural characters."[28]

Not for the first time, Bateson felt, Weldon was trying to stop him dead in his tracks. In a fury, Bateson wrote Pearson, who had sent the journal in the first place, informing him of his intention "to take the earliest opportunity of dealing with the article appearing under the name of Professor Weldon."[29] When Pearson expressed dismay at the "truculent tone" of Bateson's letter, Bateson hastened to make amends, writing, "I respect you as an honest man and perhaps the ablest and hardest worker I have met, and I am determined not to take up a quarrel with you if I can help it." It was his fondest wish to be able to sit down and share the pleasure in the "beautiful phenomena" that were being uncovered. "If through any fault of mine you were to be permanently alienated from the work that is coming," Bateson wrote, "I should always regret it."

But the situation with Weldon was entirely different. Bateson had felt "disgusted" when he read Weldon's article, he confessed to Pearson, exasperated "that he should so utterly have missed the point of Mendel and mangled a great man's work beyond recognition." As a biologist and naturalist, Weldon ought to have known better than to dismiss Mendel. Not only had Weldon misunderstood Mendel's work, Bateson added, but he had vastly underestimated its significance, and Bateson predicted that the "gravity of the issue cannot be long unknown." That Mendelism ranked amongst the most important discoveries in the history of theoretical biology, he flatly asserted, was "not a matter of opinion but certain."

A few days later Pearson wrote to Bateson, assuring him that he had an open mind about Mendelism but that his loyalty to Weldon made it difficult for him to accept some of the things that Bateson had said about him. Bateson must remember, Pearson warned, that his defense of his friend was unlimited: "I have tested the strength of his affection in the graver matters of life, and am prepared to do for him and to accept from him anything that one human being can or will do for another."[30] Nonetheless, Pearson offered to save space in the next issue of *Biometrika* for Bateson's defense of Mendel, which Bateson promptly declined. He had a far better option: Cambridge University Press had agreed to publish a short monograph that would be entitled *Mendel's Principles of Heredity: A Defence.*

As had been the case at several critical junctures in his career, Weldon's criticisms served to focus and clarify Bateson's thinking. By the end of March 1902, Bateson had completed his 212-page monograph as well as a 22-page addendum to the text of the *Report to the Evolution Committee.* While the report, and particularly the late-added addendum, focused for the most part on new developments in Mendelism, the monograph focused more directly on Weldon's critique of Mendel. At the suggestion of his Cambridge editor, he had toned down a few of the more vitriolic passages in the preface. While the editor allowed that it was fair to say that it was with "regret approaching to indignation" that he had read Weldon's treatment of Mendel, he prevailed on him to cut. "Having no practical acquaintance with the principles he questioned, Professor Weldon thinks fit to [gloss] this great discovery with that deadly mixture of mild contempt and faintest praise that may blight the strongest growth."[31] He was also persuaded to eliminate his characterization of Weldon's attitude as one of "frigid courtesy of satisfied disdain."

Bateson's review of the development of the law of ancestral inheritance, which was the centerpiece of the biometrical view of inheritance, was particularly perceptive. As it was first introduced in Galton's *Natural Inheri-*

*tance* and later applied to basset hound coat color, Bateson acknowledged, the law seemed to work remarkably well. However, Pearson had given Galton's law a major overhaul in 1898, so that it now allowed for the effects of natural selection to alter the mean of a character as it was passed down the generations. In his paper, Pearson downplayed the significance of this change, which was very much at odds with Galton's views on the discontinuity of evolution. Instead, he emphasized the fact that the modified law still applied to discrete or "alternate" inheritance, as exemplified by the case of tricolor or "lemon and white" coat color in basset hounds that Galton had first studied, as well as to continuous characters like human stature. "On the whole," Pearson had written, "the confirmation obtained from stature data for the law of ancestral heredity is very striking; I am inclined to think even more convincing than obtainable from the basset hounds."[32] On January 1, 1900, Pearson published a revision of the already revised ancestral law of inheritance, which was, like the first paper on the same subject, dedicated to Galton. In his new paper Pearson backtracked, claiming that his 1898 law had applied only to the case of blending inheritance but not to the discrete case.[33]

Although mathematics had always been his weakest subject, Bateson did not fail to appreciate that there was something soft about Pearson's so-called hard mathematical analysis, pointing out that it was the discrete case of coat color inheritance that the ancestral law had so wonderfully explained in the first place.[34] While he had allowed that the law might still apply in cases of Mendelian inheritance,[35] Bateson could not resist satirizing Pearson's equivocations, writing: "The Law of Ancestral Heredity after the glorious launch in 1898 has been brought home for a complete refit. The top-hamper is cut down and the vessel altogether more manageable; indeed she looks trimmed for most weathers."[36]

Later that year, Pearson's brilliant disciple G. Udny Yule pointed out that there was no fundamental contradiction between the two points of view. While strictly speaking it was true that it was only the *particular structural characters,* or genes, as they would soon be known, possessed by the parents that could affect the offspring, it was also true that an individual's ancestry provided the best estimate of that individual's genetic composition. Yule's insight fell on deaf ears.[37]

**Karl Pearson and Francis Galton, age 87.**

From Karl Pearson, *The Life of Francis Galton,* vol. 3A (Cambridge: Cambridge University Press, 1930), plate 36, after p. 352.

CHAPTER 7

# Mendel Wars

— ALTHOUGH HE WORRIED about De Vries, who seemed to be spinning off in a strange new direction with his *Oenothera,* Bateson's main concern was Weldon, who appeared to be increasingly determined to enter a full-fledged war against Mendelism. The first attack in the February 1902 *Biometrika* was followed by a short breather, but in November Weldon struck again.

This time Weldon's critique revolved around the Mendelian analysis of the inheritance of the hairy and smooth varieties of the flowering perennial *Lychnis,* which was an ideal target for his criticism in part because both De Vries and Bateson had studied it. In fact the crosses of the two forms of *Lychnis* were among De Vries's first hybridization experiments, and his 1897 report on them had so impressed Bateson that he and Miss Saunders had repeated the crosses themselves in 1898. Both Bateson and De Vries had mentioned the *Lychnis* results in their lectures at the First International

Conference on Hybridization in the summer of 1899, a few months before De Vries was sent Mendel's paper.

Several different aspects of the *Lychnis* story attracted Weldon's attention, not least of which was his discovery that De Vries had changed his account of the results. As Weldon explained in his paper, De Vries had first reported a 2:1 ratio of hairy to hairless types in 1897. When De Vries presented the same experiment again in March 1900 (in the first of his Mendel papers), he reported that 28 percent of the $F_2$ plants were hairless, a better fit with the 3:1 ratio predicted by the Mendelian theory. Not only did the facts suggest that De Vries was an unreliable character, Weldon implied, but they presented clear evidence against the Mendelian theory of segregation. Assuming that 500 plants were observed, Weldon computed that the odds of obtaining a 2:1 ratio due strictly to random fluctuation were 17 to 1 against—that is, for any experiment for which the true expectation was 3:1, the odds that the experiment would yield a 2:1 ratio were 1 in 17. Weldon also reproached Bateson and Saunders for confirming that the *Lychnis* cross obeyed Mendelian expectations and being willing to "pass over the deviation from Mendel's law, observed during its earlier history, without notice."[1]

The real history, however, was considerably more complex. In fact, De Vries had written about the *Lychnis* crosses in several different papers, and in each case he appears to be guilty of some after-the-fact reinterpretation of his conclusions. However, Weldon had done De Vries an injustice by implying that he had concealed data or altered it. The remarkable thing was that even while he interpreted his data as supporting a 3:1 ratio, De Vries published the raw data from the *Lychnis* crosses—99 hairy to 54 hairless plants—that clearly contradicted his conclusion. The odds that a particular experiment would yield a result this extreme were 625 to 1 against. Ironically, Weldon had vastly underestimated the true odds against seeing data like De Vries's. Assuming that De Vries had observed 500 plants (as Weldon had assumed), one would expect to see a ratio as far from expectations as 2:1 in fewer than 1 in 119,000 experiments.[2]

Although Weldon went to considerable lengths to undermine De Vries and the argument for Mendelism, his primary interest was in demonstrating that the Mendelians in general, and Bateson in particular, had failed to understand the proper mode of scientific inquiry, which involved statistical analyses of large quantities of data. It was not enough to claim that one plant was hairy while another was not, Weldon insisted; rather, what was required was a quantitative study of the character of hairiness. To this end, Weldon and an assistant had used a microscope to count the hairs on the leaves of more than a thousand plants of each of the two varieties of *Lychnis*. Afterward, following proper scientific procedure, Weldon had then computed the mean, mode, standard deviation, and other constants associated with the distribution of hairiness, and by this means he documented the degree of variation within both the hairy *diurna* and the hairless *vespertina* varieties. The inescapable conclusion of the study, Weldon declared, was that the Mendelian character of hairiness was entirely ambiguous and in fact utterly meaningless. Even among the hairy parents, he noted, "it is possible to pass by a series of small steps from the glabrous condition through individuals with various numbers of hairs per square centimeter . . . up to a condition of very great hairiness." Thus, Weldon wrote, "the statements made by Mr. Bateson and Miss Saunders are seen to be utterly inadequate either as a description of their own experiments, or as a demonstration of Mendel's or of any other laws."[3]

An ordinary reader would have been hard-pressed to find the flaw in Weldon's reasoning. But the proof of the truth of Mendelian theory would ultimately be found in its ability to explain and unify a vast array of divergent phenomena, and when it came to particular cases a determined critic could usually find a variety of flaws and inconsistencies.

For the time being, it seemed, Weldon had the upper hand, and he wasn't about to let his opponent have a breather. Along with his own resoundingly critical article, the November 1902 issue of *Biometrika* contained the first of four anti-Mendelian articles by Weldon's student A. D. Darbishire on the inheritance of coat color in mice. The article reported on the results of crosses of albino with Japanese waltzing mice, which had a white coat with yellow patches. The key observation, according to Darbishire, was that the hybrid progeny, which were all a combination of grey and white, showed a great variability in the distribution of the color: some had small patches of color on the cheeks, shoulders, and rump in the pattern characteristic of the waltzing parent, while others were almost entirely grey, indistinguishable from common house mice except for a small white patch on the belly. Darbishire made no mention of the inheritance of eye color or of the strange jerking motions for which the Japanese waltzers were named.

Bateson smelled a rat, correctly guessing that Darbishire, under Weldon's guidance, had withheld some interesting facts, and he immediately applied himself to uncovering them. "I have today seen your interesting note on cross-bred mice," Bateson wrote Darbishire. "Can you kindly tell me 1. the eye colour of the cross-breds and 2. whether they 'waltzed'"?[4]

Darbishire wrote back several days later, providing the information that the hybrids had dark eyes while both of their true-breeding parents were pink-eyed, and that there were no waltzers among the hybrids. "I am absolutely unbiased about Mendel and am very keen to come to an unprejudiced conclusion on it," he added. Although there was something a little peculiar in the fact that Darbishire had felt the need to proclaim his scientific objectivity, Bateson took his statement at face value, writing back the same day to suggest that the eye color of the hybrids was likely to be of critical importance in the interpretation of the case, and that he had been surprised to learn that Darbishire had not reported it in his first paper.

In February 1903, Darbishire's second paper containing the results from crosses of hybrids appeared in *Biometrika*. Among 66 grandchildren, Darbishire noted 13 were pink-eyed albinos like their albino grandparent and 17 were pink-eyed and pigmented like the waltzer grandparent. He did not mention the remaining 36 mice, which were all dark-eyed and pigmented to varying degrees. Because the cross of two pink-eyed parents gave black eyes, Darbishire claimed pink eyes could not be dominant. The paper ended on a belligerent note: "The behavior of eye-colours is in every respect discordant with Mendel's results."[5]

"The purpose of the allusion to 'dominance' escapes me," Bateson replied in the March 19 issue of *Nature*.[6] "In what circumstances could pink-eye be dominant, or recessive, to pink-eye?" As he had indicated to Darbishire in his letter after the publication of the first paper, he was quite certain that the mystery of the eye color might provide a clue to the case, which he now spelled out. Leaving aside the tricky issue of why pink-eyed parents produced black-eyed progeny, he pointed out that the cross could nonetheless be seen to fit the Mendelian predictions. In fact, the distribution of grandchildren types—13 pink-eyed albinos (like the albino grandparent) to 36 pigmented types with dark eyes (like their hybrid parents) to 17 pink-eyed pigmented types (like the waltzer grandparent)—was a strikingly good fit for the 1:2:1 (or 16.5 to 33.0 to 16.5) distribution expected from self-mating hybrids. Dropping any pretense that Darbishire had a say in the interpretation of his results, Weldon replied two weeks later, now challenging Bateson to explain why roughly a third of the pink-eyed pigmented grandchildren had a "lilac" coat color and resembled neither true-breeding grandparent.[7]

*Nature,* which had now become the new public forum on Mendelism, ran Bateson's next letter on April 23. Debating these questions with Weldon, Bateson quipped, "is like discussing the perturbations of Uranus with a philosopher who denies that planets have orbits," but he nonetheless acknowledged that there were "genuine difficulties" of interpretation associated with the appearance of the lilacs.[8] One way to account for the appearance of new colors among the grandchildren was to allow for the existence of compound Mendelian factors that split into new simple factors in the progeny of the hybrids ($F_2$), an idea Bateson had suggested in his report to the Evolution Committee of March 1902 but would soon abandon.

Bateson's second suggestion, however, amounted to an entirely new framework in which to think about the variability that had now been observed in the segregation of many Mendelian characters. Certain traits like coat color, Bateson speculated, might be influenced by factors other than the "visible colors of the parents," and these secondary factors might segregate independently of the primary factors. Thus rather than two classes of gametes produced by the pure-breeding parents, there would be a "heterogeneity of gametes," each carrying a different set of auxiliary factors that subtly altered the expression of the primary color factors.

Weldon promptly dismissed the existence of auxiliary factors, as "a doctrine too amazing for brief treatment,"[9] and instead focused his next *Nature* letter on attacking the far more vulnerable theory of compound factors.

The final exchange, which appeared on May 14, again consisted of Bateson's efforts to explain ambiguous and difficult phenomena and Weldon's "amazed" incredulity. This time, however, Weldon proposed that the correspondence cease until the publication of further results, and the *Nature* editors agreed, declaring: "This correspondence must now cease."[10]

Darbishire's third article appeared two weeks later in the June 1903 *Biometrika.* As in his first article, the subject was the variability of the pigmentation in hybrids resulting from the cross between waltzers and albinos. There was a new twist, however. Rather than crosses between a pure-breeding waltzer and a pure-breeding albino, which had been procured from professional breeders and presumably descended from a long line of pure-breeding ancestors, Darbishire now replaced the pure-breeding albino parent with albinos of two different parentages. In the first series of crosses, the albino parent was descended from a cross between a highly pigmented heterozygote and a pure-breeding albino, and in the second series the albino parent was the offspring of a lightly pigmented heterozygote and a pure-breeding albino. Darbishire reported that there was a nearly perfect correlation between the characteristics of the pigmented grandparent and the hybrid grandchildren, disposing entirely of "any theory which involves the gametic purity of the recessive parent."[11] The possibility that gametes might contain a heterogeneous mix of auxiliary factors, which Bateson had suggested in his letters to *Nature,* was not even considered. A harbinger of trouble to come for the biometricians, however, was contained in a brief reference to results of the Harvard biologist William Ernest Castle, who had been conducting his own extensive studies on the inheritance of coat colors in mice and rabbits and concluded, on the basis of Darbishire's first two papers, that his "interpretation of his results is clearly unsound."[12]

In addition to Darbishire's article, the June 1903 *Biometrika* contained Weldon's third and most scathing attack so far, entitled "Mr. Bateson's Revisions of Mendel's Theory of Heredity." With hindsight, it is remarkable how nearly exactly wrong Weldon was about each of the major issues he raised. As he had in each of his previous articles, he attacked Bateson for insisting that dominance could be "defective or irregular," although it was already apparent that many, if not most, alternative characters were inconstant in their expression.[13] Secondly, Weldon bitterly objected to the idea that "reversion to remote ancestors" might have a Mendelian explanation. And lastly, winning the trifecta of bad judgment, Weldon was utterly contemptuous of Bateson's suggestion that certain characters might be sex-linked, a phenomenon that would soon prove to be the key to unraveling the connection between Mendelism and the chromosome theory, which was even then rapidly becoming the hottest new topic in biology.

The following month, Darbishire, who was preparing to leave Oxford for an independent position as a zoology demonstrator at Victoria University of Manchester, sounded his first note of contrition about his behavior in a letter to Bateson. "I am very sensible of the kindness you have shown me," he wrote, "and I wish I could feel that I had done something to merit it."[14] In the same letter, he informed Bateson that he was preparing to publish what would turn out to be his final paper in *Biometrika*.

Darbishire's fourth and final *Biometrika* article was published in January 1904. In it, he insisted that waltzing did not behave as a classic recessive, reporting that only 97 out of 555 progeny of albino-by-waltzer hybrids waltzed, a deviation from Mendelian expectation (138 of 555) that he claimed would occur by chance fewer then once in 50,000 trials. However, he was at long last forced to concede that the inheritance of albino coat color among the progeny of albino-by-waltzer hybrids supported a "purely Mendelian theory of albinism regarded as a simple unit character."[15] Because he had expended much ink in claiming the opposite, this was a large concession, but Weldon had hardly given up the fight. On the contrary, he was now convinced that Darbishire's latest results would eclipse all previous doubts and settle the debate once and for all in favor of the ancestrians.

To his credit, Weldon had finally come up with a direct and elegant experimental test of the effect of the grandparental generation on the outcome of a cross between hybrids that were alleged to be heterozygous for the pigment gene. The new experiment, which was a refinement of the design introduced in Darbishire's third paper, involved monitoring the frequency of albinos among the progeny of hybrids of varying ancestry. In the new design, three different types of matings of hybrids with differing num-

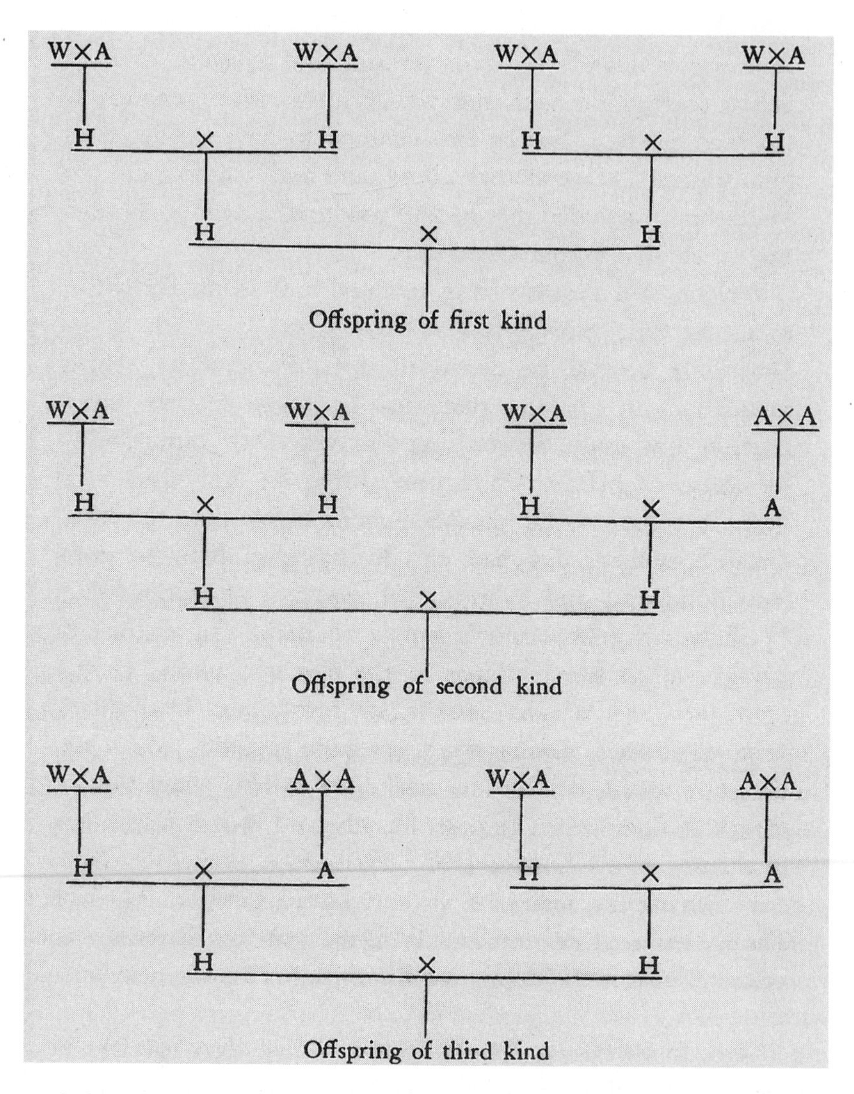


**Darbishire's illustration showing the varying ancestry of the grandparental generation.** A. D. Darbishire, "On the Result of Crossing Japanese Waltzing with Albino Mice," *Biometrika* 3, no.1 (1904): 24–25.

bers of albino parents were considered.[16] In the first series of crosses all four member of the grandparental generation were hybrids, products of crosses between a pure-breeding pigmented mouse (W) and an albino (A). In the second series, one of the four grandparents was a pure albino and the other three were hybrids as in the first series; and in the third series, two of the four grandparents were pure albino and two were hybrids.

The ancestral and Mendelian theories made two entirely contradictory predictions. Under Mendelian assumptions, a cross between hybrid parents, each of whom was heterozygous for the pigment gene, would be expected to produce a 3:1 ratio of pigmented to albino offspring, regardless of the fraction of albinos in the grandparental generation. Under the ancestral view, the greater the number of albinos among the grandparents, the higher the percentage of albinos that would be expected among the grandchildren. The data showed a clear increase in the percentage of albino grandchildren from 10.75 percent to 18.69 percent to 24.79 percent, corresponding to having zero, one, or two albino grandparents. The result, the paper triumphantly asserted, was a decisive confirmation of the ancestral view and a repudiation of Mendelism.[17]

Bateson read Darbishire's *Biometrika* article in the last week of March 1904 and found it "about the most insolent yet published," as he put it in a letter. After devoting two days to the analysis of the "tangle" of facts, Bateson concluded that the paper was replete with errors and inconsistencies.[18] Not only did some of the data on crosses presented in the fourth paper contradict the data reported in earlier papers, but also the data within the fourth paper was itself internally inconsistent. Hoping to avoid a protracted argument over the factual basis of the interpretations, Bateson wrote a carefully worded letter to Darbishire in which he proposed that they meet in person to clear up "a number of essential questions of fact."[19]

But Darbishire was now in an impossible bind. If he met with Bateson, he risked further alienating Weldon and Pearson, on whom his future career depended, and if he ignored Bateson, and the increasingly obvious fact that Mendelism was a reality that could not be ignored, he risked sacrificing whatever vestige of self-respect he may have possessed. The best he could do, he explained to Bateson, would be to answer Bateson's queries in writing, and that way both sides would have access to the same information. He hastened to assure Bateson that his ideas about Mendelism were "far from being fixed by prejudice" and "not yet completely formed," and he begged that they might remain on good terms when this was over and hoped it would be possible to "clear up the difficulties and then start a discussion in a friendly spirit."[20] A few years earlier such an accommodation might have been possible, but the rift between the two sides was now too large. Darbishire's paper was now only a new weapon in a much larger war. Bateson and Weldon, who had once been best friends, now considered each other the worst of enemies, and it was clear neither would give up until one or the other was totally defeated or, as it would turn out, dead.

In April, Bateson wrote expressing his sincere regret that Darbishire found it impossible to meet in person. This, he remained convinced, would have been the only way to arrive at a reliable record of the experiments. Nonetheless he enclosed a list of "preliminary questions" in which he noted discrepancies between the way 29 individual crosses yielding several hundred progeny had been reported in the fourth paper and the way they had been reported in the previous three.[21]

Nearly three weeks later, Darbishire provided a thorough exegesis of the 29 inconsistent crosses. His letter made clear that the inconsistencies were a result of his inexact method of scoring crosses rather than of intentional fraud, but at the same time, as Bateson wrote in his reply, his explanation had raised new questions. In particular, it was now clear that the "List of all Mice Bred during the Experiment," which had appeared as a table in the fourth paper and presented the results of more than 300 crosses producing more than 2,000 progeny, was far from complete. "It would be a great convenience," Bateson observed, "if the whole of the facts could be made available."[22] With a complete data set, he could begin the arduous task of reconciling the results from the earlier papers with those of the final one, as well as clear up a large number of internal inconsistencies in the fourth paper. A few weeks later Darbishire wrote back, providing the additional information that Bateson had requested and at the same time pleading with Bateson not to publicly discredit him.[23]

Bateson sensed that the moment was now ripe for winning Darbishire to the cause, and his reply was a small masterpiece of persuasion. He began by reminding Darbishire of his deep regret that he had rejected his offer to meet in person to straighten out the data, and at the same time assured him that he fully understood his feeling in declining it. After thoroughly reviewing the evidence, Bateson continued, he had reluctantly come to the conclusion "that to attempt to unravel the facts by correspondence would be an almost interminable task." However, he was careful to hold out some hope that the affair might not end nearly as badly as Darbishire feared. In fact, he hinted, if Darbishire handled himself properly, the entire episode might bring him considerable glory: "There are occasional evidences of orderliness in the results of the experiments, which suggest that after thorough revision and analysis, indications of positive results of great significance may well be found."[24] As for going public, Bateson was noncommittal, but even on this point he revealed that there was room for negotiation. His decision, Bateson explained, would have to depend on the degree of importance that was attributed to the paper. In the meantime, the only concrete help he could offer was to renew the offer to go over the records in person. Taken as a whole, it made a nearly irresistible case to induce Darbishire to meet and make the final break with Weldon.

Darbishire wrote for the final time on May 27, expressing the wish that the letter could remain "a private one between man and man."[25] It was not possible for Bateson to understand the "exact reason" he had been unable to meet in person to go over the records, Darbishire explained, hinting that Weldon had applied pressure. In desperation, he suggested a secret meeting at Cambridge and inquired, "Do you think that would be fair? Or even possible?" Alternatively, he pleaded with Bateson to simply ignore the paper until the two of them had a chance to review the data together, or to allow him to rewrite all the tables, this time including a complete and accurate account of all crosses. "To have my records discredited," he wrote, "would be heart-breaking and render it useless and a waste of time for me to go on with the costly experiments I am carrying out now."

Bateson wrote back a few days later to say that although he appreciated Darbishire's dilemma, he saw nothing he could add to his previous letters: he could not agree to ignore the paper or to allow Darbishire, who had already proven himself to be capable of blind adherence to Weldon's wishes, to undertake the revision of the tables unsupervised. "As for meeting in secret," Bateson wrote, "it will, I think, be obvious to you on reflexion, that any communication between us which is to serve as a basis of discussion must be of a public nature."[26]

The height of the public controversy over Mendelism was reached that summer at the meeting of the British Association for the Advancement of Science when Bateson and Weldon came face to face. As president of the Zoology Division, Bateson was for the first time on the offensive, and he had spent the better part of the June preparing the traditional presidential address in order to take full advantage of his position.

He attacked the conservatism of the British scientific community, which had utterly failed to understand the significance of Mendel. "Had a discovery comparable in magnitude with that of Mendel been announced in physics or in chemistry," he wrote, "it would at once have been repeated and extended in every great scientific school throughout the world." Returning to a theme he'd been sounding for the past fifteen years, Bateson attacked the state of contemporary Darwinism, proclaiming that research on evolution, which ought to have begun with the publication of the *Origin of Species* had in fact essentially ended there. The field was stultifying under the oppression of the "Dynasty of Selection," he insisted. It was the height of folly to assume that every feature of organisms be regarded as the result of a slow and gradual transformation; and the attempt to prove the reality of selection as a factor in evolution—as Weldon had spent the bulk of his adult life doing—"was a work of supererogation." The purpose of selection was, Bateson insisted, "to select but not to create."[27]

The real players in evolution, he insisted, were discontinuous variations, or what De Vries had called "mutations." Breeders knew that selection couldn't get you anywhere without something to start from. "Variation leads; the breeder follows," he insisted. The art of the breeder was to pick up on existing mutations and to combine them, and "we may rest assured the method of Nature is not very different." The breeder and nature itself depended on the collecting together of a dozen or so discontinuities. Contrary to the views of the gradualists, he argued that new breeds seldom went back more than a period of seventy years. The application of statistics was of little use in this arena. In the most quoted line from his address, Bateson asserted the primacy of breeders and hybridizers over mathematicians:

> The imposing Correlation Table into which the biometrical Procrustes fits his arrays of unanalyzed data is still no substitute for the common sieve of a trained judgment.[28]

The speech, which was both a literary and a scientific tour de force, was followed the next morning by shorter talks by Bateson's allies, including his first disciple, Miss Saunders, as well as the most recent convert to the cause, Darbishire, who now argued, exactly as Bateson had done earlier that spring, that the progeny of the waltzer-by-albino hybrids segregated into three groups following Mendelian proportions.

After Darbishire, Weldon rose to defend the biometricians. The epitome of health and vigor, Weldon had bicycled the eighty or so miles from Oxford to attend the meeting.[29] Rising to the cause, he spoke with his customary lucidity and passion, arguing that the Mendelian ratios were better explained by Galton's ancestral law, trotting out several of his best-rehearsed examples of fatal holes in Mendelian theory. His failure to mention Darbishire's hybrid mice, however, represented a significant setback to the cause.

The afternoon session began with Bateson answering Weldon's criticisms, comparing the ancestrians to medieval astronomers who tried to explain movements of the stars and planets starting with the assumption that the earth was flat.[30] Pearson followed for the biometricians. Attempting to strike a higher note, Pearson invoked the figure of his beloved Francis Galton, "whose introduction of methods of precision had nothing to do with Mendelism or ancestral law." He then discussed his most recent article, entitled "On a Generalized Theory of Alternative Inheritance, with Special Reference to Mendel's Laws." The remarkable thing, he noted, was that his independent work had led to principles strikingly like Galton's law of ancestral inheritance. Hoping to be able to return to his statistical studies, Pearson proposed a three- to four-year truce between the warring parties.[31]

In describing the scene, Pearson later recalled that Bateson flatly refused his offer, hoisting the issues of *Biometrika* over his head and offering them as evidence of the folly of the biometric school.[32] According to the official record, however, it was the highly revered Darwinian, Reverend T. R. R. Stebbings,[33] not Bateson, who expressed his wish to see the controversy rage on because "from it the world could only gain in light," and this view was enthusiastically seconded by the chairman, who added that "the debate had been of much value to those biologists who were still sitting on the fence."[34] Even a half century later, Bateson's Cambridge colleague R. C. Punnett could still taste the victory of the day, recalling that "Bateson's generalship had won all along the line and thenceforth there was no danger of Mendelism being squelched out through apathy or ignorance."[35]

That fall Bateson discovered a major flaw in Darbishire's experiment purporting to prove that the percentage of albinos among the progeny of hybrid parents depended on the number of albinos in the grandparent generation. Remarkably, Darbishire had failed to distinguish between pigmented homozygotes and heterozygotes among allegedly "hybrid" parents, and as a result some of them actually contained no albino gene. Properly interpreted, Darbishire's results were perfectly consistent with the Mendelian model of segregation of hybrids after all: the smaller the number of albino grandparents in a series of crosses, the greater the number of undetected homozygotes among the parents and the smaller the number of albino grandchildren.

— THE LAST MAJOR CONFRONTATION between the warring camps was again over the vexing question of the inheritance of coat color, this time in horses. Already in a paper published in March 1900, days before the rediscovery of Mendel, Pearson had taken his first stab at the problem. With the aid of Weatherby's studbook, which gave a complete record of the coat color and ancestry of all the registered thoroughbred horses in England, he had computed the degree of similarity in coat color between relatives of different degrees and concluded that the inheritance of horse coat color did not obey either Galton's ancestral law or the revised form that he had introduced in 1899.[36]

In 1903 Pearson had again returned to the coat color problem, introducing a third law of ancestral inheritance, which he now claimed described the inheritance of both eye color in humans and coat color in horses. Furthermore, he now asserted that "nothing corresponding to Mendel's principles appears in these characters for horses, dogs or men."[37] To prove his point, he pointed to cases in which the coat color of a foal could not be predicted from the coat color of its parents. In one of several examples, he traced three different matings between a grey stallion and a grey mare and showed that each pair gave rise to a different color filly—bay, brown, and chestnut.[38] With the certainty that was one of his hallmarks, Pearson wrote:

> It was the same story with every coat-colour taken, its relative constancy depends largely on the extent to which it has appeared in the ancestry, and one by one black, bay, chestnut, grey must be dismissed by the Mendelian as neither "recessive" nor "dominant," but as marking "permanent and incorrigible mongrels."[39]

While the coat color of the parents was not a sufficient guide to the coat color of the progeny, Pearson pointed out that the extended ancestry could often provide a clue. Raising the stakes for all future work on horse coat color, Pearson argued that horse pedigrees, which provided coat color data from thousands of crosses, some of which had been carried on for five to ten generations, were a far more reliable guide than the data from "five or ten individual crosses followed for two or at most three generations," that were characteristic of most of the Mendelian studies.[40]

When Pearson's article appeared in February 1903, Bateson had been preoccupied with the analysis of Darbishire's mouse experiments and did not have the time to devote to these new cases. Furthermore, as he later explained, it had not occurred to him that Pearson might be bluffing. However, with the disintegration of Darbishire's results, Bateson was once again free to consider new problems, and the final act in the Mendel Wars was set to begin. The cast of characters was still much the same, including Bateson, Weldon, and Pearson, and Galton in the wings, but there was one new central player, the indomitable Charles Chamberlain Hurst.

Hurst was the son of the owner of a century-old family nursery, which included a large and important collection of orchids.[41] When an attack of TB prevented him from studying at Cambridge, Hurst had instead devoted himself to producing exotic hybrid orchids, which was then a common practice. It was a sign of the respect he'd won that he was included, along with Bateson and De Vries, as one of six principal speakers at the opening session of the first International Conference on Hybridisation held at Chiswick in 1899,[42] where he had met Bateson for the first time. Several years later their shared interest in Mendelism brought Bateson and Hurst together, and by 1903 they were close friends and collaborators. In addition to his keen mind and a remarkable aptitude for hard work, Hurst brought to Bateson's crusade for Mendelism two qualities that would prove to be extremely useful in the battle against the biometricians: the perspective of a professional breeder and a healthy skepticism about the high-toned world of British academia.

In October 1904, Bateson learned that the Evolution Committee had rejected his second report on Mendelism after a five-month delay, and Bateson suspected that Weldon, who was chairman of the Royal Society Zoological Committee, which had jurisdiction over Evolution Committee reports, was behind the action.[43] The referee's report infuriated him:

> The literary composition of these bulky memoirs seems to me below the standard that might reasonable be exacted by the Royal Society. An unnecessary burden is thrown on the reader, by their cumbrous, confused and imperfect presentation of objects, arguments and conclusions. The hand writing shows that some part of these faults is due to off-hand composition without adequate revision."[44]

"The paragraph is almost unmistakable Pearson," Bateson wrote Hurst, whose investigation of inheritance in poultry had constituted a third of the report. He regarded the action "as part of the general conspiracy to suppress us," Bateson explained, and was tempted to withdraw the manuscript.[45]

Hurst, who had seen Pearson in action for the first time earlier that summer at the British Association meeting, wrote back, "I can quite imagine his indulgence in such superior sneers," but he urged Bateson to hold the line. To withdraw the manuscript, he argued, would be playing into their hands. Hurst swore, "A pox on all their machinations and I would see them damned before I would withdraw a single letter!"[46] showing the fiery spirit that would later prove invaluable.

Pearson's and Weldon's arbitrariness incensed Hurst and fueled his desire to punish them for their arrogance. His work on human eye color in the county Burbage, where his family had long resided, had convinced him that blue eyes behaved like a classic Mendelian recessive to brown, and that Pearson's assertions were fatuous.[47] In February 1905 he sent Bateson the eye color data showing that blue eyes were recessive to brown. "Your figures are extraordinarily interesting," Bateson replied, but he urged caution because the biometricians would pounce on any anomaly.[48] Furthermore he didn't like the idea that they would go after the eye color and leave the horse coat color problem untouched.

Over the following weeks, Hurst applied himself to the study of horse pedigrees and by March he had begun to suspect that chestnut was recessive with respect to bay. As Bateson pointed out, Pearson's 1903 paper had not mentioned a single case in which two chestnuts produced blacks or bays, which supported Hurst's theory. Furthermore, Bateson wrote, it was clear that they had purposefully withheld the data.[49]

In mid April, Hurst had found that chestnut-by-chestnut crosses produced chestnut foals almost without exception, and Bateson was dumbfounded. "After Pearson's statement," he wrote Hurst, "I had no conception there was a clear case like that." However, he doubted it would be possible to work out the color formulas based on the information from the stud records, which provided too little data. Still, the possibility of catching Pearson up in so grave an error was too tempting to resist, and he saw just how to do it. By writing a letter to *Nature* politely asking Pearson to point to the cases he had in his mind when he made that statement regarding chestnut, perhaps they could get him to reveal the exceptional matings of chestnut by chestnut. "It wants doing with great care," he cautioned.[50]

Buoyed by Bateson's support, Hurst now threw himself at the problem. The first step was to check all the recorded foals produced by two chestnut parents, but the unraveling of the pedigrees from Weatherby's famous studbook, which filled twenty volumes, was a painstaking and tricky business. To properly interpret the data in a single volume, Hurst explained to Bateson, required at least six other volumes. Adding to the difficulty was the fact that sires of the same name often had different colors, and it required special "horsey Knowledge" to get the pedigree right.[51] By April 25, with the help of his sister, Hurst had checked 180 foals from chestnut-by-chestnut crosses, and had found only two exceptions.[52] By May 9, the Hursts had looked at over a thousand chestnut-by-chestnut matings and found nine that were reported to give other than chestnut foals. Although Bateson doubted Pearson would know about the exceptions, he thought it better not to risk a public challenge, and suggested delaying the *Nature* letter. "In the great game you must assume your opponent has overlooked nothing," he wrote Hurst.

In the meantime, he hoped they might be able to eliminate at least five or six of the exceptional cases, which he had begun to suspect were caused by the common practice of mating a mare to more than one stud at a time and the owner's inclination to report that the best sire was the true father. In order to seal the case, Bateson had also begun to investigate the way the hair color pigments interacted to produce the chestnut and bay coats. Based on the histological analysis performed by his assistant, Miss Durham, he was now convinced that the bays and chestnuts had essentially identical red coats, and the only thing that distinguished bays from chestnuts was that bays possessed black points, including black manes and tails and sometimes black markings on the leg. Over the summer every horse on the street had become an opportunity for further study; in September Bateson had worked through his reservations, and it was agreed that Hurst would write up a short account of the horse coat results.[53] Bateson, as a member of the Royal Society, would submit them for publication in the society's influential *Proceedings*.

The paper, which was submitted to the Royal Society in early November, proposed that the difference between bays and browns, on the one hand, and chestnut horses, on the other, was controlled by one Mendelian factor pair, for black pigment. Both bay and brown were either homo- or heterozygous for the dominant allele, and chestnuts were homozygous for the recessive. In conformity with Mendelian expectations, Hurst did not find a single chestnut foal among 370 progeny of pure-breeding bay or brown sires. Furthermore bay and brown heterozygotes (that is, studs that themselves were the cross of brown and chestnut parents) produced very close to half chestnut and half bay and brown progeny (355:347) when mated to chestnut mares. The foals of chestnut mated to chestnut were more than 99 percent chestnut (1,095 of 1,104 chestnuts). Less than 1 percent were exceptions, which he attributed to breeders' or printers' errors.[54] "I think the horse paper a very remarkable piece of work and I am proud to communicate it," Bateson wrote on November 1.[55]

Hurst's paper was scheduled to be read at the Royal Society on Thursday, December 7, at 4:30 p.m. In the days preceding the climactic event, Bateson sent a flurry of cards and letters with last-minute suggestions and ideas. The day before the meeting, Bateson suggested that he and Hurst meet at the Austro-Hungarian Café at 1:15 and leave from there for the Royal Society. When he arrived, Hurst found Bateson in a state of high nervous excitement over the upcoming battle, and suggested that they order a bottle of champagne.[56] This idea horrified Bateson, who was in no mood for celebration or levity of any kind. After lunch, the two men hailed a hansom cab, which was drawn by a chestnut horse. As the two men climbed into the seats, which were located in front of the driver directly behind the horse, Bateson noticed that its tail was exceptionally dark, a darker shade than he had ever before encountered in a chestnut. Hurst explained that the horse was in fact a very dark form known as a liver chestnut, which had a "chocolate" colored tail, but Bateson, who had never seen such a horse before and was convinced that the tail was black like the tail of the bay, grew increasingly agitated. When the ride was over, Bateson prevailed on the cab driver to give him a sample of tail hairs to bring back to Miss Durham for analysis.

Once the meeting was under way, Bateson managed to get through the reading of his first paper, which documented the newly described phenomenon of "gametic coupling," a principle that would turn out to be of immense importance. The paper was a joint one with Saunders and Punnett, who were seated in the audience. Hurst was last on the program, and as soon as he had finished his presentation the bickering began.[57] Unlike Hurst, who was enlivened by the thrill of battle, Bateson had a tendency to come undone in the face of confrontation. Even under the best circumstance, Bateson would have found the prospect of direct confrontation with Weldon draining and debilitating, but the last-minute jolt of anxiety provided by the hansom ride seemed to push him over the brink. Weldon listed a host of exceptions, far more cases than Bateson had anticipated, drawn from a large number of volumes of the studbook. In the heat of the argument, Bateson brought out a poultry skin to illustrate a point, and his opponents objected on the grounds that it was a discussion about horses. The chairman of the meeting concurred. This was the final straw for Bateson, who slammed the skin back into his bag, announced that he was withdrawing the paper, and returned to his seat.

Hurst, who was far more sanguine, stepped up to assure Weldon that the "alleged exceptions were mere errors of entry."[58] Later Hurst recalled that the shocked, stricken faces of Miss Saunders and Punnett in the audience encouraged him to hold on. To refute the specific claims would take a great deal of study of each individual case and was clearly impossible on the spot, but Hurst's direct denial and stout defense seemed to quiet the opposition, and before he had finished, Hurst had extracted an agreement from Weldon that both sides should put the facts down on paper.[59] In the coat room on the way out, Hurst encountered Pearson, who said in a fury, "One thing, you shall never be Fellow here as long as I live."[60]

After the meeting, Bateson went out to dinner with Punnett and Miss Saunders. Punnett recalled the walk to a restaurant where they had planned to celebrate afterward as "grim and silent" and dinner that followed as "glum." Later that night Bateson wrote Hurst that he felt "utterly collapsed" and that Weldon had inflicted "a bad wound, that won't heal for a long time." Nonetheless, he had been "filled with admiration" at Hurst's pluck, and his own fundamental conviction that they were in the right had not been shaken. It was now incumbent on Hurst to become half horse himself and "get back into the collar and pull us out of the mud."[61]

In his return letter, Hurst apologized for getting them all into the mud in the first place, but he expressed his doubt that Weldon would be able to make good on his promise to produce the figures he had presented on Thursday. His cases would simply not hold up. "The Stud book is full of pitfalls for the unwary," he explained, "and records got out in a hurry will not stand the test of time."[62] For his own part, he had already begun a thorough analysis of the studbook and vowed, "If it takes me 20 years, I will now see it through!"

Hurst's letter heartened Bateson, who now regretted having withdrawn Hurst's paper without at the same time extracting in exchange a promise for a written list of the cases. "He will no doubt send in nothing and leave the onus with us," Bateson predicted, and he promised Hurst he would buy a complete set of studbook volumes in order to assist in the arduous work ahead. In addition he wrote the Royal Society, attempting to reframe the withdrawal of the paper in such a way that it could be used to pressure Weldon to make good on his promise to put his claims in writing. Addressing the secretaries of the Royal Society, he wrote that he had no doubt as to the validity of Hurst's findings. "Naturally I regret that a paper communicated by me would have been open to criticism of a seemingly destructive character," he continued, but now his greatest wish was to see the unsavory business resolved as quickly as possible. To that end, he wrote, "A definite issue will be reached more quickly, if Professor Weldon carries out his expressed intention of communicating his independent analysis to the Society which we trust he may be persuaded to do."[63] As he'd advised Hurst, one must "always take hold of a rascal by his full title."[64]

Hurst was now going full force, making separate lists of foals born of mares and foals born of sires and double-checking them one against the other. By December 19 he reported to Bateson that he had classified 223 mares and it had taken him 56 hours to do it. It was highly complex and time-consuming work, but he was convinced that only by scouring the records was it possible to avoid mistakes, and without such intense effort the results simply would not stand up. "I doubt if [Weldon] can spare the time," Hurst gloated, "and second hand will not do at all."[65]

In fact, Hurst had already gotten to the bottom of the case of the thoroughbred sire Ben Battle, an alleged example of a homozygous chestnut that produced a large number of bays when mated to chestnut mares. In his research, Hurst had discovered a record in which Ben Battle had been mated to a homozygous bay mare and produced a filly. Under the assumption that Ben Battle was homozygous for the chestnut allele, the filly was, by necessity, a heterozygote. However, when this filly was crossed with a chestnut sire, she had produced nine dominants and no recessives at all, suggesting that filly was almost certainly homozygous dominant and thus that one of her dominant factors had been inherited from the allegedly homozygous recessive Ben Battle.

On December 30, Bateson telegrammed Hurst with an independent confirmation that Ben Battle had indeed been misidentified as a chestnut. Over the Christmas holiday Bateson had contacted an entomologist friend, Verrall, who was also a handicapper at the Jockey Club, and made inquires about Ben Battle. Although Ben Battle appeared as a chestnut in the studbook, Verrall wrote back, he was always listed in the Racing Calendar as a bay or a brown and never, as far as Verrall knew, ran as a chestnut.[66] Bateson wrote Hurst the next day again, ending his letter with an exuberant, "Isn't B Battle ripping."

— ALTHOUGH WELDON HAD ALWAYS lived life at a fast pace, beginning in 1904—when the Darbishire affair had begun to unwind—his friends began to worry that he might be overdoing it. Not only had he taken over the full burden of the mice work after Darbishire left Oxford, but he had begun work on a book on heredity as well as pursuing his various interests in natural history along the way. "The torpedo boat was being run at full speed," as Pearson described him just after the 1904 British Association meeting. In addition to his frantic pace, Pearson noted that Weldon had showed occasional signs of depression, a "lack of joyousness in life."[67]

During 1905, work on the heredity book continued, and Weldon spent the Easter holidays in Ferrara because it had an ancient and famed university and was certain to have a library in which he could work. Despite the fact that the contents of the library and the building itself were medieval, Weldon managed to continue working on his book. "I feel out of the work, an absolute blank with only a slight interest in newts' tails and an even slighter in a statue of Savonarola which looks at me all day through the window," he wrote Pearson in March 1905.[68] Though he could not appreciate the irony, the talented philosopher and scientist Savonarola, who had renounced the world to join the Dominican Order at Bologna in 1474, was Weldon's soul mate, writing treatises denouncing the immoral practices of the court of Lorenzo de Medici, which was flourishing in the dawn of a new renaissance of art and science.

In November 1905 Weldon had interrupted his work on the heredity book in order to study horse color in preparation for the December 7 reading of Hurst's paper at the Royal Society. His success in quelling the opposition in the December 7 meeting seemed to fuel him. Nine hours a day, he threw himself into the study of Weatherby's studbook. As he had pledged to do, Weldon had completed a preliminary written account of his findings by January 1906, which he read at the Royal Society on January 18. The chance of obtaining a foal that was not a chestnut from the matings of two chestnuts, Weldon now reported, was 1 in 60, clear evidence of the fact that chestnut could not be regarded as a Mendelian recessive. Furthermore, he reported on the crosses of heterozygotes (bays that had one chestnut parent) with pure-breeding chestnuts. Weldon found that 1,118 of 2,430 progeny were chestnuts, a deviation from the Mendelian expectation (1,215 of 2,430) that would arise by chance less than once in a thousand times.[69]

After the January 18 meeting, Bateson had no doubt that Weldon was bluffing, and together he and Hurst drafted a short note to go after Hurst's paper in which they pointed out that Weldon's entire argument was based on the "alleged existence of exceptions," and they illustrated the point with several examples, including the ripping Ben Battle. In addition they pointed out that errors sometimes arose through the incorrect return of the sire's name, as they had long suspected, and that such errors could radically skew the number in Weldon's crosses of heterozygote by homozygotes. To illustrate this point they pointed to a case in which one allegedly heterozygous dominant bay sire that had been misidentified as a homozygote had produced 43 bays and browns. Although Hurst had never doubted it, Bateson was also now convinced that Weldon's case would not stand up. It was a testament to the power of wishful thinking that Pearson and Weldon continued to believe.

In February, Weldon retreated again to Italy, this time to Rome. Though he felt he needed a holiday, he wrote Pearson that he had felt compelled to bring along the volumes of the studbook. "To sit here eight hours a day or so, doing mere clerk's work, seems rather waste of life?" he wrote Pearson from his hotel room, wishing he could partake in the pleasures of Rome. From February to April he examined hundreds of pedigrees, and his letters were filled only with crosses.[70] Over Easter the Pearsons and Weldons met for a joint vacation in Woolstone about twenty miles from Oxford, and Weldon was still hard at work on the studbook. On Sunday, April 8, Weldon biked into Oxford to develop photographs, and complained of being tired after the trip, which was extremely unusual for him. On Monday he took a long walk over the Downs and returned home late. He got up for breakfast on Tuesday but returned to bed afterward. Pearson visited that afternoon, and Weldon insisted on smoking while he questioned Pearson on the solution to a new problem he was working on. The next day Weldon's wife begged him to stay in bed, but he refused and instead went to town to visit an art gallery. On Thursday he had a dentist appointment from which he was taken first to a doctor and then to a hospital. He died of pneumonia the following day, Good Friday, April 13. It was shocking news to everyone that the most vigorous of men had been cut down in his prime. For Pearson it was a devastating loss, of his dearest friend and his partner in the founding of a new school of biology. The immensity of the loss he felt is evident throughout the fifty-page obituary he wrote immediately following Weldon's death.

In early May, Hurst sent Pearson a reprint of the horse coat-color paper, which had appeared in print on the day of Weldon's death, along with a brief note of condolence. Although Hurst had been moved by true regret, Pearson refused to accept his condolences. In Pearson's imagination, Weldon was a soldier who had died in battle, and Pearson was not prepared to take succor from the enemy. "Only a few days before his death he condemned in stronger language than I have ever heard him use of any individual the tone and contents of the *Note* added to your paper," Pearson lashed out. "It is a judgment of which I believe everyman who has the interests of science at heart will concur."[71] Although it was true that Hurst and Bateson had each made scathing attacks in the heat of battle, the late added postscript to Hurst's paper was an entirely dispassionate scientific statement.

"I could have warned you," Bateson wrote an astonished Hurst.[72] After the death, he had considered carefully whether to write himself, he explained, and had realized that it would be futile. But for Bateson, Weldon's death involved a far more complex mix of emotions. On the one hand, he clearly saw that the death of Weldon cleared the way for the broad acceptance of Mendelism. Furthermore he was bitterly resentful at the unfair treatment that he had been subjected to, convinced that Weldon had been determined to destroy his work.

But at the same time Bateson mourned the loss of his oldest friend. "I was more intimate with him than I have ever been with anyone but you," Bateson wrote his wife, who had been out of town the week Weldon died.[73] In trying to put into words what Weldon had once meant to him, Bateson made clear how far things had gone wrong and how much good had been squandered:

> To Weldon I owe the chief awakening of my life. It was through him that I first learnt that there was work in the world which I could do. Failure and uselessness had been my accepted destiny before. Such a debt is perhaps the greatest that one man can feel towards another; nor have I been backward in owning it. But this is the personal, private obligation of my own soul.[74]

Edward Beecher Wilson (1890).

Marine Biological Laboratory Archives.

CHAPTER 8

# Cell Biology

— THE PRODUCT OF AN ASTONISHING clarity of thought and analytic penetration, Mendel's abstract system of heredity had seemed entirely ungrounded to his contemporaries. In fact it would take four decades of intensive investigation of the cell, the nucleus, and its contents to discover that there were physical entities in the nucleus, the chromosomes, that behaved exactly as Mendel's theory predicted they should. Unlike Mendel's solitary effort, the development of an understanding of the material basis of heredity was a vast communal effort that drew on the insights and observations of many investigators in Europe and later in America.

The new understanding of the cellular basis of life began in Germany in the 1840s, with the recognition by Theodor Schwann and Matthias J. Schleiden that cells were the fundamental building blocks of life and that each cell contained a nucleus. Although the Schleiden-Schwann cell theory, as it would come to be known, was a major advance, both Schwann and Schleiden had believed that new cells were born spontaneously, either in the cellular fluid surrounding the cells or from within.[1]

The next crucial insight, which was pivotal in understanding inheritance, was that cells did not arise *de novo* but instead one cell divided to give birth to two daughter cells. In particular, unless there was a connection between the cells of the parents and the sex cells that combined to produce their child, it was extremely difficult to conceive how parents could transmit their traits to their children. Already in 1841 the brilliant microscopist Robert Remak was convinced, based on his observations of the proliferation of red blood cells, that cells arose from other cells by cell division. A decade later in 1851, after intensive research on the division of embryonic cells and muscle cells, he published a classic paper in which he proclaimed the "Law of Multiplication by Cell Division." In the same paper, he also proposed that two new nuclei were derived from the mother nucleus by nuclear division.[2]

Reflecting the nearly unanimous consensus among the leading figures of European science, Rudolfe Virchow, then the rising star of cell biology, rejected Remak's theory in his 1854 *Handbook on Special Pathology*, proposing instead a theory of cell creation from within the mother cell that he had borrowed (without attribution) from Schleiden. The following year Virchow reviewed Remak's own definitive work on the embryology of vertebrates, which included a chapter reviewing the history and current status of cell theory. The publication of Remak's masterly review of cell theory coincided with the creation of a new Berlin professorship, and the two leading contenders were Remak and Virchow. Afraid that Remak would beat him out, Virchow rushed out a new paper containing a revisionist history of cell biology. Although in his 1854 book he had credited Remak with what he had then regarded as a flawed theory of cell origination by binary fission, he made no mention of Remak in his new book and instead claimed the idea as his own. In the flowery romantic style perfected by the German nature philosophers, Virchow wrote, "Omnis cellula e cellula" (All cells from cells), ensuring that he would be forever celebrated as the originator of Remak's theory.[3] Despite the fact that Virchow's dictum is one of the hallmarks of nineteenth-century biology, it was oddly imprecise, failing to make clear the crucial fact that new cells were created by the process of binary fission.[4]

The next great leap forward in the cellular study of heredity was taken by Virchow's student Ernst Haeckel, who would become a leading disseminator of the theory of evolution, perhaps more influential than Darwin himself, as well as a great believer in the importance of German racial purity.[5] Drawing on the knowledge he had gained as Virchow's student, in 1866 Haeckel made the astonishingly perceptive speculation that it was the cell nucleus that controlled heredity. Without providing a shred of evidence, he boldly declared: "The inner nucleus is responsible for the transmission of the inheritable characteristics, while the outer 'plasma' is responsible for the accommodation of or adaptation to the conditions of the environment."[6]

Haeckel's insight was all the more remarkable in light of the fact that biologists of the time believed that the protoplasm was responsible for all the basic functions of life, including those of heredity. Furthermore, there was little to support the idea that the nucleus played the role Haeckel assigned it and much to recommend against it, the main problem being that the nucleus seemed to disappear at the beginning of the cell division process, dissolving into the cytoplasmic soup without a trace.

In 1874 Leopold Auerbach, who had studied histology, embryology, and microscopy with Robert Remak at the University of Berlin, produced a series of striking drawings of the large, clear germ cells of a threadworm of horses *(Ascaris),* which gained worldwide recognition.[7] With remarkable clarity and detail, Auerbach's drawings depicted the fusing of two nuclei in the formation of the fertilized egg and the subsequent cell division.

Although Auerbach had carefully observed the fusing of the two nuclei in the fertilized egg, he missed the crucial fact that the two nuclei were the sperm and egg nuclei.[8] But Auerbach's paper was soon read by Oscar Hertwig, who together with his brother Richard was part of Haeckel's circle in Jena. Immediately after reading the paper, Hertwig dropped everything to study fertilization in the sea urchin, whose large translucent egg cells were ideally suited to the purpose.[9] Five to ten minutes after spermatozoa were added to a suspension of ripening eggs, Hertwig observed, a second

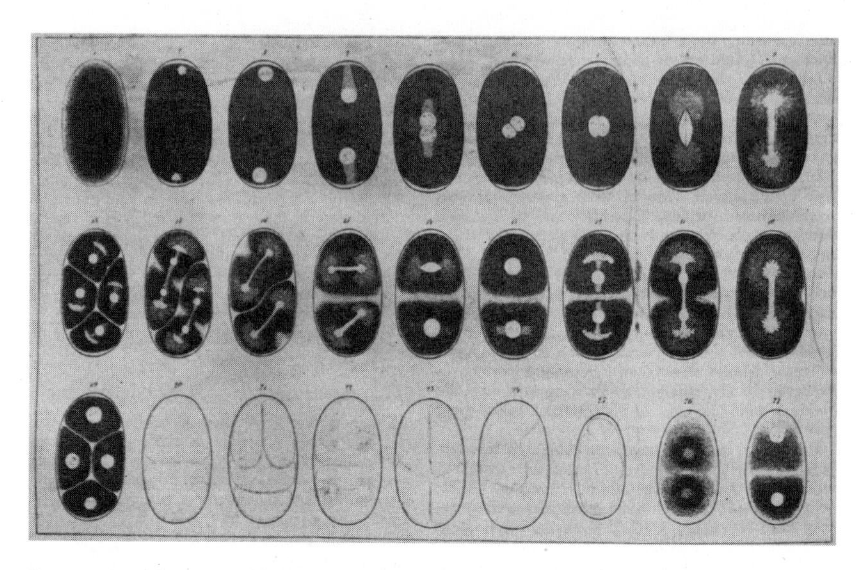

**Progressive changes of cells and nuclei in fertilized egg of *Ascaris.* Second line reads from right to left.** From Leopold Auerbach, *Organologische Studien,* vol. 2 (Breslau, 1874).

nucleus appeared in the egg. Although this observation did not constitute direct proof that the second nucleus had come from the sperm, the fact that the second nucleus appeared inside the cell at exactly that point nearest to where the sperm had approached the outside of the egg provided strong circumstantial evidence. In the course of his observations, Hertwig had also been able to see that the putative sperm nucleus approached the other nucleus in the cell and that the two were fused, which supported the idea that one and only one sperm was required for fertilization. Although there were some doubters, the idea that fertilization consisted of the fusing of one sperm nucleus with the egg nucleus was recognized as a revolutionary advance.

While Hertwig had brilliantly interpreted the fusing of the two nuclei in Auerbach's pictures, it still remained to properly understand the next set of slides depicting the first division of the newly fertilized egg. Auerbach himself had adopted the still-standard view that the new nucleus that appeared in each of the daughter cells bore no relation to the nucleus of the mother cell. To support this interpretation, he suggested that the two spheres with outward-pointing rays that appeared when the cells underwent their first division were the visible manifestation of the nuclear material dissolving into the cell plasma.[10]

Unknown to Hertwig, the Breslau zoologist A. Schnieder had just published a paper in which he solved the mystery of the disappearing nucleus. By treating cells with vinegar Schnieder had seen that the nucleus did not disintegrate and re-form in the course of cell division—instead it had transformed itself into a heap of fine curled-up threads. In the course of the transformation, he observed the threads thickening and giving rise to thicker rods (the chromosomes), which were first aligned along the equator and then distributed to either end of the cell, where they gave rise to new daughter nuclei. The complex rearrangements of the nuclei persuaded him that the nucleus did not disappear and that there must be some larger significance to his findings. In his own words, Schnieder had seen

> for the first time clearly, how elaborate a metamorphosis the nucleus (the little germ bubble) can undergo during the cell division. Apparently, this metamorphosis is not necessary in every cell division, but—very probably—it happens always when the nucleus seemingly disappears. If here the nucleus did not happen to be large, and the cell not transparent, one would probably also assume that the nucleus disappears as in other cases.[11]

Published in an obscure journal, Schnieder's report received little notice, but the same phenomena were soon noticed in other organisms. In 1879 Walther Flemming added a new piece to the description of the nucleus during cell division. Using newly available Pekonic dyes,[12] he was able to see that the nuclear rods, which had first been visualized by Schnieder, appeared to be split down their lengths into two sister strands that appeared to be identical in every respect.[13] The following year, Flemming further demonstrated that the matching daughter strands were actually separated in the process of nuclear division, for which he coined the term *mitosis* in 1882.[14] In addition, he was now also able to see that the radiating spheres first noticed by Schnieder actually formed a system of delicate strings emanating from the centrally located chromosomes to each of two poles located at either end of the cell, the whole figure forming a pair of inverted cones joined at their bases. (These structures are visible in the top row, fourth and fifth cells from the left, in Auerbach's depiction of the divisions of cell and nuclei.) In the process of cell division, the two members of each pair of matching daughter strands were drawn down these strings to opposite sides of the cell. His discovery had major implications, as Flemming recognized, for if the chromosomes were in fact true copies of each other, longitudinal splitting could be the cell's mechanism for accomplishing an exact division of the hereditary material.

In 1883 the French microscopist Edouard van Beneden published an extraordinary 350-page study of the chromosomes of the sperm and egg nuclei in *Ascaris* that added a significant new element to the picture of nuclear division described by Flemming.[15] First, van Beneden noted, the sperm and egg cells in *Ascaris* had only two of the four chromosomes characteristic of the non-sex cells, and further careful study of the process of egg cell formation revealed that the two missing chromosomes were expelled in a series of divisions giving rise to three small vestigial eggs (the so-called "polar bodies"). Furthermore, he showed that in the process of fertilization the two male chromosomes donated by the sperm were added to the two female chromosomes contained in the egg nucleus, giving back the full complement of four chromosomes. Van Beneden's results were a striking confirmation of the breeding results that had long pointed to the equivalence of maternal and paternal contributions to the characteristics of the offspring, as well as a definitive confirmation of Hertwig's idea that fertilization consisted of the joining of the egg and sperm nuclei.

By the early 1880s the idea that the nucleus was the repository of the hereditary traits had been elaborated by several writers independently, most importantly by the master of German theoretical biology, August Weismann. Steeped in evolutionary theory, embryology, histology, and cytology, Weismann was best positioned to synthesize the many threads of knowledge that were then converging. In his 1883 essay "On Heredity," Weismann coupled a penetrating critique of the doctrine of acquired traits with the idea that the hereditary material, for which he introduced the term *germ-plasm,* was isolated in special germ cells, where it remained "uninfluenced by that which happens during the life of the individual which bears it."[16] Furthermore, he declared, germ-plasm was immortal—while the bodies that possessed it might perish, the unaltered germ-plasm survived in the bodies of their offspring.

A storm of voices objected to the "germ theory," but the most telling critique came from Haeckel's protégé Edouard Strasburger, who argued that the fact that leaves or virtually any part of a *Begonia* plant could be used to propagate a new plant showed that all of its cells contained germ-plasm and there was no need for specialized germ cells. In response to Strasburger's salient criticism, Weismann revised his theory in 1885, proposing that the germ-plasm was contained in the nuclei of cells, and more particularly, in the nuclear rods whose behavior at cell division had been described by Schnieder, Flemming, and van Beneden.[17] It was perfectly possible, Weismann now declared, that a small amount of germ-plasm could be passed on into the nuclear material of some, or even all, somatic cells.[18]

His interest in the germ-plasm led him deeper into investigations of the process of sex-cell formation. Having undertaken a series of experiments on egg production in the asexually reproducing zooplankton in 1885, he became convinced that the theory of "mitotic" division of cells was not sufficient to explain the production of all reproductive cells. Though he believed that parthenogenetic eggs, like those of zooplankton, were produced by a process of mitotic division resulting in the production of an egg that contained a full complement of hereditary units, he now proposed that the production of egg cells in sexual organisms involved a second kind of mitosis. This second kind of mitosis, which he called a "reducing division," resulted in the halving of the chromosome number, as van Beneden had reported in his 1883 paper on egg production in *Ascaris.*[19] But a reduction in the number of hereditary determinants could have been predicted on theoretical grounds alone, Weismann insisted. Unless there was a halving of the hereditary material, the fusion of the male and female germ cells would result in a doubling of the quantity of the hereditary units in each successive generation. This was the very same argument Galton had first put forward in 1871, and Weismann would later acknowledge his priority.[20]

There were, however, troubling reports issuing from Flemming and others indicating that the second division was, like the standard mitotic division, a separation of longitudinally split daughter chromosomes, and thus could not result in a reduction of the nuclear material of the egg-cell by half. Despite evidence to the contrary, the theoretical necessity for a halving of the genetic material was so compelling to Weismann that he was willing to disregard the empirical evidence: "Even if I am mistaken in this interpretation, the theoretical necessity for a reduction . . . seems to me to be so securely founded that the process by which it is effected must take place, even if they are not supplied by the facts already ascertained."[21] Although he would be reviled in some quarters for his failure to subordinate theory to empirical evidence, he defended his theoretical approach. "To go on investigating without the guidance of theories," he replied to his critics, "is like attempting to walk in a thick mist without a track and without a compass."[22]

Weismann's prediction that two distinct types of cell divisions would be found to be involved in the maturation of the sex cells—in the first type of division, which Weismann termed "equal division," each daughter nuclei inherited a full complement of hereditary material by the process of longitudinal splitting of chromosomes, and in the second, a reductive division, each daughter nucleus received only half the number of hereditary determinants—proved to be prescient. In the words of the Princeton geneticist Edwin G. Conklin, it was "as brilliant an example of scientific prophecy as was the prediction of the existence of the planet Neptune."[23] The second process would later be termed "meiosis," from the Greek for "that which is reduced."[24] Although Weismann got some of the details wrong—the reductive division turned out to precede one final mitotic division in the formation of the sex cells—he had grasped an elemental fact.

— HOWEVER FASCINATING the speculations brought forth about the role of the nucleus and the behavior of the chromosomes in heredity, they were not satisfactory to the young German biologist Theodor Boveri, whose lifetime ambition was to establish the facts of heredity experimentally. The 22-year-old Boveri began work in Richard Hertwig's Zoological Institute in Munich in 1885, just at the moment when the idea that the hereditary material was located in the as-yet-unnamed chromosomes began to gain currency. Through Hertwig, Boveri received the ideal introduction to cytology, and in particular to the study of the nucleus, which was rapidly becoming a subfield of its own. Deeply impressed by van Beneden's results, Boveri spent his first several years in Hertwig's institute repeating van Beneden's work on egg cell formation in *Ascaris*.[25] In January 1888, Boveri visited Naples, where as fate would have it he was a guest in the same pensione where Weismann was staying.

Although few men were better prepared than Boveri to accept Weismann's views on the role of the reduction division in the formation of sex cells, Boveri and Weismann had a fundamental disagreement over the nature of the chromosomes. Whereas Weismann believed that the chromosomes were all essentially identical, each containing multiple copies of "ancestral germ-plasms," Boveri's work on the chromosomes in the egg cells of *Ascari* had convinced him that the individual chromosomes differed in form and function, and each evening after returning from the Zoological Station the two men argued.[26] That spring, Boveri began to collect a variety of marine forms from the Sea of Naples and made detailed renderings of carefully prepared chromosome arrays isolated from their sex cells, documenting the differences in the sizes and shapes of the individual chromosomes, as well as the way they changed from one generation to the next. He soon found that in each case the sperm and egg cells contained identical sets of chromosomes;[27] this was consistent with the idea that the chromosomes are the repository of the hereditary material, but it did not yet constitute a proof.[28]

Boveri continued the work on chromosomes when he returned to Munich in April 1888, and later the same year he presented the fruits of this new line of inquiry in the second of six "Cell Studies," a unique amalgam of genetics, cytology, and embryology that immediately established him as a world leader in the study of heredity. In this paper Boveri clearly showed that the chromosomes, which disappeared during the resting period between nuclear divisions, reappeared in roughly their old positions, and retained their sizes, shapes, and number. "The chromosomes," Boveri now asserted, "are autonomous individuals that retain this autonomy even in the resting nucleus."[29] In the conclusion of the paper he took the argument one step further, suggesting that each of the chromosomes of *Ascaris* might contain "different qualities."

The remarkable sea urchin, which had played the leading role in the development of the Hertwig's fertilization theory in 1875, took center stage once again at the end of the century when Boveri saw how it could be used to provide definitive experimental proof of the role of the nucleus and chromosomes in heredity. As Oscar Hertwig had first noted, sea urchin eggs were occasionally fertilized by two sperm simultaneously, and these doubly fertilized eggs bypassed the normal two-cell stage of development and instead went directly from one to four cells on the way to becoming fully formed larvae.[30] However, these tetrafoils, or *simultaneous fours* as they were more commonly called, never became healthy larvae, instead suffering from a variety of developmental irregularities and, most often, premature death.

Around 1901, Boveri saw how his early work on the mechanism of fertilization and cell division in the sea urchin could be used to prove that each chromosome in an organism was a unique individual containing different qualities. Already in 1888 he had shown that in addition to injecting its nucleus into the egg, the sperm donated a small circular body known as the centrosome, and that, independently of the nucleus, the sperm centrosome divided into two parts, which moved to opposite sides of the cell where they acted as attractive centers. Furthermore, he observed that threads emanated from each of these centrosomes and attached to opposite sides of each of the recently divided chromosomes, which were then dragged to opposite ends of the cell. While the newly fertilized egg containing one divided centrosome was an exquisitely well-designed machine to evenly distribute one set of longitudinally divided chromosomes to two poles, Boveri realized that the doubly fertilized egg, which received two sperm centrosomes that gave rise to four attractive centers, would not be nearly so effective. When four attractive centers were competing to attach to each split chromosome, Boveri conjectured, it would simply be a matter of chance which two of the four made the attachment and thus which two cells would receive the chromosomes.[31] When he computed the odds, Boveri found that the likelihood of getting four cells each of which contained a complete set of chromosomes was less than 1 in 3,800.[32]

Although Boveri's study strongly supported the idea that each chromosome (in a haploid set) was a unique individual and that an organism required at least one complete haploid set of them to function, it did not yet constitute a definitive proof. Once again the miraculous sea urchin provided precisely the tool needed to advance the science of heredity. While in residence at Naples, a young American, Thomas Hunt Morgan, made the seemingly insignificant discovery that by shaking dispermic sea urchin eggs he could sometimes produce three-cell embryos (called trefoils) in the first division rather than the usual tetrafoils. The beauty of Morgan's trefoils was that they allowed Boveri to formulate a testable hypothesis. For while the chance that each of four cells received a complete set of chromosomes was less than 1 in 3,800, the chance that each of three cells received a complete set was better than 1 in 25, therefore if the poor distribution of the chromosomes was truly the cause of the premature death of the dispermic embryos, then the trefoils ought to survive better than the tetrafoils.

With the expert assistance of his new wife, Marcella O'Grady, who was a former student of Edmund Wilson's, Boveri began the double-fertilization experiment in the winter of 1901. By the spring of 1902, four and a half months after they'd begun, the Boveris had the result in hand: Of 1,170 tetrafoils examined, none developed into healthy larvae, while 58 of 695 (or 8.3 percent) of the trefoils yielded perfect larvae.[33] It was a beautiful solution of the riddle of the double-fertilized eggs and a resounding confirmation of the theory of the individuality of the chromosomes.

— IT WAS A LONG WAY from the sophisticated world of European biology to the sun-bleached wheat fields of Kansas where Walter Sutton spent his summers off from college working on his parents' ranch. Standing in the "header box" on the horse-drawn wheat cutter, trying not to get pricked, stabbed, or buried alive by wheat was strenuous work.[34] In addition to spreading the fast-coming wheat, Sutton took the opportunity to collect insects to study in the lab of his mentor, Clarence McClung, where he had been working for his sophomore and junior years at the University of Kansas. The insect collecting proved to be far easier than the wheat spreading. Even in a nonplague year like the summer of 1899 when Sutton first discovered them, there were thousands of lubber grasshoppers hopping in with the wheat, crawling into overalls and down shirtfronts.

A letter from Sutton describing the great lubber grasshoppers elicited an urgent request for a sample from McClung. As a prank, Sutton enclosed a live specimen in his next letter, which managed to escape upon arrival, but the grasshopper was eventually corralled and brought to the lab for analysis. Upon learning of the outcome of McClung's observations, Sutton made the following prophetic prediction: "From what you say of the 'immensus' I infer that the gentleman's cells are about the largest that have ever been discovered, and if they are so our department may derive a little fame from that fact." The great cell biologist Edmund Beecher Wilson, who had been McClung's teacher, later described the lubber as "one of the finest objects thus far discovered for the investigation of the minutest details of cell structure."[35] Recognizing the potential windfall that had almost literally fallen into their laps, McClung wrote again urging Sutton to get as much as possible of the material, which was not a difficult task.

Early in college Sutton had decided to pursue the study of medicine, but the excitement of McClung's 1898 discovery of a new structure in the nucleus of the testicular cells of the locust *(Xiphidium fasciatum)* had caused him to drop his medical ambitions and take up research full-time. A similar darker-staining element had been seen in the fire wasp (*Pyrrochoris*) by the German cytologist Hermann Henking in 1891, and he had called it the "X element" because, although it was chromatic (stain-absorbing) like the ordinary chromosomes, unlike ordinary chromosomes it failed to dissolve into the nuclear soup during the period between cell divisions.[36] The fact that both Henking's X element and the structure McClung had noticed in the locust looked and behaved exactly like chromosomes in the period when the chromosomes were visible, however, was enough to persuade McClung that they might actually be chromosomes. Because his locust structure existed apart from the other chromosomes, McClung proposed to call the new element an "accessory chromosome."[37]

Over the next year, while Sutton began his analysis of the lubber chromosomes, McClung expanded the search for an accessory chromosome in a variety of different insect groups. In January 1900 he reported that he had found it in every case, noting that it came in a great variety of sizes and shapes and that it was this, he believed, that had been responsible for the great confusion that had surrounded it.[38] With growing confidence in the reality of the accessory chromosome, McClung mentioned his suspicion that it might behave strangely in sperm cell formation. At a division in which the ordinary chromosomes appeared to go to both poles, the accessory chromosome went to only one. "I hope by the aid of more favorable material," he wrote still withholding final judgment, "to reach a conclusion concerning the matter."[39]

In April 1900 Sutton published his first study on the development of the testicular cells in the lubber, providing strong confirmation of the reality of the accessory chromosome.[40] His paper, though, dealt only with the early stages of the maturation of the sperm cells and did not shed any new light on the idea that the accessory chromosome was distributed to only half of the maturing sperm cells (spermatids).[41]

One year later, on January 1, 1901, McClung submitted the third and most remarkable of his papers on the accessory chromosomes to the *Journal of Morphology*.[42] After weighing the evidence, he was now fully persuaded that the accessory chromosome was passed to only half of the daughter cells in the final stages of sperm formation. Not only had McClung reassessed the facts surrounding the accessory chromosome, but as suggested by the title of his paper, "The Accessory Chromosome: Sex Determinant?" he was now prepared to speculate on its significance. Assuming the chromosomes were functionally distinct, he deduced that the two kinds of sperm must have a different influence on the offspring produced from them. But what was there that might divide a generation into two even parts? The solution to this problem, when it occurred to him, was inescapable: "A careful consideration will suggest that nothing but sexual characters thus divides the members of a species into two well-defined groups, and we are logically forced to the conclusion that the peculiar chromosome has some bearing upon this arrangement."[43] This was the first example in which a particular chromosome had been shown to determine a definite combination of characters, those for female or maleness.

Although he made a truly inspired deduction, McClung then went wrong. Assuming that the function of the extra chromosomes was to "carry the transformation beyond the production of ova to sperm," he was convinced that the X chromosome contained factors that facilitated this transformation and therefore would be found only in males.[44] More than two years after its initial submission, McClung's tour de force "The Accessory Chromosome: Sex Determinant?" was finally published in the *Biological Bulletin* in 1903. In the period between the first submission of the paper and the publication of the final version, Sutton had become convinced that the female lubber grasshopper had 22 chromosomes while the male had 23.[45] Such a finding, had it been true, would have greatly supported the theory that the accessory chromosome was male-determining, and it was undoubtedly partly in response to Sutton's finding that McClung added an addendum to his original paper. "Regarding the theory of its function advanced in this paper," he now added, "I can say only that it has, if anything, been strengthened by later researches."[46] But it turned out that Sutton had been wrong, and he had missed one chromosome pair in female ovicular cells, which were far less plentiful and more difficult to work with than male testicular cells. Later analysis would show that females had 24 chromosomes and males had 23, which suggested that males had one copy of the accessory and females had two. Contrary to McClung's speculations, it appeared that maleness was determined by the *lack* of a second copy of the X chromosome.

In the meantime, it had become clear to McClung that Sutton could not realize his full scientific potential in Kansas. Although McClung was deeply attached to his star student, McClung urged him to pursue a Ph.D. degree with Edmund Wilson at Columbia and helped him secure a fellowship. McClung recalled helping Sutton pack on his last night in Russell and then falling asleep on a lab bench. "In the morning with a heavy heart I said good-bye," he later wrote describing their final night together, "and saw him pass from under my halting and immature guidance into that of the best men the country afforded—and the pleasure was near as great at his good fortune as if it had been my own."[47]

With Wilson's expert guidance, Sutton now continued his study of the male grasshopper chromosomes. Having already noted that there were

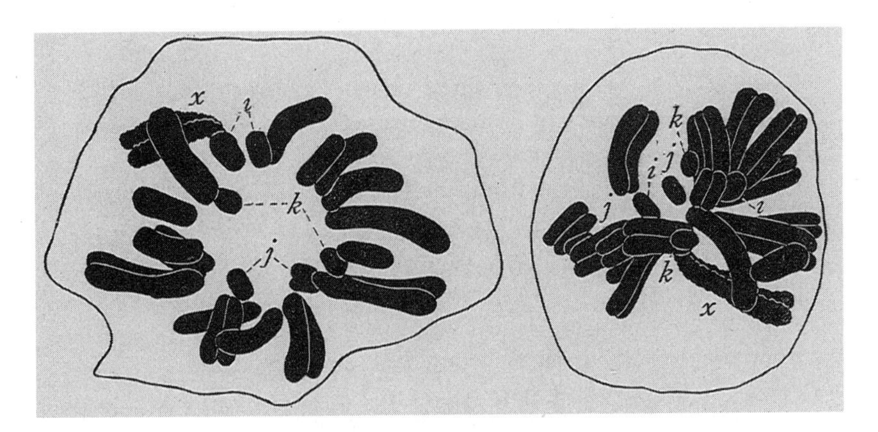


**Sutton's drawing of eleven different-size chromosome pairs.** From Walter S. Sutton, "On the Morphology of the Chromosome Group in Brachystola Magna," *Biological Bulletin* 4 (1902): 26.

clear differences in the sizes of the individual chromosomes while working in McClung's lab, Sutton now succeeded in displaying the chromosomes in far more revealing way. Leaving aside the odd accessory chromosome, Sutton saw that there were 22 chromosomes that could be grouped into 11 distinct pairs forming a nearly smooth progression from smallest to largest.

While his first paper written under McClung had reported exclusively on observations of the early maturational divisions of the spermatogonia, where the chromosome number remained constant, Sutton extended his investigation at Columbia to include the later stages of maturation. Sutton's key observation was that the 22 chromosomes found in each of eight or so generations of spermatogonia suddenly, in the first generation of spermatocytes, appeared as 11 chromosomes. Although the chromosomes in these early stages of the forming nucleus were less regular in shape and therefore more difficult to measure than the more condensed, solid chromosomes characteristic of the later stages of nuclear division, it was still possible to see that they represented a series of increasing length.

The fact that the spermatocytes contained 11 unpaired chromosomes of graded length instead of the 22 chromosomes seen in spermatogonia, convinced Sutton that each of the 11 was in fact a double chromosome made up of two members of a matching pair.[48] Close observation showed that each member of a matched pair was itself longitudinally split, forming

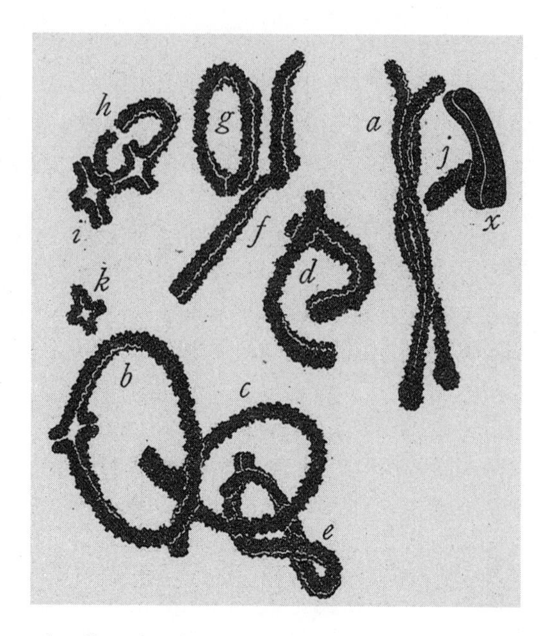


**The eleven longitudinally split chromosomes that appear in primary spermatocytes.**

From Walter S. Sutton, "On the Morphology of the Chromosome Group in Brachystola Magna," *Biological Bulletin* 4 (1902): 29.

part of a tightly packed structure known as a *tetrad* (see the diagram illustrating spermatogenesis). As the nucleus divided, each of the 11 tetrads came apart, sending each of the already divided chromosomes that made up the matched pairs their separate ways into the two forming nuclei. As a result of the pairing of the identical partners and their subsequent separation, each of the resulting daughter cells–the "secondary spermatocytes"—contained only one of the two matched chromosomes of the original spermatogonia.[49] In the final division, each of the longitudinally split chromosomes in the secondary spermatocytes was distributed to the spermatids (the direct precursors of the sperm cells), each of which now contained exactly half of the full complement of chromosomes present in the other cells of the body.

As McClung had first noticed, Sutton confirmed that the accessory chromosome seemed to march to its own beat. Unlike the normal chromosomes, which dissolved into the nuclear soup between cell divisions, the solitary accessory chromosome remained intact and apart throughout the

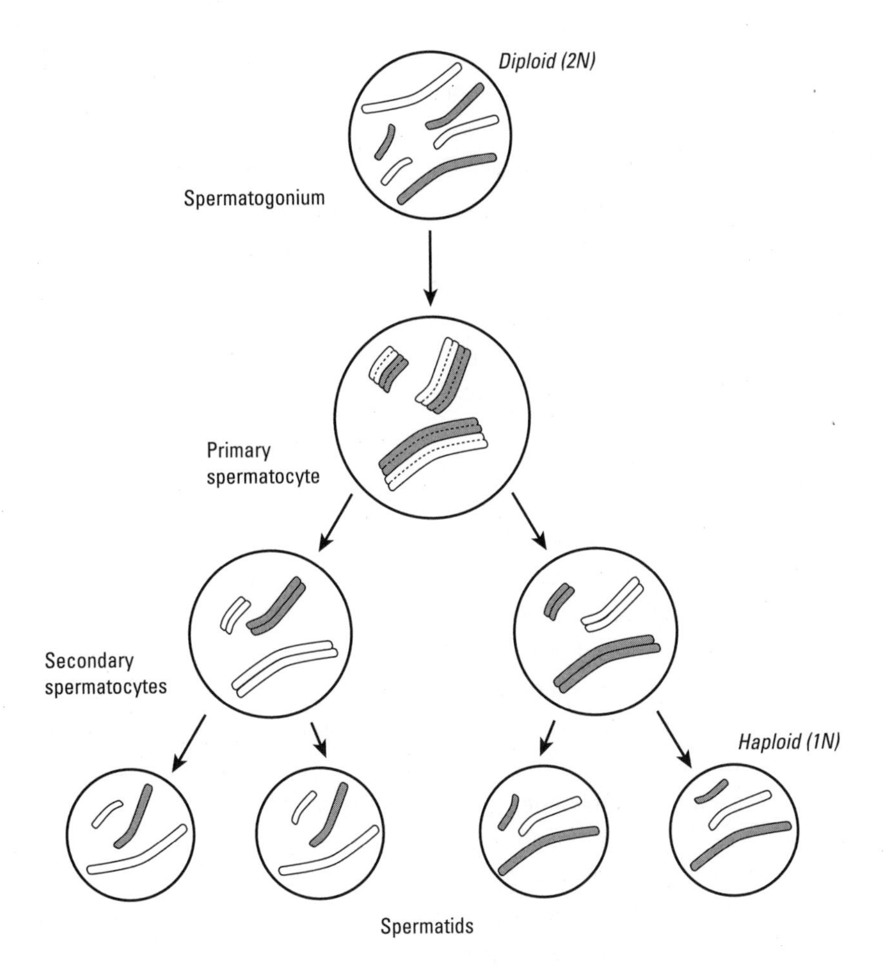


**The process of spermatogenesis.**

Illustration by Tamara L. Clark.

transformation of the spermatogonia into spermatocytes, only joining the other 11 tetrads at the moment before the first division. Unlike the other chromosomes, however, each of which was separated into two halves that were sent to two daughter cells, the accessory went undivided into only one of the two secondary spermatocytes. In the final stage, the accessory chromosome was divided and distributed to two (of four) daughter spermatids.

The clear depiction of the processes of the formation of the "reduced" sperm cells, containing only half of the chromosome number of all the other cells of the body, as well as the details of inheritance of the accessory chromosome, were major accomplishments, but Sutton had deduced something of even greater significance. As Wilson would later recall, in the early spring of 1902 Sutton came into his office and told him that he had finally understood why "the yellow dog was yellow." With great humility, Wilson would later confess that he did not at first "fully comprehend his conception or realize its entire weight."[50] Over a summer filled with swimming, sailing, and science in North Carolina and Maine, Sutton gradually brought Wilson around.

As Sutton acknowledged in the first sentence of his extraordinary paper written in the fall of 1902, he was building on Boveri's recent demonstration that each chromosome was a unique individual containing a unique set of hereditary determinants.[51] Applying Boveri's principle, Sutton declared that each of the different size groups in the lubber must correspond to a definite set of characters, and that a given size chromosome in the sperm nucleus corresponded to the same size chromosome in the egg nucleus. When the two members of each chromosome pair were joined together in the spermatocyte, the maternal and paternal hereditary qualities were localized in a single continuous chromosome mass for the first and only time in the lifetime of an organism, and when, during the process of sex cell formation, they were again separated, so were the corresponding qualities.[52] This then was the basis for the combining of the factors in Mendel's hybrids and the segregation into pure types seen among the gametes. A model of understatement, Sutton's paper concluded with one of the central insights of biology:

> I may finally call attention to the probability that the association of paternal and maternal chromosomes in pairs and their subsequent separation during the reducing division as indicated above may constitute the physical basis of the Mendelian law of heredity.[53]

The first paper Sutton wrote in Wilson's lab, which was submitted for publication on October 17, 1902, was in many ways a direct outgrowth of Boveri's work, but the second one, which was completed two months later, bears the unmistakable stamp of Bateson. By a wonderful twist of fate it happened that Bateson lectured for ten days in New York City in late September. According to McClung, Sutton had been fortunate enough to attend the lectures and had been deeply struck by them.[54] Bateson spoke at the Biological Laboratory, which would later merge with the Station for Experimental Evolution to become the Cold Spring Harbor Laboratory, to an audience of several hundred in a whirlwind tour that included Boston, Woods Hole, Chicago, and Washington. "It has been *Mendel, Mendel all the way,* and I think the boom is beginning at last," Bateson wrote home after his presentations were over.[55] The timely review of the basic facts of Mendelism is clearly evident in Sutton's next paper, which was submitted for publication on January 25, 1903.[56]

Contrary to a widespread belief that the male passed on only the chromosomes he had inherited from his father, and likewise the female passed on only those she received from her mother, Sutton now maintained that the selection of the male or female homologue during the reducing division was an entirely random affair. From this it immediately followed that the segregation of the pairs of homologous chromosomes into sperm and egg cells followed exactly the same rules as the segregation of Mendel's factors. In particular, if a pair of contrasting characters (*A*/*a*) was located on one pair of homologous chromosomes, and another pair of contrasting characters (*B*/*b*) was located on a different homologous pair, then one would expect to see equal numbers of each of the four combinations of factors, the fact from which Mendel's law of independent segregation followed.

Sutton also realized that each chromosome must contain more than one factor, otherwise there could be only as many characters as chromosomes, which was clearly absurd. Furthermore, he suggested that traits that were located on the same chromosome must be inherited together, and thus he gave the basis for the linkage of genes, an idea that would be thoroughly explored in the decade ahead.[57] With the publication of Sutton's remarkably lucid paper, the results of nearly three decades of intensive investigation of the nucleus and the chromosomes had finally come into accord with the facts of heredity that Mendel had elucidated in 1865.

**Thomas Hunt Morgan (1891).**

From the 1891 yearbook at Johns Hopkins University.

CHAPTER 9

# Sex Chromosomes

EDMUND WILSON HAD SOON appreciated that Sutton truly had understood why the yellow dog was yellow, but others were not so quick to sign on to the idea that chromosomes and Mendelian factors bore an intimate relationship or even that chromosomes played a decisive role in the transmission of the heredity. Ironically, among the most stubborn skeptics was Wilson's close friend Thomas Hunt Morgan, who would, almost against his will, take the first crucial step in the experimental verification of the modern chromosome theory by proving that particular genes resided on the X chromosome. But it would be a long, slow journey, hampered by Morgan's extreme empiricism and disdain for speculative theories, including Darwin's theory of natural selection.

To those who knew him well, Morgan's opposition to the chromosome theory, and more particularly to McClung's suggestion that the accessory chromosome was involved in sex determination, would have been almost a forgone conclusion. True to form, Morgan considered the idea that there were two kinds of sperm and that the difference between them was connected with the determination of sex to be ungrounded speculation—"nothing more than supposition," as he wrote in a 1903 review of the subject.[1] In the autumn of 1903, he was more than ever convinced that chromosomes had very little to do with sex determination, and he believed he'd found the perfect organism in aphids—the bane of rose gardeners—with which to prove his point.

Part of what made the insect so troublesome to gardeners but desirable to Morgan was its unique ability to reproduce in two modes. Hatching from a winter egg in the spring, a wingless female laid unfertilized eggs that gave rise to a burst of new daughters, all of whom continued to reproduce asexually, wasting no time or energy on the production of males. But then, as the weather turned cool in the autumn, these parthenogenetic females suddenly gave birth to a generation of sexual females and males, which in turn mated to form a new winter egg that would give rise to a parthenogenetic female the following spring.[2] Although there was a certain beauty to the closing of the circle, it was the middle step, the ability of parthenogenetic females to produce *males* and a different kind of female, that so excited Morgan about the aphids. If it could be shown that the "sex change" was caused by a drop in the temperature or a change in the food supply, for example, and there was much anecdotal evidence to indicate that this was so, this would be a significant blow to the notion that the chromosome reigned unchecked over the affairs of the cell.

That fall Morgan divided up the aphid work with his star student, Nettie Stevens, who had just that summer completed her Ph.D.[3] Morgan would study the effects of a changing environment on sex determination in aphids while Stevens, who specialized in cytology, conducted the analysis of the chromosomes of the germ cells, and particularly the chromosomes of the male and female eggs.[4] By the end of 1904, Morgan concluded that the effect of food and temperature on reproductive mode was highly exaggerated, if it existed at all. In many cases, parthenogenetic females went on reproducing asexually into late autumn and, in one case, into the winter freeze, while the sexual forms often appeared long before the weather changed. A separate analysis showed that there was no significant difference between the nutrients provided by the June leaves on which newly hatched parthenogenetic females emerged and the autumn leaves on which the sexual types later appeared.[5]

Although the failure to find external factors that affected the sex change was discouraging, Stevens's chromosomes analysis, which she submitted for publication in December 1904, provided some hard facts against the chromosome view of sex determination. As expected, Stevens found that all of the early-arriving female parthenogenetic eggs were identical, containing five pairs of chromosomes of differing sizes. Likewise the eggs that gave rise to sexual females were in no way different from those giving rise to the earlier generations of parthenogenetic females. But the most exciting discovery to Morgan was that the male and female eggs were absolutely identical in chromosome content, each containing the same five pairs of chromosomes found in the female and males to which they gave rise.[6] In her precise and cautious prose, Stevens wrote:

> There is no evidence in my material of any difference between the maturation of the female parthenogenetic egg and that of the male egg and until I can procure more material and examine the point further, I shall assume that the method of maturation of all parthenogenetic aphid eggs is the same.[7]

As it would turn out, Stevens's caution was more than warranted, and she had in fact overlooked a crucial difference in the maturation of male and female eggs, but for the moment the assumption that the male- and female-producing eggs contained identical chromosomes stood, and Morgan continued his hunt for the elusive external conditions that would change females into males. In January he quipped to Charles Zeleny, another former student, "My aphids poke along parthenogenetically and refuse to adapt more modern methods of reproducing themselves," but, he hastened to add, he had "not exhausted all the possibilities by a long shot."[8] That summer Morgan wrote his close friend and former mentor Hans Driesch that "Wilson is wild over chromosomes, says so himself—they do very funny things," but he was still confident that Wilson was barking up the wrong tree.[9] In fact, as he explained to Driesch, he had some results that indicated that the male hereditary qualities were inherited via the cytoplasm, and that the chromosomes were not even involved.[10]

The following month, Wilson's groundbreaking new study of sex determination in Hemiptera—a large class of insects that included cicadas, leafhoppers, and aphids—changed everything.[11] In his paper Wilson announced the discovery of a hitherto unnoticed chromosome pair, which he dubbed "idiochromosomes." The prefix *idio-,* he explained, was meant to emphasize the strange nature of the new chromosome pair, which manifested itself in several ways, but most noticeably in the fact that the two members of the matching pair were of unequal size. In subsequent years Wilson would suggest that the idiochromosomes could also be called "sex determinants" or "sex chromosomes," and later he would denote the larger member of the pair the X chromosome and the smaller one, when it existed, the Y chromosome.[12] As in the case of McClung's accessory chromosome, Wilson found that the sperm differed with respect to the idiochromosomes, half of them containing the larger X and half containing the smaller Y. "Their marked difference in size suggests a corresponding qualitative differentiation," wrote Wilson, reflecting his adherence to Boveri's doctrine of the individuality of the chromosomes, "and this inevitably suggests a possible connection between them and sexual differentiation."[13]

The fact that both the accessory chromosomes and the idiochromosomes seemed likely to play a role in sex determination led Wilson to consider whether there might be some relationship between them. One clue was provided by the observation that there was a more or less continuous range of size differences between the larger and the smaller of the idiochromosomes across the various species he had examined. At one extreme, the bigger chromosome was six times as large as the smaller one, and at the other the two were the same size. Wilson boldly conjectured that the smaller of the two idiochromosomes might be in the process of disappearing, and that in more evolved species, such as grasshopper, it had disappeared all together. This view, he wrote, "obviously suggests the view that the single accessory of other forms may have arisen by disappearance of one of the idiochromosomes."[14]

The conjecture that the accessory chromosome represented the larger of the two idiochromosomes that remained after the Y had disappeared from the species led Wilson to a grand unifying theory of sex determination. In this view, all reduced egg (or oöcycte) nuclei contained a set of chromosomes that included a single X chromosome. Sperm nuclei, on the other hand, were of two types: in some species they contained an X chromosome exactly like the one found in the egg, and in others they contained a smaller X (Wilson's Y) or no X at all. The sex of the egg, which was determined at the moment of fertilization, depended on the nature of the fertilizing sperm. Eggs that were fertilized by sperm containing the X chromosome produced females, while those that were fertilized by sperm lacking the X altogether, or containing the smaller X chromosome, produced males.

A more beautiful and consistent theory would be hard to imagine, but nonetheless one recalcitrant fact seemed to contradict it. The problem concerned the parity of the chromosome number in species containing an accessory chromosome. If the chromosomes come in pairs, as had been definitively demonstrated by Sutton, and the "accessory chromosome" is truly the vestigial member of a chromosome pair, then the theory predicted that the male germ cells prior to reduction must contain an odd number of chromosomes. The fact that the spermatogonial cells of the squash bug, *Anasa tristis,* as well as several other species, had been found to contain an even number of chromosomes represented a nearly mortal blow to the theory.[15] Reluctantly Wilson was forced to admit that *Anasa tristis* presented, "a formidable if not insuperable difficulty" to his theory, and, in the end, he reluctantly conceded defeat. In the final draft of the paper he submitted for publication on May 5, he wrote, "Apparently therefore the hypothesis I have suggested must in the present state of our knowledge be considered untenable."[16]

However, some time after the submission of his manuscript, Wilson made a last-minute discovery that saved his theory. Over the summer he devoted himself to an extended review of the cytological slides that had caused him to doubt his theory in the first place. In a footnote added to the paper after it had already been sent to press, Wilson reported that a careful reexamination of the male squash bug chromosomes revealed, contrary to all previous reports, that males contained 21, not 22, chromosomes. "This wholly unexpected result perhaps justified a certain skepticism in my mind in regard to the accounts of other observers, who give an even spermatogonial number for forms possessing an accessory chromosome," he explained, now confident that his original theory had been correct after all.[17]

What Wilson failed to mention was that on May 29, twenty-four days after sending off his completed manuscript, he had received a paper for review from the Carnegie monograph series written by Nettie Stevens, who had recently returned from a visit to Boveri's lab at the University of Würzburg. Stevens's paper on the sex chromosomes of the yellow mealworm, *Tenebrio molitor*, provided a striking confirmation of Wilson's findings in *Hemiptera*. Like Wilson, she had found that in the early stages of sperm development there was one pair of unequally sized chromosomes among the 10 chromosome pairs characteristic of the species. After the reduction division, half the sperm contained the larger member of the unequal pair, and half contained the smaller one. *Tenebrio* eggs, on the other hand, contained 10 pairs of equal-size chromosomes, and all the eggs that resulted from the reduction division contained an identical set of chromosomes.

But Stevens had taken her analysis one step further than Wilson had by analyzing the chromosome content of the somatic cells in males and females as well as of the reproductive cells, and this analysis provided a crucial additional piece of evidence: All of the female somatic cells contained 10 matching pairs of chromosomes, while the male somatic cells contained 9 matching pairs and 1 uneven pair. The conclusion was inescapable:

> Since the male somatic cells have 19 large and 1 small chromosome, while the female somatic cells have 20 large ones, it seems certain that an egg fertilized by a spermatozoon which contains the small chromosome must produce a male, while one fertilized by a spermatozoon containing the 10 chromosomes of equal size must produce a female. The small chromosome itself may not be a sex determinant, but the conditions in Tenebrio indicate that sex may in some cases be determined by a difference in the amount or quality of the chromatin in different spermatozoa.[18]

Not only had Stevens independently discovered a theory that was strikingly like Wilson's idiochromosome theory, but she had provided a crucial piece of confirming evidence that Wilson had lacked, and immediately after receiving her manuscript he devoted himself to an extended review of the cytological slides that were the basis of the claim that males in *Anasa* contained an even number of chromosomes.[19] Although he did not directly acknowledge that he had read her results and that they had influenced his thinking, Wilson did add a second footnote acknowledging Stevens's independent discovery of the fact that there were two types of sperm distinguished by the presence or absence of the small chromosome, as well as what he referred to as "the significant fact" that the small chromosome was found in all the somatic cells of the male, while the female somatic cells contained only matched pairs of larger chromosomes.[20]

In October 1905, Morgan wrote Driesch that Wilson's new findings "make it look at first sight as though the chromosomes were the thing." But upon reflection he had realized that the logic of the paper "lands you in an absurdity."[21] Though he did not spell out the details, Morgan had indeed discovered a flaw in Wilson's argument that followed from his assumption that sex chromosomes conferred specific male and female qualities.[22] When he realized the flaw in Wilson's theory, Morgan continued, he had felt a great relief. Once again comfortable in his conviction that chromosomes were not the sine qua non, Morgan argued that the simplest way to account for the fact that there were two kinds of sperm "was to assume that there were two kinds of protoplasm."[23]

In the meantime, however, Morgan had forged ahead with his own study of sex determination on a different track. Already in the spring of 1905, Morgan had begun collecting various species of an aphid-like insect group known as the *Phylloxera*. As was the case with aphids, a colony was begun by a single insect, which hatched from a winter egg. This "stem mother" then reproduced asexually, giving rise to a generation of winged migrants, also parthenogenetic, that laid male and female eggs. Finally, the union of the resulting male and female insects gave rise once again to a winter egg, out of which would hatch a new parthenogenetic stem mother the following spring. But what excited Morgan about the phylloxerans, and set them apart from aphids, was that the male eggs were smaller—sometimes thirty times smaller—than the female eggs, and the males were also significantly smaller than the females. Morgan was convinced that in this strange asymmetry he would find a clue to the mystery of sex determination.[24]

This time Morgan himself undertook the "painful cytological work,"[25] as he described it to Driesch, finding that even the very different-appearing male and female eggs of phylloxerans appeared to be endowed with an identical set of chromosomes.[26] Thus, Morgan concluded, the difference between male and female eggs was not nuclear nor was it due to external conditions, as his previous work in aphids had finally convinced him. Having exhausted all other options, he next turned to the cytoplasm to look for clues to sex determination. It was immediately evident that male and female eggs followed radically different developmental paths. Almost from the moment they began to differentiate, male eggs began to devote a shocking percentage—nearly half—of their embryonic cells to the production of sperm. Not only were the embryos packed with sperm, but the sperm were ready to swim the moment the egg hatched. If early embryonic development was under cytoplasmic control, a point of view that was being championed by his friend Driesch, then it was possible to conclude that the immediate cause of sexual differentiation of eggs was indeed cytoplasmic.[27]

In a *Science* review in December 1905, Morgan once again argued that "sex determination may not be the result of differential *nuclear* divisions that locate sex determining chromosomes in different cells."[28] The assumption that there were preformed male or female qualities and that these qualities were separated in the formation of egg and sperm cells, he argued, was reminiscent of the concrete thinking of the early spermists, who believed that the sperm was a miniature replica of the adult form.[29] He called this view "preformist." In the more dynamical view, the "epigenetical view" as Morgan termed it, every egg contained the potential to be either male or female. The sex of the embryo was determined by an internal condition that was present in the egg or sperm, which led to the dominion of one or the other of the alternatives.

The same kind of reasoning that made him doubt the chromosome theory of sex determination had also led Morgan to begin to question the basic assumption of Mendelism. In particular, the recent finding in mice that it was impossible to isolate a pure-breeding yellow-coated mouse led him to question the entire notion of the purity of the germ cells.[30] He began to doubt that it was possible to cleanly extract the dominant factor from a recessive, and that the idea that factors combined to give a 3:1 ratio was only a convenient fiction. His dictum, as he explained it to his friend Davenport, was "once crossed always mixed," and he insisted that there was "no evidence to show that pure extracted races do not carry latent characters, and a good deal of evidence to show that they do."[31] As in the case of the two sexes, he believed that the contrasting characters coexisted, and local conditions determined which was expressed.

Swimming directly against the ever-strengthening tide of Mendelism, he also attacked the stability of the Mendelian factors in his address to a conference on sex determination in December 1906.[32] As he had suggested in regard to sex determination, the view that characters resided fully formed in the embryo was an anachronism. It did not make sense to treat them as "entities that could be shuffled, but seldom get mixed." Continuing with the metaphor, he rejected the idea that with "each new deal the characters are separated, one germ cell getting one character and another the contrasted characters."[33] With an uncanny precision, he put his finger on the formal properties of Mendel's factors that made possible an exact theory of heredity—that they could live together for countless generations without producing any influence on each other—and then proceeded to reject them without a pang of doubt. In fact, it hardly seemed to faze him that he now stood opposed to the three major developments in contemporary biology—Darwinism, the chromosome theory of sex determination, and Mendelism. It was this immunity to the prevailing opinions and, in certain cases, to the power of logic to persuade, that would lead some of his colleagues and students to despair.

The following year Driesch was slated to be the lead speaker in the session on experimental zoology at the Seventh International Zoological Conference, which was held in Boston, but he had withdrawn at the last minute because his wife had fallen ill. "Embryology and Experimental Embryology had rather a slim attendance, but the men present were just those who are especially interested in these matters, so that our audience was select if not large," Morgan reported wistfully to Driesch. "The crowd went mainly to hear papers on chromosomes and Animal Behavior." But the really big news of the conference was Bateson, who "made a great hit, both personally and with his address."[34]

After years of battling against the anti-Mendelian biometric school in England, Bateson was swept away by the enthusiasm of the Americans for the new religion of Mendelism. "It was rather fun being lionized so much," he confessed after giving the obligatory lecture at Davenport's Station for Experimental Evolution at Cold Spring Harbor, "but it will make the prophet's 'own country and his own house' rather flat!"[35] At Woods Hole the following week, the frenzy continued. While he complained bitterly about "the tyranny of religion and temperance" among the New Englanders that prevented him from getting a drink, the scientists were wild with enthusiasm for Bateson and the new gospel he was spreading: "I only retain faint control over my movements. I am everyone's prey, being torn in pieces by my admirers!" he wrote from the Marine Biology Labs.

At Woods Hole, Bateson met E. B. Wilson, whom he liked, as well as his wife, whom he found to be "a particularly nice girl, rather young for him," and his mother, "the great lady of Woods Hole, Europeanized and inclined to religion, as so many here are." Concerning the other great Woods Hole luminary, Bateson was succinct and to the point: "Morgan is a blockhead. Here he counts for a good deal all the same."[36] Furthermore, he noted, "Morgan supplies the excitement in the place. He is in a continual whirl—very active and inclined to be noisy."[37]

— LOUD AND SLIGHTLY OBTUSE, the whirling dervish Morgan moved into high gear in the fall of 1907, and most of all, he was busy with the phylloxerans. New material he'd collected that spring and summer made it possible to study spermatogenesis and the entire cycle of cytological phenomena. That November Morgan confessed to Driesch, with whom he had often shared his doubts about reductive science, that Wilson seemed to have "some fine things." However, he hastened to add the caveat that he might be "too near the light to see sharply."[38]

Indeed, Wilson's new work was blinding in its clarity. He had now found two classes of sperm, differing in one chromosome, in more than sixty insect species. Furthermore, he had shown that in each case these differences in the sperm chromosome sets corresponded to precisely parallel differences in the chromosome groups of the male and female somatic cells. The chromosome theory of sex determination was definitely coming of age.

By the beginning of the new year, 1908, after months of eye-straining chromosome work on the tiny cells of *Phylloxera,* Morgan made the breakthrough that finally resolved the paradox of Nettie Stevens's observation that male and female aphid eggs seemed to contain identical chromosome sets. He made a preliminary announcement at the Society for Experimental Biology and Medicine in February,[39] and a detailed treatment followed a year and a half later. Morgan's 113-page treatise on sex determination in aphids and phylloxerans was unmistakably a masterpiece of scientific reasoning and technical virtuosity.[40]

Contrary to his previous report on phylloxerans and Stevens's finding in aphids, Morgan now reported a difference in the maturation of the male and female parthenogenetic eggs of the winged migrants. While female eggs retained a complete set of paired chromosomes, the chromosomes of eggs destined to give rise to males underwent a subtle pas de deux in which they rid themselves of a single member from each of two pairs of chromosomes. The working out of the details, which would have been an impressive feat even for an expert cytologist, led Morgan to the remarkable conclusion that "the egg as well as the sperm has the power of determining the sex by regulating the number of its chromosomes."

In male phylloxerans, the two unpaired male chromosomes appeared to play the role of the solitary X chromosome in other insect groups. During the reductive division, the *two* unpaired male chromosomes went into one daughter cell while the other daughter received neither, giving rise to two different types of haploid sperm. While the daughter cell that inherited the two extra chromosomes gave rise to two "female-producing" sperm, the cell that failed to inherit the extra chromosomes withered and died. Thus the only functional sperm were "female-producing." These sperm combined with the eggs laid by sexual females, which contained a full haploid set of chromosomes, producing a new parthenogenetic egg, and the cycle began anew.

After so brilliantly divining the secrets of the phylloxeran chromosomes, there was something magnificently consistent in Morgan's refusal to abandon his cytoplasmic theory of sex determination. Because the transition from parthenogenetic female to sexual female did not involve a change in chromosome number, Morgan now argued, the loss of chromosomes need not necessarily be the cause of the sex of the males. In fact, he suggested, it could be just the other way around: the loss of the two sex chromosomes might actually be a consequence of the cells being already male.[41] Though his argument was defensible, his heart no longer seemed to be quite so much in opposing, and in the end Morgan granted, "The chromosomal relations, however, remain still a fact and a very extraordinary fact. It is hard to conceive that these relations have no connection with sex determination."[42] Contrary to his deepest wishes, Morgan had finally succeeded in clearing the path for the chromosome interpretation by reconciling a set of seemingly irreconcilable facts.[43]

— WHILE PURSUING the *Phylloxera* study, Morgan had been moving forward on several other fronts, most importantly with research on De Vries's mutation theory, in which he saw the potential for creative evolutionary change. He had visited De Vries's experimental garden in 1902 and been greatly impressed by the mutating *Oenothera,* but he had missed De Vries himself.[44] In his 1903 *Evolution and Adaptation,* Morgan had expressed his excitement over the mutation theory, arguing that natural selection could not possibly have played the central role in evolution that Darwin claimed for it. As Morgan saw it, it was the mutations themselves that dictated the direction of evolution, not the selection of minute differences.[45] While De Vries had granted that selection had the ability to pick and chose among the new mutant forms, Morgan denied it any formative role at all, allowing only that selection might eliminate mutations that were positively harmful.

The two men seem to have met for the first time in California during the summer of 1904 when De Vries was lecturing in the Berkeley summer school, and they had hit it off immediately during a hike along the coastal brush on the Monterey Peninsula. Among the thrilling array of new plant species he had never before seen, De Vries also found one of his beloved evening primroses, *Oenothera cheiranthifolia,* to show Morgan.[46] They met again on several occasions in California, and at the end of the summer at the International Conference on Science and the Arts in St. Louis. In the fall, Morgan invited De Vries to stay with him in New York City, where Morgan had just been recruited to take up the position of professor of experimental zoology at Columbia. Shortly after De Vries returned to Amsterdam, Morgan wrote Driesch that De Vries was "a most interesting and agreeable man,"[47] and it was clear that from De Vries's account of his visit to America that the feeling had been entirely mutual. De Vries recognized in Morgan a valuable advocate for his mutation theory, and in De Vries's theory of the origin of species, which seemed to lend itself to experimental verification, Morgan found an antidote to what he saw as the ungrounded, polemical writings of Darwin and his followers.

In January 1905, shortly after De Vries left New York, Morgan wrote his former Ph.D. student Charles Zeleny: "If you come across any animals that are mutating send me a few of them. It seems to me there must be many forms that occasionally 'jump' and with encouragement they might be kept on the jump."[48] In April 1906, De Vries visited America for a second time and once again stayed with the Morgans, and Morgan was more impressed than ever. "Not only was he personally charming," he wrote Driesch shortly after De Vries left, "but far more level headed and at the same time original than most biologists one meets."[49]

That summer in Woods Hole, Morgan took up the study of mutation in earnest, hunting for mutations in chickens, ducks, mice, guinea pigs, and rats and flirting with raising butterflies, hoping to find some animal that was jumping or that could be induced to jump. Following many of the suggestions that De Vries had laid out in his 1904 inaugural address to the newly created Carnegie Station for Experimental Evolution in Cold Spring Harbor,[50] Morgan tried to induce mutations by injecting a variety of salts and other chemicals into the pupae of insects in the region of the reproductive tissue. Although the injections did not seem to do anything, the next summer he found that centrifuging insect eggs seemed to have produced a fly with abnormal wings, and, moreover, the deformed wings were passed to the next generation. In a letter to Driesch, he expressed a near certainty that it was the centrifuging that had resulted in the altered fly. While he could not rule out other causes, the significance of the result had not escaped him: one way or another he had managed to isolate a mutation in a fruit fly.[51]

A year after the isolation of the mutant with the deformed wings, Morgan started to work more with the fruit fly *Drosophila,* which was in many ways the ideal animal for mutation work. It was easy to grow and easy to test, the two conditions that De Vries, who took it as an article of faith that all organisms were subject to brief but intense periods of mutation, had considered essential for mutation studies.[52] It could be bred all year round in small milk jars, lived on mushed up bananas, and gave rise to a new generation every twelve days. Beginning around the summer of 1909, Morgan, again paying strict attention to De Vries's emphasis on external conditions, subjected the flies to wide temperature ranges and various chemicals, as well as to radium and X-rays.[53]

Following De Vries's injunction that "no differentiating marks, however slight, should be considered as insignificant,"[54] Morgan painstakingly examined each fly with a jeweler's loupe. In the first months, Morgan discovered that he could select for flies with a dark marking that looked like a trident across the normally amber-colored thorax, but it was only in this single character that the flies differed, and the color markings were sometimes subtle and hard to read. In early 1910 Ross Harrison, who had been a fellow graduate student with Morgan, recalled visiting Morgan at Columbia and his waving his hand at a wall full of vials filled with *Drosophila* and exclaiming: "There's two years work wasted. I've been breeding flies all that time and have got nothing out of it."[55]

After Harrison's visit, Morgan's prospects suddenly improved when out of the blue his flies seemed to start sporting. In a stock that had been breeding true for the absence of the dark trident marking for four generations, Morgan suddenly discovered that 32 of 199 members of the fifth generation showed some dark specks just under the wings. Granted it wasn't the creation of a full-blown species, but it was a definite and clearly defined mutant type, which Morgan dubbed *speck.* In the next generation, he found that 1 in 15 flies were of the mutant type. This was more like it. Not only did the strain seem to be jumping, but it was keeping on the jump.

Sometime in May 1910, Morgan spotted a single mutant white-eyed male fly in a culture filled with normal red-eyed brothers and sisters. When the white-eyed male was mated to his sisters, 1 of 163 progeny was again white-eyed. The white-eyed fly seemed to be yet another example of a spontaneous mutation, and Morgan became more convinced than ever that the flies had entered a mutating period. He believed he may have finally found the animal equivalent of De Vries's *Oenothera* mutants, and he announced his new results in an abstract entitled "Hybridization in a Mutating Period in Drosophila" on May 18, 1910.[56] The mutation for white eyes, which would change the course of modern genetics, merited only two sentences tacked on at the end of the abstract.

Although further crosses of the white-eyed male to his red-eyed sisters continued to yield a small but appreciable number of new white-eyed "sports," by June, Morgan's thinking about "white" had undergone a radical transformation.[57] If he ignored the additional white-eyed sports, he saw that the cross of the white-eyed male with his red-eyed sisters behaved like a standard Mendelian cross of a recessive by a dominant, giving exclusively red-eyed hybrids in the first generation and the classic 3:1 ratio of red- to white-eyed when the white-by-red hybrids were interbred. But the truly arresting thing was that recessive white eyes appeared only among the grandsons, behaving exactly like color-blindness in humans, which was passed from the affected male, through his daughters, who were unaffected, to the grandsons. On June 15 he wrote exuberantly to his friend Goodale from Woods Hole, "My white eyed fly gives a splendid case of sex limited inheritance: the $F_2$ gives white eyes only in the males,"[58] and he included a brief sketch of a model that could be used to explain the results.

As it turned out, it was Morgan, not his flies, who had entered a mutating period in the summer of 1910. His first great inspiration was that the sex-limited inheritance pattern of white eye color could be explained if he assumed that the color factors were linked to the X chromosome. As he first explained it in his paradigm-shifting *Science* paper of July 11, the eye-color factor (red or white) and the X chromosome were "combined, and have never existed apart."[59] Later that fall, in a graduate seminar, Morgan was more explicit, commenting that the red color factor sticks to the X chromosome and that it may therefore be regarded as being in the X chromosome.[60]

In particular, the hypothesis that red was stuck to X accounted perfectly for the fact that white eyes were found in half of the male grandchildren and in none of the females. As he explained in his *Science* paper, when the white-eyed male was crossed to his red-eyed sisters, the $F_1$ daughters inherited one red X from their mothers and one white X from their fathers (see the illustration). The $F_1$ sons, on the other hand, contained a single copy of the red X chromosome inherited from their mothers.[61] When these flies were interbred, the $F_2$ females inherited one red X from their red-eyed fathers and either a red or white X from their mothers. Because red was dominant over white, all the granddaughters were expected to have red eyes. On the other hand, the $F_2$ males, which carried only one copy of the X chromosome, would inherit the red X roughly as often as the white, and thus half the grandsons would be expected to be white eyed. Of 4,252 grandchildren examined, Morgan found 2,459 red-eyed granddaughters and 1,893 grandsons of whom 1,011 were red-eyed and 782 white-eyed.[62]

Meanwhile, Morgan pressed forward with the search for new mutants. Taking his inspiration once again from de Vries, who had suggested that radiation might cause mutations in pangenes, Morgan began in May 1910 to subject flies—pupae, larvae, and eggs—to radium rays.[63] In the first batch of treated flies, one mutant was produced that showed an unusual pattern of beads along the edge of its wings. When the "beaded mutant" was bred to his sisters, the mutations reappeared in 1 in 60 flies of the first generation. By selecting out the beaded types and mating them, Morgan found he could increase the frequency of the mutation to 1 in 35 in the third generation and 1 in 12 in the fourth. Finally, after many generations, he succeeded in isolating a pure-breeding beaded stock. Not only did selection increase the percentage of abnormal wings, but also it resulted in more extreme forms.

Although the beaded mutants, which seemed to defy the rules of Men-

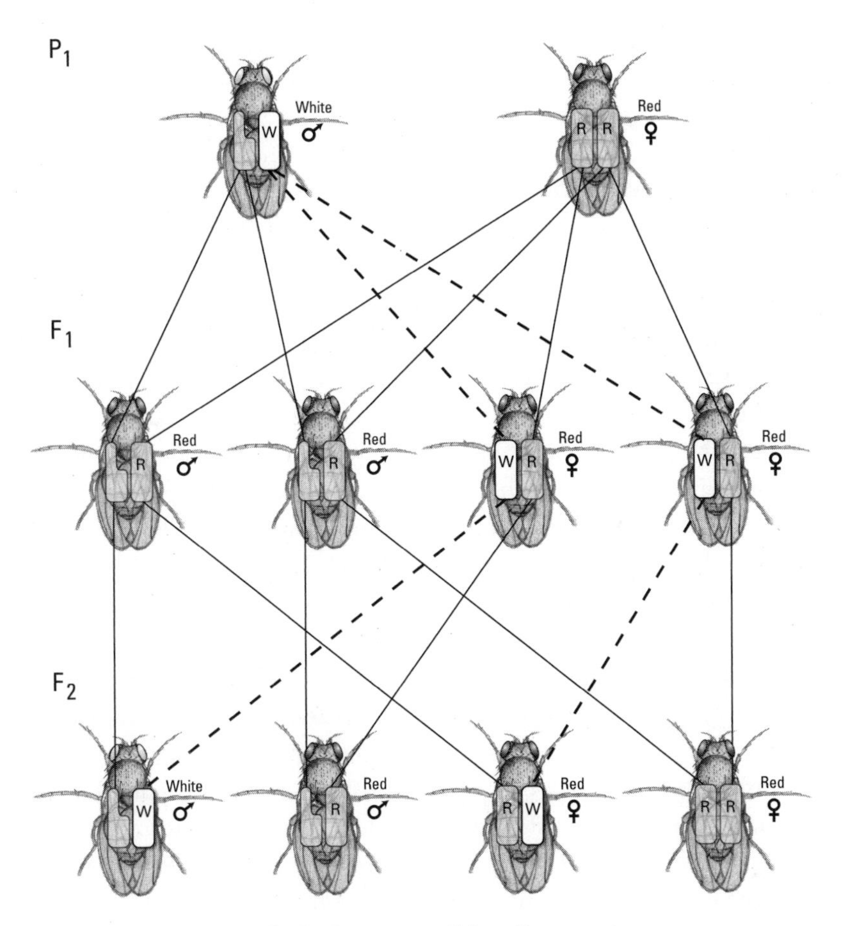


**Sex-linked inheritance of the *white* mutation.**

Illustration by Tamara L. Clark.

delian segregation, clearly required further analysis, in the rush of new results Morgan set them aside.[64] Instead he turned his attention to another mutant type with wings of half the normal length. Like the white-eyed mutation, the mutation for short wings, which Morgan dubbed "S," also seemed to stick to the X chromosome.[65] By crossing, Morgan constructed a female fly containing the white mutation on one copy of the X chromosome and short on the other. He expected half of her sons to inherit the white X chromosomes and the other half to inherit the short X chromosomes, giving rise to half white-eyed sons and half short-winged sons. Astonishingly, Morgan found that a small fraction of the sons had both white eyes and short wings and another had neither.

In a brief paper submitted on October 19, 1910, he listed all the different types of males and females that he'd seen in the second generation, including the two new types, but he was still confused, writing that "the results are difficult to explain fully."[66] By the end of the year, he had found a way to understand them. The new combination was possible, he wrote, "because in the female of the hybrid ($F_1$) a shifting of the gene for long and that for short wing (both carried by the X chromosomes) takes place."[67] Boveri had first suggested the idea on purely theoretical grounds in 1904, writing, "the traits localized in a chromosome can go independently of each other into one or the other daughter cell which would point to an exchange of parts between homologous chromosomes."[68] It was fitting that Morgan, who was deeply wary of speculation, would be the one to provide the experimental verification.

By July 1911, Morgan found further confirmation of the process of exchange of parts of chromosomes, which he dubbed "crossing over," in the work of the Belgian cytologist Frans-Alfons Janssens.[69] Studying the formation of sex cells under a microscope, Janssens observed that during the period just prior to the nuclear division, after the matching chromosomes derived from the mother and father formed pairs and began to move apart, they appeared to be stuck together at certain points. Under the microscope the strands appeared to form a crosslike figure, which Janssens called a *chiasma*.[70]

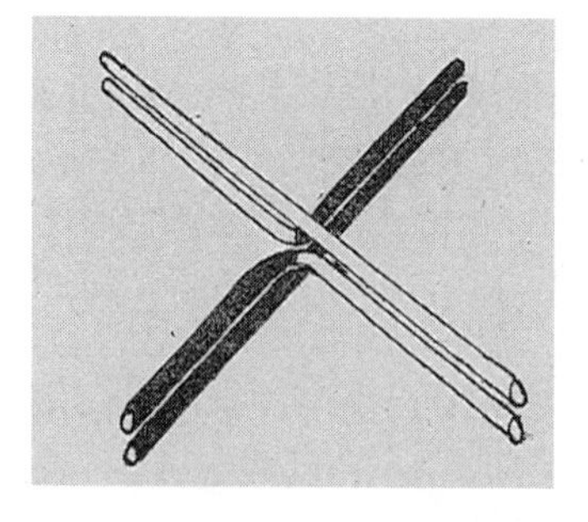

**Chiasma.** From F. A. Frans Alfons Janssens, "La Théorie de la Chiasmatypie: Nouvelle interpretation des cinèses de maturation," *La Céllule* 25 (1909): 400.

These strange X's, Janssens speculated, might be the result of an exchange between chromosomes, giving rise to hybrid, partly maternal, partly paternal chromosomes. Morgan saw that he could connect the physical evidence of crossing over presented by Janssens with the results of his breeding data. Factors on matching pairs of chromosomes could be exchanged, Morgan suggested, and the chiasmata that Janssens had seen under the microscope were the physical manifestation of crossovers between these matched pairs. The result of such a crossover was the creation of two new hybrid chromosomes.

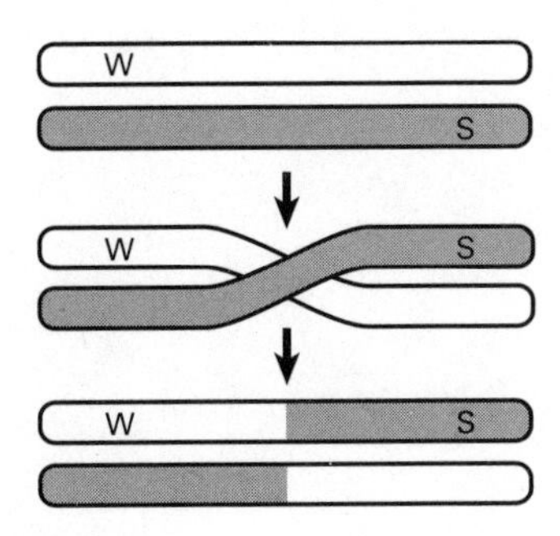


**Crossover between** *white* **(W) and** *short* **(S).** Illustration by Tamara L. Clark.

Furthermore, Morgan hypothesized, the closer to each other two genes were located in the chromosome, the less likely it was that a crossover would take place between them. "Instead of random segregation in Mendel's sense," Morgan asserted, "we find *associations of factors* that are located near together in the chromosomes."[71] Conversely, genes farther apart would have more crossing over between them. It was thus that Morgan, one of Mendel's most vociferous doubters, supplied the first significant refinement of Mendelism in forty-five years.

**The Fly Room, circa 1912.**

Curt Stern Papers, American Philosophical Society, image 4986.

CHAPTER 10

# The Fly Room

IN A WELL-KNOWN and perhaps apocryphal story, Edmund Wilson is reputed to have once quipped that Morgan's graduate students Alfred Sturtevant, Calvin Bridges, and Hermann Muller were his three greatest discoveries.[1] Although there are two radically different accounts of the atmosphere and dynamics of Morgan's lab at Columbia, there is nearly universal agreement on the view that it was these three young men working together in Morgan's lab in the years between 1912 and 1915 (when Muller left Columbia to take a job at the recently formed Rice Institute in Houston) who were responsible for the integration of Mendelism and the chromosome theory that is the basis for modern genetics.

Five years before his death in 1970, Sturtevant wrote an account of the history of genetics that has for several generations been taken as the definitive, objective account of the founding of classical genetics. His *History of Genetics* paints a portrait of a kind of scientific utopia, where Thomas Hunt Morgan and his group of enthusiastic young protégés at Columbia University worked together in an atmosphere of harmony and mutual respect to solve the great mystery of modern genetics. Sturtevant's evocation has an almost surreal quality:

> This group worked as a unit. Each carried on his own experiments, but each knew exactly what the others were doing, and each new result was freely discussed. There was little attention paid to priority or to the source of new ideas or new interpretations. What mattered was to get ahead with the work. There was much to be done; there were many new ideas to be tested, and many new experimental techniques to be developed. There can have been few times and places in scientific laboratories with such an atmosphere of excitement and with such a record of sustained enthusiasm.[2]

Despite Sturtevant's bold attempt to airbrush the history, there was something deeply troubled about the state of affairs in the Morgan lab. A few years before his death in 1967, Muller granted an interview to Columbia sociologist Harriet Zuckerman on the subject of collaboration in science, during which he made it clear that he believed Morgan to be guilty of a serious breach of professional ethics in the fly room, and that it was his example that led to problems among the students.[3] As he explained it, "What happened in this case, it seems to me, is that the major man—if I may speak out—stole from the younger men, and then some of the young men stole from each other and were able to get away with it." After the first announcement of the new chromosome theory in the 1915 textbook *The Mechanism of Mendelian Heredity,* which was coauthored by Morgan, Sturtevant, Muller, and Bridges, Morgan attempted to take credit for the work. In the highly prestigious Vanuxem Lectures given at Princeton in 1915 and 1916, which were published as a book, Morgan made little reference to the participation of any of the others, presenting the new developments as if they were the natural outgrowth of his early work without making any explicit mention of who had formulated the ideas and the experiments to test them.[4] In 1919 Morgan wrote a second textbook, *The Physical Basis of Inheritance,* but this time he wrote the book entirely under his own name. The second book was so similar to the original and brought out so soon after it (in 1919), that Holt and Co., the original publisher, considered bringing a lawsuit against Morgan. This book, which was widely translated into foreign languages, brought Morgan worldwide fame, and in Russia, Morganism became synonymous with Mendelism.[5]

Although both Morgan and Sturtevant went to considerable pains to convey the impression that the problem in the fly room was caused by Muller's competitiveness and obsession with priority, the atmosphere in the lab did not improve with Muller's departure. According to the Russian cytologist Theodosius Dobzhansky, who joined the lab in 1928 and would go on to gain renown for his work in experimental population genetics, Bridges and Sturtevant "had this intense resentment and jealousy, largely to Morgan and also to each other." Unable to express himself to Morgan, with whom he'd always played the role of favored son, Sturtevant became deeply embittered. Dobzhansky recalled that a "conversation about Morgan having taken the credit for what has been made by him, Sturtevant, was repeated again and again and again and again."[6] In the end Sturtevant's loyalty to Morgan prevented him from ever asserting himself or achieving independence. Years of deferring to the boss had instilled in Sturtevant a fundamental passivity and indecisiveness, "the boss was the boss, and Sturtevant deferred in gloomy silence," as historian of science Rob Kohler summarized Sturtevant's attitude in the latter years.[7]

Filled with rage and disappointment over his own failure to achieve independence from Morgan, Sturtevant tried to make sure that Muller—the disfavored son—did not end up outdoing him in history's eyes. In a biographical memoir of Morgan written for the National Academy of Sciences in 1959, Sturtevant made a bold, possibly unconscious, attempt to expunge Muller from the history of genetics as well as from his otherwise prodigious memory. In his description of the early fly room, Sturtevant left Muller off of a list of persons who had been given desks in the fly room, relegating him instead to a group of "Others, who did not have desks in the *fly-room,* but worked actively with the group and were often in and out." Among this secondary group, Sturtevant acknowledged that "H. J. Muller must be specially indicated, since his share in the early developments was especially important."[8]

While his friends and supporters had for years urged him to ignore the perceived slights of Morgan and Sturtevant, Muller, who was by then a world-renowned Nobel laureate, refused to leave Sturtevant's account unchallenged. "I want to thank you for having singled me out for the special compliment," he wrote Sturtevant with ironic detachment, "but at the same time to point out that I did occupy a regular desk in the 'fly-room', for a longer time than I think any of those mentioned . . . occupied one, namely, from Sept. 1912 until Sept. 1915 . . . My desk was in the southwest corner of the room, just south of Bridges' desk. Do you not remember that?"[9] Sturtevant's reply, if he ever wrote one, has been lost, but he was apparently chastened, for a few years later he revised his account of the fly room for inclusion in his *History of Genetics,* now allowing that Muller had had a place in the inner sanctum: "Beside the three of us," Sturtevant wrote, "others were always working there—a steady stream of American and foreign students, doctoral and postdoctoral. One of the most important of those was H. J. Muller, who graduated from Columbia in 1910." In the next sentence, he allowed that Muller had taken "a very active part in the Drosophila work."

For Muller, unlike Sturtevant, the discoveries made in the early days of the fly room had been a prelude to discoveries of his own of equal or even greater significance. Even in the height of the turbulent 1930s, when he lived and worked first in Hitler's Germany and then in Stalinist Russia, he managed to continue to keep his research going at a feverish pitch and play a central part in the continuing development of genetics; however, he did not have energy left over to defend himself against efforts by Morgan et al. to discredit him. As a result, Morgan and Sturtevant largely succeeded in shaping the perception of Muller for much of the twentieth century, but in the end, Muller's boundless supply of new ideas and energy could not be suppressed.

— ALREADY AS A YOUNG BOY Hermann Joseph Muller was steeped in the scientific view of the world. Forty years later, he still vividly recalled the day his father taught him the theory of natural selection:

> When I was about eight years old, my father took me to the American Museum of Natural History, and, as I well remember, made clear to me, through the simple example of the succession of fossil horses' feet shown

> there, how organs and organisms became gradually changed though the interaction of accidental variation and natural selection (those which happened to have stronger middle toes, better adapted to escaping from the carnivores, tending to survive and so to leave more offspring, through numberless generations).[10]

On the night of August 29, 1900, Hermann Sr., who was then only 47 years, died of a stroke, leaving his wife with two young children and little money. For the 9½-year-old Hermann the loss of his father meant the premature loss of his childhood. Already a serious boy, Hermann exacted an even higher standard from himself after his father's death, excelling at all things academic and collecting prizes and notice from the many adults who found in him a diligence and intelligence that was truly remarkable.

In his senior year at Morris High School in the Bronx, Hermann began a lifelong friendship with Edgar Altenburg, who would remain his best friend, defender, collaborator, and scientific critic for the next half century. As Muller would later recall, it had not taken them long to establish the pattern that would characterize the next fifty years of their relationship, immediately entering into intense dialogue, "argu[ing] out vehemently and to the bitter end all questions of principle" until they'd resolved all differences.[11] But in many ways, Muller and Altenburg were a study in contrasts. While Muller had fine strong features, he was extremely short, just over five feet tall, prematurely balding, and lacking in social confidence. Edgar, on the other hand, was tall and confident with movie-star good looks, which many women seemed to find irresistible. If Hermann may have felt physically intimidated by Edgar with his good looks, height, and ease with women, he also basked in Altenburg's admiration. Edgar believed that Hermann possessed a rare genius, and never ceased to be amazed by his friend's quickness of mind.[12] In the late 1930s and early 1940s, when Hermann's reputation was at a low ebb, Edgar was Hermann's relentless advocate and defender.

After scoring second on a citywide college entrance exam, Muller was awarded a fellowship to attend Columbia College in 1907, and he was joined there by Altenburg, who transferred from the City University of New York in the spring of his freshman year.[13] At the end of the summer of 1908, in preparation for E. B. Wilson's undergraduate course, Muller read *Recent Prog-*

*ress in the Study of Variation, Heredity and Evolution* by R. H. Lock, a remarkably perceptive integration of Darwin's theory of natural selection, the recently disinterred laws of Mendel, and the new understanding of the nucleus and the chromosomes revealed by cytology.[14] Muller was thrilled by Lock's prophetic suggestions that genes lay in chromosomes and that there might be a possibility of exchange between them, and by the time he had completed Lock's book he was a confirmed Mendelian.[15]

Lock's ideas were further reinforced in Wilson's course during his second year. By his third and final undergraduate year at Columbia, Muller was so intensely interested in biology that he began to present his own ideas about the future direction of biology to an undergraduate biology club that he founded. In addition to Altenburg, the club's regular members included two sophomores, Alfred Sturtevant and Calvin Bridges, and this core group would go on to form the core of the famous fly room. At 21, Bridges, who had been orphaned at 3 and brought up by his grandmother with very little money in upstate New York, was several years older than the ordinary sophomore.[16] Sturtevant, by contrast, was only 17 and lived in Edgewood, New Jersey, with his older brother Edgar, who was a professor of classics at Barnard and had persuaded Sturtevant to attend Columbia in the first place.

By a remarkable stroke of good fortune Bridges and Sturtevant were given their first introductory biology course by Thomas Hunt Morgan, who taught the course for the first and only time in his twenty-four-year career when they were freshman in 1909–10. Sturtevant, in particular, was immediately drawn to Morgan's enthusiasm and his unpretentious southern manner. Like Morgan, Sturtevant was born into an old American family that had known better times, and both men had grown up in the South. As a boy, Sturtevant had constructed a pedigree of the horses on his father's farm in Mobile, Alabama, already then showing signs of the remarkable ability to organize and retrieve data that would prove invaluable in later years. Later, when he went to college, he followed up on this interest in genealogy by using the New York Public Library to research his family pedigree. At his brother's urging, Sturtevant applied himself to the task of learning more about heredity in order to pursue his interest in pedigrees. Morgan, who still had serious reservations about Mendelism, did not mention the new genetics at all in his undergraduate course, and Sturtevant instead found the basics in a sixty-page primer entitled *Mendelism,* written by the Bateson disciple R. C. Punnett.[17]

Taking his cue from Hurst's 1906 Royal Society paper, Sturtevant saw that he might be able to clarify the case of coat color inheritance in horses using the data provided by American harness horses, close relatives of the English thoroughbreds studied by Hurst. Although Hurst had proved beyond a reasonable doubt that chestnut was recessive to bay and brown, he did not have enough data to explain the other coat colors, but the American breeds, which presented a far greater range of colors, allowed Sturtevant to formulate a complete model. When his analysis was done, Sturtevant had introduced a total of five genes that together accounted for the complete range of coat colors, and he had thereby succeeded in finally bringing closure to the savagely contested problem of coat color inheritance.[18]

After much agonizing and prodding from Edgar, Sturtevant wrote up his results in the spring of 1910 and submitted them to Morgan, who was duly astonished.[19] Not only did Morgan urge Sturtevant to publish the paper in the *Biological Bulletin,* but also he immediately offered him a research position in his lab to begin the following fall when Morgan returned from Woods Hole. At the same time Morgan hired Bridges, who in his early twenties already had a wife and children to support, to maintain the fly stocks. When, a few months later, Bridges picked out a mutant fly with vermillion eyes from a discarded bottle, Morgan immediately offered him a desk in the lab.[20] Meanwhile, Muller, who had been heartbroken over his failure to receive funding from the zoology department, accepted a fellowship to work on nerve-impulse transmission.

That fall, shortly after Sturtevant and Bridges joined the lab, Morgan made his paradigm-shifting observation that mutations (for white eyes and short wings, respectively) that were located on different X chromosomes in the mother occasionally appeared on the same chromosome in the sons, and a few months later, in February 1911, he proposed that there was a physical exchange of chromosome parts to account for the doubly mutant males. In September 1911 he made the further radical suggestion that the association of factors, which had been first described by Bateson without reference to chromosomes, could be explained by the fact that they lay near one another in a linear series along the chromosome, adding that "the proportions that result are not so much the expression of a numerical system [as Bateson had suggested] as of the relative location of the factors in the chromosomes."[21]

Sturtevant took the next big step later that year.[22] Half a century later, he still vividly recalled the afternoon that he hit upon what was without doubt the greatest idea of his life. He and Morgan had been discussing a 1909 paper by the American geneticist W. E. Castle in which Castle seemed to be suggesting, contrary to the inspired deductions of the great European cytologist Wilhelm Roux in 1883 and Morgan's recent elaboration on the same theme, that genes were not arranged in a line along the chromosomes after all. Morgan and Sturtevant agreed that Castle's idea was not worth further consideration—the genes were almost certainly arranged in a line along the chromosomes, and the association between them, as Morgan had recently pointed out, was a function of their distance apart. It was then that Sturtevant had his insight, suddenly seeing that the ratio of crossover to noncrossover types among the progeny could be used as a measure of the distances between two genes on a chromosome and that the use of this measure made possible the construction of a linear map of the genes.

While Morgan had been vague about the relationship between association of factors and their placement on the chromosome, Sturtevant proposed a concrete way to quantify the distance between genes. Furthermore, he claimed that if one knew AB (the distance between genes A and B) and BC (the distance between B and C), then AC (the distance between A and C) would equal their difference or their sum.

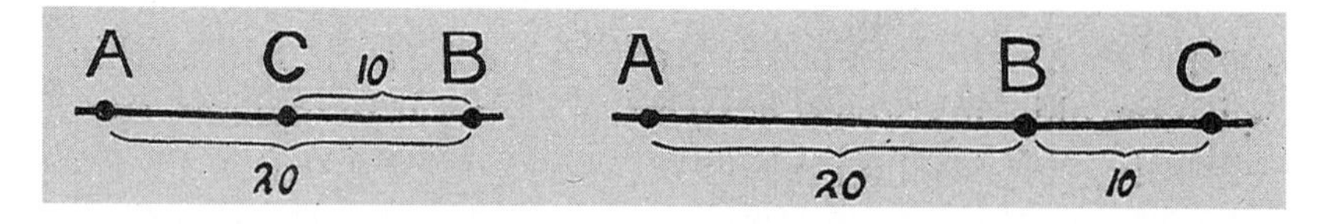


**Gene mapping.** From Hermann J. Muller, "The Mechanism of Crossing Over," *American Naturalist* 50 (1916): 200.

By applying the same tests to strategically chosen combinations of genes, he could make a map of the genes showing their linear order and the relative distances between them. In his *History of Genetics* Sturtevant described this banner day with his characteristic dry understatement:

> I went home and spent most of the night (to the neglect of my undergraduate homework) in producing the first chromosome map, which included the sex-linked genes y, v, m, and r, in the order and approximately the relative spacing that they still appear on the standard maps.[23]

In fact, Sturtevant's proposition that AB + BC = AC (when B lay between A and C) corresponded only approximately to the actual data based on the numbers of crossover and noncrossover types, and AC was typically somewhat less than AB + BC. In order to account for the discrepancy, Sturtevant suggested that a chromosome occasionally underwent two crossovers simultaneously, the first between genes A and B, and the second one between B and C.

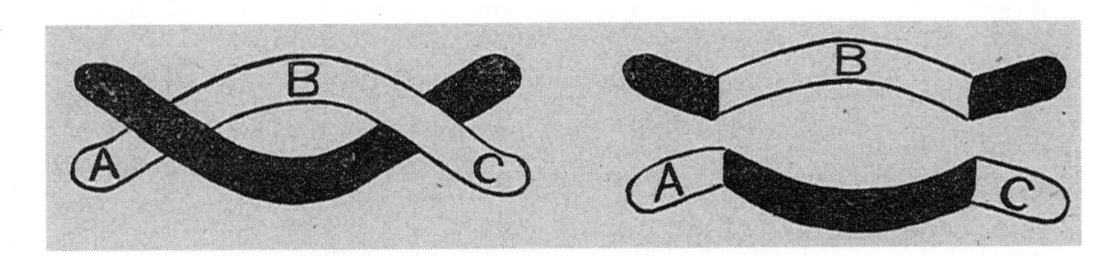


**Double crossover.** From Hermann J. Muller, "The Mechanism of Crossing Over," *American Naturalist* 50 (1916): 202.

In the cases of a *double crossover,* as Sturtevant called it, A remained linked to C (though no longer to B). Since each double crossover actually represented two single crossovers, the number of crossovers between two genes A and C (and consequently the distance between them) would tend to be underestimated.

Muller, who was in the midst of his second fellowship year in physiology at Cornell Medical school, first learned about Sturtevant's idea at one of the Friday night get-togethers at Morgan's house where Morgan prepared fried squid Neapolitan style and a small group of students read recent papers (of Morgan's choosing) out loud. Though it was Morgan's policy to restrict the discussion those evenings to topics other than genetics, he made an exception for Sturtevant's gene-mapping technique. Muller, who would often leap up from his chair and pace madly back and forth across a room even in ordinary conversation, found the news of Sturtevant's discovery so exciting that he began literally to jump up and down.[24] A decade later, in a lecture given at Cold Spring Harbor in the summer of 1921, Muller was still brimming over with enthusiasm about Sturtevant's achievement:

> And I have seen in the space of only one year a student—Sturtevant—when only 19 years old—first in the world get it into his head that it might be possible to make a real map of the inside of a chromosome, and then proceed actually to do it. Just one boy, in one year, could do that. What then cannot be done by the friendly cooperation of many people in the course of time?[25]

While the original idea behind gene mapping was due to Sturtevant, the working out of the general theory was in large part done in collaboration with Muller. Among other significant insights, Muller realized that distance would be better measured as the fraction of flies that had undergone recombination among all progeny rather than the ratio of recombinant to nonrecombinant types,[26] the measure that Sturtevant, following Bateson and Punnett, had used up to that time.[27] When distance was measured this way, it became clear how to adjust for the double crossovers, resulting in a much better fit with AB + BC = AC.[28] Ironically, Muller's unit of distance would come to be known as the "centimorgan."[29]

In the first mapping paper of 1912, Sturtevant generously acknowledged that he had been "greatly helped by numerous discussions of the theoretical side of the matter with Messrs. H. J. Muller, E. Altenburg, C. B. Bridges, and others," and he even went so far as to single out Muller for making suggestions that had been "especially helpful during the actual preparation of the paper."[30] The following year, in his thesis, Sturtevant once again graciously acknowledged his debt to Muller, writing that he "had been greatly helped by numerous discussions with Mr. H. J. Muller and others," but he did not credit Muller with his specific contributions to the theory. Although Muller's contributions to mapping were relatively insignificant compared to those he would later make, a precedent was being established, and it soon became the unspoken policy of the lab that while published results and unpublished data were to be acknowledged, general theoretical contributions and specific ideas for the design of experiments to test them were common property.[31] For Muller, who provided an inexhaustible flow of new ideas, many of which would turn out to be absolutely crucial to the continuing development of the subject, the policy was grossly unfair.

While Muller had wanted nothing more than to join the *Drosophila* work after college, he had been forced to support himself as a graduate student in physiology. During this period in exile, he later recalled that the only time he had to think about the *Drosophila* work was on the New York subway while he commuted to and from his night job teaching English to foreigners in the New York public schools, a job he needed to supplement his meager income. While he struggled to support himself, Muller could not help resenting the fact that Morgan lavished support on Sturtevant, paying out of his own pocket in his final semester of Columbia after his undergraduate fellowship ran out and then supporting him on his Carnegie grant, freeing Sturtevant of the teaching obligations that would have otherwise diverted a great deal of his time and energy from his fly work.[32]

But the bond between Morgan and Sturtevant was more like father and son than professor and student. Unlike his own father, who had cut him out of his will, Morgan was unfailingly solicitous and generous toward his surrogate son Sturtevant, and Sturtevant, who brought fame and attention to Morgan and himself with his first genetic map, more than repaid Morgan's fatherly devotion.[33] Sturtevant was every professor's dream student: self-confident, independent, and productive. He had a prodigious memory, and was the one to whom all questions concerning the scientific literature were referred. In his later years he found relaxation in reading through the *Encyclopedia Britannica,* but he complained that it was difficult to find an article that he had not already read. His encyclopedic grasp of the facts of genetics, essential and arcane, was a marvel to all who knew him, and he was an ace problem solver, capable of completing the most difficult Sunday crossword puzzle in a single sitting.

Muller's first in-depth exposure to Morgan was in the year 1910–11 when he took Morgan's graduate course in experimental zoology in his first year as a graduate student in physiology, and he had quickly grown weary of Morgan's sloppiness and lack of precision. In one particularly ill-prepared lecture, Morgan used *W* and *R* for the white and red eyes of *Drosophila* and the same letters for wrinkled and round peas. Charming and self-deprecatory, Morgan wrote a poem for the next class:

> Wrinkle, wrinkle little pea
> How I wonder what you be
> Whether Round, or whether Red,
> In the pod, or on the head?[34]

But for Muller, Morgan's confusion was a far more serious issue. Unlike Wilson, whose careful synthesis and presentation of the facts Muller revered, he found Morgan's thinking "chaotic and undeveloped."[35] But it was not only Morgan. The entire field was being suffocated by a host of antiquated ideas, "vague and even metaphysical conceptions," as Muller later described them, that held back progress.[36] The time had come to reexamine the foundations of genetics and to put them on a sound logical basis, and while he was completing his second master's degree, in neurobiology at Cornell, and otherwise struggling to keep himself and his mother and sister afloat, he began a systematic reexamination of the relationship between genes and characters.

In "Erroneous Assumptions regarding Genes," as he called the lecture he delivered to the Biology Club during the winter of 1911–12, Muller not only displayed his gift for analytical reasoning and abstraction, but he insisted on the importance of theory and speculation in the biological sciences. It was critical to give free rein to one's imagination and follow an argument to its logical end, Muller argued, "to conjecture freely and fully as to all the possibility involved." Those who thought otherwise, "those who would try to reach *natural* conclusions but avoid *unbridled theorizing*," Muller said in a conscious echo of Morgan, would in the end, themselves be operating under a host of unexamined assumptions, which were tantamount to the worst destructive form of wild speculation. Though he did not use Morgan's name in the article, no member of the Biology Club would have missed the reference.

"Erroneous Assumptions" was both a merciless deconstruction of the often illogical and fallacious reasoning of Morgan (and others) as well as a bold attempt to reconceptualize the relationship between genes and characters. The idea that there was a gene that became "incarnated" in a character of the organism was far too simplistic. While Mendel's identification of factors with clear-cut characters had been absolutely essential in the early stages of working out of the theory, the identification of factors and characters was in reality considerably more complex. To say that a particular gene was responsible for color, for example, was merely a "convenience." Not only were many other genes responsible for producing the color, but also the color factor itself was also undoubtedly involved in other entirely unrelated processes. It was only an illusion, a product of human psychology, that characters were "direct materialization of metaphysical entities," as he put it in his lecture. While genes were highly stable, separate material particles that strictly obeyed Mendel's laws, the assumption that characters were actually distinct and that they would remain constant during the process of segregation was true only in special cases.

One consequence of this new point of view was the realization that most characters must be due to the combined effects of several, and sometimes many, genes, and that a single gene, in turn, must influence more than one character.[37] Not only did the genes and their products and reactions due to them influence each other, but also they would be influenced by the chemistry of the cell where they acted. The effects of these genes, and the interaction of the genes with the environment, were necessarily "mingled or superimposed" in a vast complex network of chemical processes. Furthermore, he argued, one could not ignore the effects of gene dosage, and a far more nuanced understanding was required of dominance and recessiveness.

In the process of his critique, Muller suggested that the nomenclature for describing mutations be revised, explaining that "a clear agreement as to method of symbolization is important for clarity of thought regarding the objects of the symbols."[38] Following the discovery in 1911 by Morgan of a partial reverse mutation (from white to eosin)[39] it became clear that it was necessary to abandon the then-prevailing notion that recessive mutations were always by loss.[40] As a consequence, it now made sense to adopt the far more convenient system of nomenclature proposed by Muller in which a gene was to be named after its mutant function or form, with a capital letters for dominant mutations and lowercase for recessives. Additional mutations of the same gene would be distinguished with subscripts rather than by giving different letters for each new form, and the normal (or wild-type) allele would always be denoted with a "+" (this last simplification suggested by W. E. Castle, who would otherwise contribute mightily to the muddle over characters and their relation to genes). Lastly, genes would be listed in a horizontal line in the order in which they were linked, and homologous chromosomes would be stacked one on top of the other, separated by a horizontal line.

$$\frac{a\ b + D\ e\ f}{a + c + e +}$$

**Gene nomenclature.**

— IN THE FALL OF 1911, Altenburg was finally given access to the kingdom in the form of a bench in the fly room on the sixth floor of Columbia's Schermerhorn Hall.[41] He and Bridges were seated on one side of the small lab, Morgan was usually shown in pictures standing, in the center, and Sturtevant on the opposite side.[42] Despite the fact that he was the next student after Sturtevant and Bridges to work in the fly room, and remained there for two years, from 1911 to 1913, Altenburg is hardly mentioned in the accounts of the early years. While Sturtevant and Bridges were busy collecting and mapping new mutants, Altenburg was assigned what seemed to be the most unpromising problem of the lab, a mutant race called "truncate," named for the stunted development of their wings.

Truncate, the bugbear of the laboratory, was one of two major problem cases among a host of well-behaved classically segregating Mendelian mutations. It was a direct descendant of the other problem case, the first of the wing-shape mutants, known as "beaded," so named for the small bead-like nodes that interrupted the ordinarily smooth flow of the outer wing edge. The transmission of both mutations was entirely erratic and seemed to defy conventional Mendelian analysis. Morgan had first isolated beaded in the summer of 1910 after treating flies with radium. For several months he had grown up bottles of beaded flies, picking out the most beaded male and female from each culture to serve as the parents of the next generation, hoping without success to isolate a true-breeding stock. Then in August, the seventh generation of the beaded stock suddenly gave rise to truncate, which would give rise to a mutant race that would prove to be every bit as strange as its beaded ancestors.

In the early months, Morgan bred truncate parents and found that the trait was inherited by about 50 percent of the offspring. By selecting the

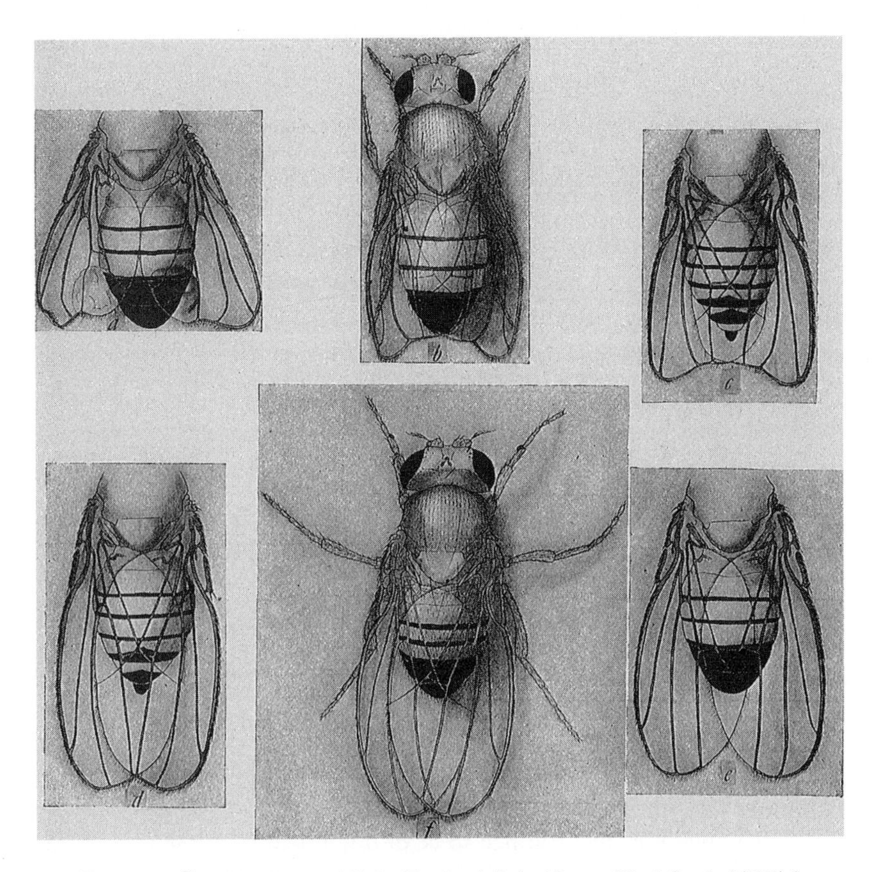

**Truncate flies.** From Hermann J. Muller, "The Genetic Basis of Truncate Wing," *Genetics* 5 (1920): 3.

most extreme cases, he "improved" the grade of the character but not the frequency of its inheritance. After a year's worth of selection Morgan managed to get a truncate stock that yielded about 90 percent truncate offspring, but the strain simply could not be induced to breed true. By then, Morgan had had enough, declaring that the inheritance of certain characters simply could not be explained by the segregation of stable Mendelian factors.

Such equivocation was entirely unacceptable to Muller and Altenburg, who were increasingly viewed as fanatical by Morgan, Sturtevant, and Bridges for their insistence that all inheritance must be chromosomal and Mendelian.[43] In particular, Muller and Altenburg refused to accept the idea that the strange and irregular inheritance of truncated wings was caused by fluctuations of the underlying genes or their contamination by other alleles.

Support for the stability of the underlying factors was given a powerful boost by the recent work of Wilhelm Johannsen, who spent a month at Columbia in the fall of 1911 as part of a several-month cross-country tour publicizing his new "genotype-phenotype theory."[44] Following Galton, "the creator of the study of exact heredity," as Johannsen wrote in the dedication of a 1903 monograph on inheritance, Johannsen studied the inheritance of size of beans, but unlike Galton, who'd conducted his experiments on beans that were selected from a bushel bought at a local nursery, Johannsen had been careful to use seeds of known ancestry. In particular, he made sure to select seeds that were the descendants of a single self-fertilizing plant, or what he called a *pure line,* that was sure to be homozygous for all its genes.[45] Although such a plant produced seeds of varying weights, some heavier and some lighter, Johannsen's striking finding was that progeny grown up from the heavier seeds gave rise to seeds of the same average weight as plants grown up from lighter seeds. Thus Johannsen concluded that the variability in size in a pure line was due exclusively to environmental effects—to the position of the bean in the pod, for example, or the position of the pod on the plant. The variability in "appearance type" or "phenotype," as Johannsen dubbed it, was therefore "a superficial phenomenon" and had to be distinguished from the underlying unchanging "genotype."[46] Galton had gone wrong, Johannsen correctly saw, in failing to consider that a population might actually be the composite of several biologically distinct subpopulations, each with its own distinct fixed genotype. If a population was properly subdivided into pure lines, one saw no evidence of regression to the population mean.

In November 1912, the Harvard geneticist E. M. East made the case for assuming that there was a fixed underlying genetic basis for varying traits on strictly logical grounds. "A factor not being a biological reality but a descriptive term, must be fixed and unchangeable," East wrote. "If it were otherwise it would present no points of advantage in describing varying characters."[47] Thus, East insisted, the inconstancy of a character, if it were not caused by changing environmental conditions, must be explained by the segregation of multiple genes. It was much the same point as Johannsen had made, and East had breeding data that proved that the inheritance of seed color in maize was influenced by three "modifying genes," but his data, which turned on the differentiation of fine shades of purple and red, were hardly conclusive. What was needed was a well-developed genetic system in which the ideas could be put to a rigorous test, and *Drosophila* with its ever-growing collection of well-characterized genes with specific locations was just such a system and truncate the perfect fluctuating character on which to experiment.

In his first series of experiments, Altenburg crossed highly selected truncate flies to normal flies and found only zero to 8 percent truncates among the progeny. When the "extracted" truncates were inbred, the frequency of truncate could be raised from 8 percent up to 20 percent among the offspring, and sometimes as high as 40 percent. As Morgan had noticed before him, not only was the transmission of the character erratic, but the degree of the truncation and the shape of the deformed wings were highly variable as well. Still there was no doubt that the inconstancy was at least in part genetic, as shown by the fact that the degree and frequency of truncation of the parents were still highly correlated with the degree and percentage of truncation among the offspring. By continuing the selection, Altenburg regenerated the 90 percent truncate stock that Morgan had created, but even after three years of continued selection he could not produce a pure-breeding truncate.[48]

Despite the fact that the conventional mapping technique only worked to the extent that a gene was determined by a clearly segregating visible difference, Altenburg initiated a series of crosses to see if he could find any sign of truncate's linkage to known mutations. Not unexpectedly, there was no clear-cut case of a 1:2:1 ratio of types that was diagnostic of closely linked genes.[49] The data seemed to show, however, a mild degree of linkage between truncate and genes in each of the three known linkage groups, which was expected of a character that was influenced by several genes. Although the results were vaguely encouraging, it was not clear how to proceed. But the linkage tests did reveal one curious fact that would turn out to be enormously useful. In the presence of *black,* a recessive mutation located on the second chromosome, truncate behaved more like a dominant than a recessive, appearing in 25 percent to 50 percent of offspring.

In the fall of 1912, Muller was finally awarded a fellowship in zoology and allowed at long last to join the fly work. By November, the 22-year-old Muller had his first great inspiration, an idea that would lead to the resolution of the stubborn nonconformance of truncate to conventional Mendelian principles and at the same time provide a new approach to the study of all complex traits. Muller's inspiration was to see how to use the growing assembly of known mutants with clearly identifiable phenotypes to map the far larger class of mutations whose inheritance was more difficult to track. The idea was prompted by the then-new observation that certain genes that underwent crossing over in females never crossed over in males. Muller saw that if no crossing over in males turned out to be the rule rather than the exception, one could follow the inheritance of all the genes contained in a given chromosome by means of tracking the inheritance of a single heterozygous gene that determined a clearly differentiating visible difference.[50] If, for example, a male that was heterozygous for *pink*—a recessive mutation on the third chromosome causing pink eyes—was mated to a female that was homozygous for *pink,* one could be sure that each pink-eyed progeny had inherited the *pink* third chromosome from its father while each red-eyed progeny carried the other copy.

By taking advantage of the absence of crossing over in males, Muller and Altenburg were finally in possession of the tool they needed to analyze the inheritance of truncate. By means of a carefully arranged cross, truncate males were obtained in which one member of each chromosome pair came from a truncate parent and the other member contained a recessive marker.[51] In particular, the recessive body-color mutation *black* was used to mark chromosome 2, and *pink* (as in the example discussed in the previous paragraph) was used to mark chromosome 3. Chromosome 1 could be followed because it was sex-determining. By backcrossing these marked males to tester females that were homozygous for both *black* and *pink,* it was possible to track the inheritance of the males' individual chromosomes in the subsequent generation. Depending on whether an individual offspring was male or female, whether its body was grey or black, and whether its eyes were red or pink, one could infer that it had or had not received the truncate copy of its father's first, second, or third chromosome, respectively.[52]

In the winter of 1912–13, it was clear to Muller and Altenburg that the experiment was working. Among the progeny of the truncate males, truncate wings were never observed in the absence of the truncate copy of the father's second chromosome, showing that the second chromosome contained a gene that was necessary for the expression of truncate. Furthermore, among the progeny that inherited the truncate copy of their father's second chromosome but the marked copy of each of the other chromosomes, truncate wings were sometimes observed. Thus, it was clear that the second chromosome contained at least one gene—a "chief" or "master" gene, as Muller dubbed it—that by itself was capable of causing truncated wings. Moreover, those progeny that contained the truncate copy of the father's third chromosome in addition to the truncate second chromosome showed both a higher frequency of truncate and a higher grade of truncation, and the same was true of flies containing the first and second of the father's truncate chromosomes, which implied that both the first and third chromosomes contained genes that intensified the effect of one or more master genes on chromosome 2.

By comparing the degree of truncation among a set of brothers that contained the identical set of truncate genes with the expression of truncate in sons that had inherited these same genes, Muller and Altenburg were finally in a position to ask if fluctuations in the genes themselves were responsible for the high degree of variability of the truncate mutants.[53] If the variability of truncate was in fact due to the fluctuation of the underlying truncate genes as opposed to some nongenetic, environmental factors, then one would expect a high degree of correlation between the degree of truncation of the brothers and the average degree of truncation of their sons. On the other hand, if the sons of highly truncated fathers (each of whom carried exactly the same truncate genes as their fathers) showed a wide range of truncation, one could safely conclude that the father's genes themselves had undergone no permanent modification that was passed on to their sons. As Muller and Altenburg had long argued must be the case, no correlation at all was found, and the proof of the stability of genes was complete.

— THE SCHISM IN THE FLY GROUP, between the intellectually aggressive and challenging Muller and Altenburg on one side, and Morgan and Sturtevant, each of whom shunned overt conflict, on the other, broke out into the open in 1913 when Altenburg approached Morgan to discuss possible thesis projects with him. In the course of their talk, Morgan led him from his office to the aquarium room across the hall. Although Morgan was by then devoting his major energies to the new *Drosophila* work, he had not altogether abandoned his interest in the study of development of marine forms. As Altenburg recalled their conversation, Morgan had dipped his finger in a stagnant tank, held it up to a light bulb, and stared at the water drop for a moment. "There are a lot of daphnia in here," he had then observed. "Why don't you work on them?"[54] To Altenburg, who rightly considered himself to be one of the founders and leading lights of the new *Drosophila* genetics, Morgan's suggestion was both a huge disappointment as well as an insult that he never forgave. He promptly quit the fly room and decided instead to do his thesis under the aegis of the botany department.

Despite years of determined resistance to the basic premises of the chromosome theory, by 1913 it had become obvious even to the remarkably resistant Morgan that the *Drosophila* work must be correct, and he proposed that he and the four founding members of the fly group, Sturtevant, Bridges, Muller, and Altenburg, join forces to write the textbook that would become *The Mechanism of Mendelian Heredity.*[55] Still bitter over Morgan's attempts to marginalize him, Altenburg refused to have anything to do with the project and thereby denied himself immortality as one of the founders of modern genetics. More self-preserving than his exceedingly hotheaded friend, Muller accepted Morgan's offer to coauthor the textbook and stayed on in Morgan's lab for the three years necessary to complete his thesis.

A preliminary account of the truncate story appeared in *The Mechanism of Mendelian Heredity,* but the full account, which filled an entire issue of the journal *Genetics,* was published in January 1920.[56] As Altenburg later recalled the experiments, the labor had been mostly his, while the ideas were nearly all Muller's.[57] When it came time to submit the manuscript to *Genetics,* Muller had insisted that the order of the authors be decided by the toss of a coin. While he had won, Altenburg would always regret allowing his name to appear first on the paper, which was a marvel, both conceptually and in the details of the experimental plan.[58] Muller anticipated with astonishing accuracy developments that would reach fruition only with the completion of the human genome sequencing project in the twenty-first century:

> Remote as the possibilities of such work may seem to be in the case of such animals as mammals, it is nevertheless difficult to conceive how the genetic bases of the more elusive and complicated characters in them can be determined adequately by other means. Even in the case of man, an attempt in such a direction would be justified, for here the most important characters,—such as the psychological ones,—are perhaps more plastic, obscure, and complicated in genetic basis, than any others in the entire animal kingdom, and it would seem next to impossible ever to give any real Mendelian analysis of most of them without studying them by the method of linked identifying factors.[59]

Nearly a century later, the "method of linked markers," as Muller dubbed it, has made possible the identification of genes that influence complex diseases ranging from diabetes to high blood pressure and to traits such as sweetness in tomatoes.

In its time, however, the impact of the truncate work was not practical but theoretical. It was a keystone to the connection of genetics and evolution, the integration of Mendel and Darwin. Years later, in 1946, Muller explained what he believed to be the true significance of their joint work on truncate:

> It has been too little realized in general how important the multiple factor theory was, first, in making it possible to show that the inheritance of all variations found at that time in Drosophila (and by implication elsewhere) was in fact stable and mendelian, chromosomal and not subject to contamination . . . and secondly, that the multiple factor basis thus revealed converted the "mutation theory," contrary to the ideas and wishes of its early proponents, into the basis for essentially Darwinian selection of "continuous variations" afterall.[60]

**Muller and Altenburg, circa 1919.**

Generously provided by Helen J. Muller.

CHAPTER 11

# *Oenothera* Reconsidered

— AFTER THREE YEARS in the lab of Thomas Hunt Morgan, Muller was dying to get away. In 1915, even before he'd officially been granted a degree, he leapt at the opportunity to join Julian Huxley (grandson of Darwin's friend and ally Thomas Huxley) in the biology department of the newly created Rice Institute in Houston. Sharing a passion for Darwin and eugenics, the high-strung, 28-year-old, Oxford-educated Julian Huxley and Muller were a natural pair, and it was a great disappointment to Muller when, a year later, Huxley decided that it was his duty to return to England and join the Britain war effort against Germany.[1] In Huxley's absence, Muller became acting head of the department and persuaded his faithful friend Altenburg to join him on the Rice faculty.

At Rice, Muller's first priority was to once and for all time put an end to the doubts about the universality of Mendel's laws. While the truncate case had been solved by the end of 1913, the inheritance of beaded wings, the other most frequently cited example of allegedly non-Mendelian inheritance, was still muddying the waters. Isolated by Morgan in May 1910, the same month Morgan found the famous white-eyed mutant, beaded was in many respects the most puzzling of all the several hundred *Drosophila* mutants that had been isolated over the subsequent five years. In the early generations, Morgan had found that beaded parents produced only a small percentage (from 3 to 10 percent) of beaded progeny, and the strange deformations of the wings were highly variable in shape and degree. By picking out the most beaded male and female from each culture to serve as the parents of the next generation, he had quickly improved both the frequency and the grade of the beaded stock, and he soon had beaded flies that produced nearly two-thirds beaded progeny of much more extreme type. Selection was continued for more than a year, over 25 generations, and just when he was about to conclude that the system had reached its limit, cultures were found that were 90 percent pure for beaded, and from these came a truly pure-breeding stock. But the sudden appearance of a constant-breeding beaded stock was just as mysterious as the variability that had been observed in the earlier days, and Morgan concluded that the erratic behavior of beaded was simply not amenable to conventional Mendelian explanation.[2]

In 1912 the beaded project was given to a short-term member of the Morgan lab, J. S. Dexter, who used the linkage method just then successfully worked out by Altenburg and Muller on truncate to identify a chief gene for beaded on chromosome 3 and an intensifier on chromosome 2.[3] But at the same time that he managed to clarify one aspect of the inheritance of beaded, Dexter introduced a new complication by showing that upon crossing with a normal fly, a true-breeding beaded fly gave rise to two very different classes of progeny, as if it were a heterozygote. The first class, which consisted of 15 to 30 percent of the progeny, were beaded and the remainder were not. Furthermore, when the beaded progeny were intermated, they gave about 25 percent beaded offspring, while the unbeaded progeny gave only 10 percent, which showed that the difference between the two classes was genetic.[4]

— IN 1915, JUST BEFORE leaving for Rice, Muller crossed one of the true-breeding beaded flies to a normal-winged fly, and selected one normal-winged daughter for further study. Much to his surprise, Muller found that there was no crossing over at all among markers on the third chromosome. This finding suggested that either the copy of the third chromosome that the daughter had inherited from her beaded father or the copy that she had inherited from her mother, the normal parent, contained a factor that inhibited crossing over. Because he knew the maternal copy behaved normally in other crosses, Muller concluded that the paternal copy—the copy inherited from the beaded father—must contain a genetic factor that prevented crossing over.[5] This in itself was unusual but not unheard of, as it was known that certain chromosomes did not participate in crossovers with their partners, and, in fact, Sturtevant had recently found a case in which crossing over was suppressed in the third chromosome and attributed the effect to the presence of a genetic factor, which he dubbed the C′ factor.[6] Following Sturtevant's lead, Muller concluded that one or both of the copies of chromosome 3 from the beaded father contained a suppressor of crossing over that he dubbed $C_b'$ (the b to designate the beaded stock that gave rise to the crossover suppressor).

The nonbeaded daughter held another surprise. Not only did she fail to produce crossovers, but she failed to produce a single beaded offspring in the next generation when crossed to a normal-winged male. Evidently, the factor or combination of factors that had caused beaded wings in the apparently homozygous father had entirely disappeared in the nonbeaded daughter. In particular, the nonbeaded daughter's paternal third chromosome seemed to have no vestige of the master gene for beaded that was known to map chromosome 3. All that was known about it was that it contained a crossover suppressor.

Muller's special gift was to see a way through just such difficulties when no one else could. His idea in this case was to create a new strain in which flies that were homozygous for the mysterious chromosome 3 could be identified. To do this he made a fly that contained one copy of the third chromosome from the nonbeaded daughter (carrying the crossover suppressor $C_b'$) and one copy marked by a dominant mutation, a well-known mutation called $D_f'$ that resulted in deformed eyes.[7] When these flies were self-mated, the progeny with normal eyes would be those that were homozygous for the "beaded" chromosome 3.[8]

To Muller's utter astonishment, the normal-eyed class was entirely absent. Furthermore, not only were the normal homozygotes absent, but the homozygotes for deformed eyes were also missing.[9] In fact the only flies that appeared were all heterozygotes like their parents. Adding to the mystery, when these heterozygotes were allowed to mate with each other, the next generation was again exactly like their parents. The flies were then allowed to mate without any intervention for six months (over 14 generations) and at the end of the period the flies were identical to those that had initiated the experiment.

In a pattern that would be repeated over and over again in Muller's scientific life, out of the ashes of a disappointing result came an entirely new way of looking at a classic problem in genetics. While the results seemed to support the view that genes were in fact fluctuating, mixing, or any of a number of non-Mendelian interpretations, Muller was determined to explain his results from a strictly Mendelian point of view. First, he argued, it was most reasonable to explain the absence of the deformed-eye homozygotes by assuming that the $D_f'$ mutation was lethal when homozygous.[10] It was then not a long step, Muller contended, to assume that the absence of the homozygous class with normal eyes, the $C_b'/C_b'$ flies, was also due to a recessive lethal factor, and that this second recessive lethal was distinct from $D_f'$. Because the two lethals were assumed to affect different genes, the heterozygote, which contained only a single copy of each of the recessive lethals, would survive and be propagated as a constant hybrid. This remarkable situation in which the action of two lethal factors prevented both types of homozygotes from surviving, Muller called a condition of "balanced lethal factors."

By analogy with the constant hybrids he'd artificially created, Muller was now in a position to interpret the baffling behavior of the beaded flies as well. The fact that dominant mutations in *Drosophila* were very often lethal led Muller to the assumption that that the chief beaded gene, $B_d$, which could cause deformed wings in a single dose, was lethal when homozygous.[11] This hypothesis provided an explanation for Morgan's early difficulty in getting pure-breeding beaded flies during the early stages of his selection.

As had been the case for truncate, Muller explained, Morgan's early success in raising both the frequency and the degree of the beaded wings could be attributed to the selection of modifying genes, but the lethality of the beaded gene when homozygous had prevented Morgan from isolating a true-breeding beaded. The appearance of a second lethal on the homologous chromosome allowed for a sudden improvement. Under ordinary circumstances, Muller pointed out, the appearance of a recessive lethal would have been selected against by natural selection and passed out of the stock unnoticed.[12] But as the goal of the experiment was to get a race that expressed as low a percentage of normals as possible, a mutation that resulted in the death of homozygous normals would tend to be selected.[13] Since all of these offspring would then contain both lethal factors, a nearly pure race would become established in a single step, and, locked in a state of enforced heterozygosity, the two factors would then persist indefinitely. The only other ingredient necessary for the creation of the pure-breeding beaded race was the existence of the crossover suppressor.[14] Without such a factor, crossing over between the new lethal and the beaded mutation would generate one normal and one doubly mutant chromosome containing both the new lethal and the beaded mutation with a frequency of 10 percent.[15]

— THE DISCOVERY of the balanced lethal factors in the beaded flies was a remarkable piece of detective work, and with it Muller provided an entirely Mendelian explanation for the most troubling of all the cases of allegedly non-Mendelian inheritance in *Drosophila*. But the relentlessly probing Muller believed that his discovery might have a more general significance, and that it could provide the key to the aberrant behavior of De Vries's *Oenothera* crosses, which still remained one of the great unsolved mysteries of the post-Mendelian era. Not only did he see a striking parallel between the behavior of the crosses of beaded *Drosophila* and the breeding behavior exhibited by *O. lamarckiana* and the other *Oenothera* varieties, but he was convinced that he had finally found the true explanation for the "mutants" of *O. lamarckiana*.

De Vries himself had believed that there was a connection between the aberrant breeding behavior of *O. lamarckiana* and its tendency to produce mutants, and that the explanation would lie in the changes of individual pangenes.[16] Beginning from the premise that *O. lamarckiana* was not of hybrid origin, he argued that the differences between *O. lamarckiana* and its mutants were caused by change in one or at most a few pangenes. He claimed that the dwarf mutant, *O. nanella,* for example, was caused by the inactivation of a pangene required for growth, and the brittle stemmed *O. ribrinervis* resulted from a mutation in a pangene required for synthesis of lignin.[17] In addition to these loss-of-function mutants, which he called "retrogressive," he postulated the existence of "degressive mutations," which resulted in the return of long-latent characters and "progressive mutations," which gave birth to entirely novel pangenes, but in each case the mutant was envisioned to be the result of one, or at most a few, mutations of the *O. lamarckiana* parent, which he presumed to be homozygous.

Likewise, the splitting of the $F_1$ generation, which was the starting point of his genetic theory, was predicated on the purity of the *O. lamarckiana.* As he had first observed in the 1890s, when *O. lamarckiana* was crossed to other related varieties of *Oenothera,* the progeny routinely split into two (and occasionally more) distinct types, each of which was again true-breeding. This splitting was the fact that had so aroused Bateson's initial suspicion and caused him to wonder if *O. lamarckiana* itself might be a hybrid. By 1907, De Vries had observed a vast number of crosses involving *O. lamarckiana* and its mutants with other closely related *Oenothera* varieties, and in nearly every case the first generation split into two distinct types, each of which in turn bred true. He had been so impressed with the phenomenon that he wrote a paper entitled "On Twin Hybrids" in which he gave names to the two distinct classes, laeta and veluntina.[18] Laeta was always like the *O. lamarckiana* parent itself, with strongly crinkled, clear green foliage and relatively slender hairless stems, and veluntina, which appeared at roughly the same frequency, was a blend of its parents, with narrower, less crinkled, greyish green foliage and hairier, stouter, and woodier stems.

In 1913 De Vries suggested that the splitting of *O. lamarckiana* hybrids upon crossing with other varieties was due to the instability of the *O. lamarckiana* parent's laeta-determining pangene. By some unspecified mechanism, the non-*O. lamarckiana* parent was thought to induce the laeta pangene to mutate into an inactive "veluntina" form. Progeny showing the laeta characters had unmutated pangenes while those that had undergone a mutation showed the veluntina characters.[19] It followed that the $F_1$ generation would breed true, the self-mated laeta progeny having nothing but laeta pangenes, and likewise for the veluntina offspring.

Struck by the exact parallel between De Vries's twin hybrids and the two distinct classes of $F_1$ progeny that resulted from a cross between a beaded and a normal-winged fly, Muller suggested that *O. lamarckiana,* like the beaded fly, was a permanent hybrid, maintained by balanced lethals. In the terminology introduced by the German botanist Otto Renner in the same year, the hybrid *O. lamarckiana* contained two "gene complexes," each containing 7 of the 14 chromosomes characteristic of *Oenothera.* Renner named the two complexes gaudens and velans. Muller and Renner independently suggested that except for crossing over there would be only two types of germ cells, each containing a distinct lethal factor. When a gaudens-containing egg was fertilized by a gaudens-containing pollen, the zygote died, and likewise for the combination of velans with velans. As in the case of the beaded fly, it was only the heterozygotes that survived.

Furthermore, different species might contain their own distinctive gene complexes, to which Renner, a classics scholar, assigned recondite Latin names.[20] When different species were crossed, each of the two distinct germ cells from one species could combine with each of the two distinct germ cells provided by the other, giving four different types of fertilized eggs. When two gene complexes containing distinct lethals were combined, a new stable hybrid would be formed. But if the two complexes contained identical lethals, as would be expected in a cross between closely related species, the resulting hybrid would die. In this way, the balanced lethal theory accounted for the puzzling fact that certain crosses split into two or more different types in the first generation, each of which went on to breed true when self-mated.[21]

— BY 1918, DE VRIES had begun to accept that some of the stable varieties of *Oenothera* might be permanent heterozygotes held together by a pair of interlinking lethals, as Muller had argued in 1917, but he continued to insist on the purity of *O. lamarckiana* and the origin of the new varieties that arose from it.[22] To yield on this point was the death toll for his entire mutation theory. Writing to his old friend Jacques Loeb, De Vries complained, "Morgan has had such nonsense published by one of his pupils."[23]

But not only was Muller insisting on the heterozygosity of *O. lamarckiana,* by 1918 he had collected sufficient evidence in *Drosophila* to be confident that De Vries's mutants were not due to actual changes in the genes but rather were the result of recombinations and anomalies in the segregation of chromosomes. Under ordinary circumstances a chromosome carrying a new recessive mutation would, over the course of generations, be paired with an identical copy of itself, and when that happened it would be subject to the action of natural selection. In the likely event that the mutation had a detrimental effect on survival, the homozygous fly would be eliminated from the population, and the frequency of detrimental or lethal mutations would slowly diminish. However, the situation was fundamentally different, Muller explained, when a new recessive mutation arose near a lethal on either of a pair of balanced chromosomes. In such flies, a member of a balanced chromosome pair could never be paired with itself, ensuring that any mutation it contained escaped the policing action of natural selection. As a result, a pair of balanced chromosomes tended to accumulate multiple recessive mutations. In the past, Muller pointed out, the infrequent but regular appearance of a recessive trait from normal parents had been interpreted as a mutational event, and organisms were said to "sport" or "throw mutants." But it was now clear, Muller argued, that what was being observed were actually recombination events in which a recessive mutation broke free from its enslaving lethal.[24]

By analogy, the transformation of *O. lamarckiana* wasn't caused by a new mutation at all; instead, it was caused by the sudden expression of a group of cryptic recessive mutations that had been accumulating for generations on a pair of balanced lethals. Every now and then a new recombinant would arise that differed in several independent characteristics from its parent, and just as it had in the parent plant, the balanced lethal system ensured that the new recombinant type would breed true. Over the subsequent years, as the nature of the segregation in *Oenothera* became better understood, it became clear that the plant had evolved a wonderfully intricate means to ensure that the maternal and paternal chromosomes formed two complementary gene-complexes that were passed on intact from one generation to the next. As in the case of beaded *Drosophila*, the complementary complexes were kept in heterozygous condition by means of a system of balanced lethals, and over generations, as Muller had hypothesized, recessive mutations accumulated and a line became increasingly heterozygous. It also became clear that De Vries's mutants were due to chromosome rearrangements, by means of recombination or the failure of homologous chromosomes to segregate, that resulted in previously heterozygous genes breaking out into homozygous condition. Though the fully elaborated theory took another decade to develop and would extend beyond Muller's original formulation, already in 1917 Muller had pointed the way by providing the essential idea of balanced lethals.

In the meantime, De Vries was forced to watch his mutation theory dwindle into irrelevance. By 1919 he had wholly retired from his position in the university to spend the last seventeen years of his life working in the private laboratory and experimental garden that he had built next to his home in Luntern, a small village in central Holland. Following his long-established practice, De Vries refused to engage in public debate, even though he was convinced that he was being wronged. To Loeb, he now explained that his *Oenothera* were "always true to their principles, despite the rather curious and vague objections of Morgan and others."[25] Until his death in 1935, De Vries, tall, thin and regal, with a great Darwinesque white beard, actively tended to his beloved *Oenothera*, unshaken in his conviction that the *O. lamarckiana* and its varieties were essentially pure species, homozygous for most traits, and that their differences were to be accounted for by rare mutations within species.[26]

— AS WORK ON THE CYTOLOGY and genetics of *Oenothera* proceeded, De Vries's account of the origin of *O. lamarckiana* began to seem less probable. De Vries had long maintained that the *O. lamarckiana* he had found in Hilvershum had escaped from a local garden where it had been planted by seeds purchased from a local seedsman who gotten them from the London firm of Carter and Company, who in turn claimed to have imported them from Texas.[27] In 1895 and in 1913 again, he had visited the Muséum d'Histoire Naturelle in Paris to examine specimens of the original species that Lamarck had grown in the museum garden and described in his famous *Encylopédie Méthodique, Botanique* in 1797. During his first visit, De Vries confirmed that one of the two remaining specimens was virtually identical in every aspect to the plants he'd found in Hilvershum and it bore a notation in Lamarck's handwriting indicating that it had come from North America. When he visited the second time, he located other examples in the museum collection, the most interesting of which he reported was a nearly perfect specimen of *O. lamarckiana* collected from the eastern United States.

But the American cytologist Bradley Davis, who was an ardent opponent of De Vries's mutation theory, claimed to be unable to find any evidence that *O. lamarckiana* existed in the wild in Texas or anywhere else in North America.[28] Instead, he speculated that *O. lamarckiana* was a hybrid, and to prove it he set out to produce an artificial hybrid by crossing two true-breeding, native American *Oenothera* species (*O. grandiflora* and *O. biennis*). By the following year Davis had succeed in creating a stable hybrid that differed from De Vries's *O. lamarckiana* in certain vegetative characters, but was virtually indistinguishable from it in terms of its flowers, fruits, and inflorescence.[29] To further support the view that *O. lamarckiana* first arose as a hybrid, Davis found pictures of an *O. lamarckiana*-like plant that had characteristics of both the American *O. grandiflora* and a European *O. biennis* in an 1806 encyclopedia of English plants.[30] According to the accompanying text, the plant came from an extensive tract of sand hills a few miles north of Liverpool where millions of the same species grew wild and covered large tracts of land. One hundred years later, Davis pointed out, the hills were covered by *O. lamarckiana* plants, and thus, he argued, it was entirely plausible that modern *O. lamarckiana* was a descendant of the plants pictured in the 1806 encyclopedia.

By 1914 Davis had found a stable hybrid, a cross between an *Oenothera* species native to the western United States *(O. hookeri)* and an *O. biennis* that was for all practical purposes indistinguishable from De Vries's *O. lamarckiana.* In addition to possessing all the physical characteristic of *O. lamarckiana,* the new hybrids, which Davis called *neo-Lamarckiana,* produced twin hybrids when outcrossed and generated *O. lamarckiana*-like "mutants."[31] With the simultaneous publication in 1917 of Muller's theory of balanced lethals and Renner's theory of the gene complex, Davis's claim that *O. lamarckiana* was a permanent hybrid was given a solid theoretical foundation.

In the early 1930s, Davis's student Ralph Cleland saw that it was possible to measure the degree of relatedness of any two Renner complexes by means of a simple but clever cytological test.[32] Using his method, Cleland was able to divide the more than 400 Renner complexes found in American *Oenothera* species into ten groups, each of which had a definite geographic range.[33] Cleland's work made it clear that Davis, who had made his speculations before the concepts of balanced lethals and Renner complexes had been introduced, had not been far off. *O. lamarckiana* was a stable hybrid containing one complex derived from an *O. hookeri* variety indigenous to the far western states, and another that was a close relative of an *O. biennis* native to the Northeast.[34]

*O. lamarckiana* most likely originated in Europe sometime in the late eighteenth century, and Carter and Co.'s claim to have received the seeds from Texas was false. In all likelihood, *O. hookeri* and *O. biennis* seeds had been mixed in the dirt that was used as ballast in the empty ships returning from America and piled up in large heaps near Liverpool and other European ports. Once ashore, the two varieties proliferated madly in the sandy coastal hills where they crossed and formed the true-breeding hybrid, which eventually became known as *O. lamarckiana.* Smitten by their beautiful large flowers, Carter and Co. put the seeds up for sale in 1860, and thus they found their way to a private garden in Hilversum where they escaped and took over a corner of an abandoned potato field and were eventually spied by Hugo De Vries in the summer of 1886.

**Hermann Muller in Austin, Texas, circa 1926.**

Generously provided by David Muller.

CHAPTER 12

# X-Rays

— WITH THE RESOLUTION of the beaded and truncate cases, the last serious doubts about the universality of Mendelism and the chromosome theory were extinguished. However, genetics, as William Bateson had defined the word in 1905, was the study of two central topics, heredity and variation, and while the problem of inheritance was more or less solved, the study of variation had barely progressed at all in fifty years. Furthermore, Muller had succeeded in discrediting De Vries's mutation theory, which had been the only promising lead. But having cleared the field, Muller was now free to make a new mutation theory.

As he wrote to Julian Huxley, almost nothing was known about the topic: "We simply know that mutation occurs, and occurs 'rarely', whatever that means, though in fact on its rate and mode of incidence depend evolution."[1] In fact, the rate of appearance of new mutations was so low—the best estimate was that 1 in 50,000 flies contained a new mutation—that it was difficult to conceive of how one might study the problem at all.

Guided by the idea that mutations were generally variations for the worse, one of his earliest and most profound insights into genetics, Muller refused to be stymied.[2] The beaded work and the discovery of the balanced lethal system had strengthened his belief in the importance of deleterious mutations, and more particularly, he was now convinced that recessive lethals were likely to be the most common class of mutation.[3]

Although Muller was convinced on theoretical grounds that lethal mutations would greatly exceed the rate of visible and other classes of mutations, he had no idea what the actual frequency of lethal mutations might be. Of course, it was a tricky matter to detect the presence of newly created lethal mutations. A dominant lethal mutation would often kill the embryo and escape detection in that way, while a recessive lethal would be carried along invisible through many generations until the mating of two individuals who happened to be heterozygous for it. The detection of X-linked recessive lethals in females, however, was a special case, as Muller pointed out in his 1918 paper on beaded.[4] Although an equal number of sons and daughters would inherit an X-linked lethal from their mothers, daughters were protected by a good copy of the gene contained in the X they had inherited from their father, while the unlucky sons would die from it. Thus females carrying a X-linked lethal mutation would give rise to twice as many females as males. The problem was that sex determination was a random process, and therefore the ratio of males to females would be expected to fluctuate, the values following a bell-shaped distribution with a mean of one-half.[5] For this reason, it was sometimes difficult to distinguish those sex ratios that were a reflection of the natural variability of the process of random sampling from those due to the presence of a lethal.

In the spring of 1918, Muller designed a genetic screen that would allow him to detect the creation of X-linked recessive lethals without relying entirely on the sex ratio method. The experiment, which he assigned as a lab project in his undergraduate genetics course, involved following several generations of multiply marked females and recording the frequency of different recombinant classes.[6] As a class, Muller's students examined a total of 600 X chromosomes and failed to find a single new lethal.[7] Although he believed it likely that many of the cultures were not properly carried through, due to the inexperience of the students, nonetheless the result seemed to suggest that large numbers of flies would have to be examined to find a rate for the appearance of new lethal mutations.

In the fall of 1918, Muller returned to Columbia, where he'd been offered a temporary appointment as an instructor that he hoped would lead to a permanent position, and his hunt for a mutation rate was put on hold. Meanwhile, Altenburg, who had remained in Houston, took up the project, counting lethals on the X chromosome using the more straightforward sex ratio method. He began by breeding 90 females and counting the sex ratios of their progeny, eliminating those flies that carried a preexisting lethal mutation. Daughters from normal females were then bred and screened for the presence of new lethals by the sex ratio test, and this process was repeated for five generations. As expected, there were several ambiguous cases, which were subjected to further tests, but when the uncertain cases were resolved, Altenburg had found an astonishing 13 lethals in a total number of 385 females tested, which was equivalent to a rate of 1 lethal in every 60 X chromosomes. At this rate, an X chromosome would be expected to accumulate one recessive lethal mutation every four years. Although it was clear that humans could not possibly accumulate mutations at the same rate, otherwise a girl would be likely to possess several X-linked lethal mutations before she was of reproductive age and would therefore be incapable of producing sons, nonetheless Altenburg's result was mind-boggling.

In the end, Altenburg's simple and direct approach yielded the first major breakthrough in mutation theory in thirty years, though the results of his experiment so defied expectations and met with such skepticism among other *Drosophila* workers that he did not rush to publish them.[8] Instead, he and Muller met the following summer at the Marine Biology Lab at Woods Hole and undertook a joint experiment to verify Altenburg's initial findings. To avoid the uncertainty associated with the sex ratio test, they followed the method Muller had invented for his undergraduate class at Texas the previous year. At the same time they looked for an effect of temperature on mutation rate by dividing the cultures into two lots, one grown at 80°F. and the other, kept cool in shallow pans of running seawater, at 66°F.[9] Among the warm-grown cultures, the frequency was 1 mutation per 80 X chromosomes, as compared to 1 in 180 for the cooler series. Although the average rate (1 in 106 chromosomes) was slightly lower than that found by Altenburg, the two numbers were in the same ballpark.[10]

That fall Altenburg suggested that he and Muller coauthor a paper in which they presented both his results using the sex ratio method and the results of the joint work done at Woods Hole. "It would be a privilege," Muller wrote back, "but oh so nervy."[11] Because Muller and Altenburg were coauthors on the paper, Altenburg was not singled out for making the first successful determination of a mutation rate, but in a later paper authored by Muller alone he set the record straight, writing, "The first experiment intentionally giving a quantitative result concerning mutations in a group of loci was carried out by Altenburg in the winter of 1918–1919."[12]

The following summer Muller and Altenburg met again to resume their joint mutation experiments. The new goal was to repeat the previous summer's experiment using a more sophisticated genetic stock that had been engineered by Muller over the winter at Columbia.[13] Literally thousands of cultures of flies were bred that summer before it became apparent that there was a problem with the reengineered stock that prevented the males from thriving in the Woods Hole climate and that hundreds of hours of work had been in vain.

Once again, however, Muller's special gift for seeing an opportunity in the most unpromising results saved the day. As it had been designed, the experimental protocol called for a female fly carrying *Bar*—a dominant X-linked mutation that caused a pronounced narrowing of the eyes—to be crossed to a normal male, and their progeny to be grown up. As expected, about half of the resulting offspring—those that inherited the *Bar*-containing copy from their mothers—showed bar-shaped eyes. However, among the last of the cultures to be examined before the experiment was aborted, Muller noticed one anomalous culture.

More than forty years after the fact, Altenburg still vividly recalled the moment late in the summer of 1920 when Muller found the mutant that would soon be known around the world. They were working together in a basement lab examining the glass vials in which their cultures were grown.

Suddenly Muller jumped up and called him over to look, his entire being alight with excitement, as he held up a vial for Altenburg to examine. In this culture there were no bar-eyed male progeny. Not only that, it was also clear, as in the case of the beaded chromosome, that there was no crossing over between the maternal X's.[14] Instantly, Muller had surmised that a novel mutant condition had arisen in the *Bar*-containing X of the mother. Cℓ, as he called the new mutation, simultaneously conferred two new properties on the *Bar* X, acting both as a crossover suppressor and a recessive lethal, resulting in the death of half of the grandsons.[15]

Not only could he explain the absence of *Bar* grandsons in the anomalous culture, but Muller had also seen how the new chromosome carrying Cℓ and *Bar* (soon to be known worldwide as the CℓB chromosome) could provide a simple new way to screen for spontaneously arising X-linked lethals in male sperm; one would simply cross a male to a female carrying one normal and one CℓB chromosome. In the next generation, the $F_1$, those daughters containing the CℓB chromosome, which could be identified by the presence of the dominant *Bar* eye mutation, were individually mated to their brothers, each in its own vial. Any $F_1$ daughter that failed to produce sons contained a new lethal mutation in her non-CℓB chromosome, the one inherited from her father.

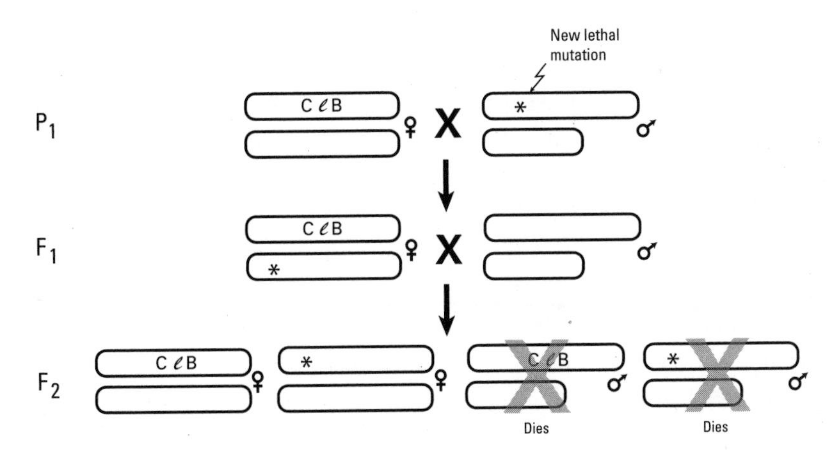


**Method for detecting creation of new lethals using the CℓB chromosome.**

Illustration by Tamara L. Clark.

New lethals could be identified simply by checking each culture for the absence of males. One would no longer have to worry over the ambiguities of the sex ratio test. Altenburg never forgot that moment in the lab when Muller held the vial up to the light. Muller had seen it in a flash, he proudly recalled.[16]

— THAT FALL MULLER took a position in the zoology department of the University of Texas in Austin, disappointed that he had not been offered a permanent position at Columbia when his two-year appointment had run out. According to Sturtevant, Muller mistakenly believed that Morgan was responsible for his not being offered a job at Columbia, but in fact it had been Muller's hero, E. B. Wilson, who as chairman of the department had actually made the decision.[17] If it were true that Wilson had blocked Muller, it would have been gratifying to Sturtevant, who had long resented the fact that Wilson had a very high opinion of Muller and a far less favorable impression of him.[18] In any case, it was with regret that Muller left "the happy swirl of civilization, friends, etc. for what [he hoped was] a better chance to work."[19]

In the early days, Muller was happy with the atmosphere in the zoology department in Austin, which was run in a highly democratic fashion by John Thompson Patterson, or "Dr. Pat," as he was known to his students, a short, pear-shaped man who was legendary for his use of profanity. The other powerful figure in the department was the cytologist Theophilus Painter. Like Muller, Painter was extremely ambitious, but he was in other ways Muller's antithesis. Highly reserved and self-controlled, Painter was an expert furniture maker and sportsman who loved tennis, golf, fishing, and most of all, hunting, especially deer and antelope.[20]

During Muller's first winter in Austin, he measured the lethal mutation rate using the new CℓB method and was shocked to find only 4 lethals in a total of 3,935 flies examined. While Altenburg had seen a rate of close to 2 percent and the joint experiment had yielded a rate of 1.26 percent, Muller now found a mutation rate of 0.1 percent.[21] The new results were puzzling, for there did not on the face of it seem to be any genetic difference between the CℓB strain and those that had given the higher rates, nor was there any obvious difference in the growth conditions in the

CℓB experiments and those in the earlier joint experiment. The mystery of the high variability in the mutation rate required further investigation, and as it would turn out, Muller's attempts to resolve the inconsistencies in the lethal mutation rate would consume the better part of the next six years.

As usual, Muller's powerful imagination raced ahead of the experiments, and despite the difficulties he was now completely convinced that the study of the nature of mutation was a key to understanding the gene. As he had done ten years earlier for complex traits, he was now prepared to begin to lay the conceptual groundwork for a real mutation theory, which he introduced in September 1921 at the Second International Congress on Eugenics in New York City. "Beneath the imposing building called *heredity* there has been a dingy basement called *Mutation,*" he dramatically introduced his subject. "Lately the searchlight of genetic analysis has thrown a flood of illumination into many of the dark recesses there."[22] Several distinct phenomena had been grouped together under the name of "mutation" because they all involved the appearance of a new genetic type, he explained. It was unfortunately now necessary to recognize that the results with *Oenothera,* which had once formed "the backbone of the earlier mutation theory," as he put it, were no longer strictly relevant to the modern problem. It was now time to replace the "elaborate system of conclusions" derived from *Oenothera* and build a theory based on the study of changes in the individual hereditary units.

Although he granted that almost nothing was known about the mechanism of mutation or about the nature of the genes, other than that "they behaved like material particles existing in the chromosomes," it was possible to begin to put down the basic outlines of a new theory of mutation. More than a hundred examples of changes in individual genes had been found in *Drosophila*, and these would provide the basis for the new theory. The most important new principle, he asserted, gleaned from the quantitative study of mutations conducted by Altenburg and himself, concerned the relatively high stability of an individual gene. Even at the higher rate found in their first summer at Woods Hole, Muller computed that a typical gene underwent mutation no more frequently than once every 2,000 years.[23] A corollary of the stability of genes was the principle that external agents could not in general be shown to significantly affect the rate of production of mutations, a rule to which Muller himself would soon find a remarkable exception.

Furthermore, despite the large number of changes in individual genes that gave visible phenotypes, the new evidence suggested that the visible mutations were only a small fraction of the "spectrum" of mutations, and in particular that there were many more lethal than visible mutations.[24] More generally, he now had the data to support the previously unsupported hypothesis that mutations were far more often detrimental to an organism than beneficial. Among the other principles that could be gleaned from recent empirical work was that reverse mutations occurred that restored mutant to normal types, which made it clear that mutations were not exclusively losses, although there were cases in which losses clearly did occur. Furthermore, a single gene locus could give rise to many different phenotypes. The eye color mutant *eosin,* for example, was a partial reversion of the *white* mutant, and the same locus had given rise to *buff, coral, blood, tinged,* and *ivory.*[25]

That December, while en route to the annual meetings of the American Zoological and Naturalist Societies in Toronto, Muller met William Bateson at Morgan's house in New York City. Muller was unimpressed by Bateson, feeling he was out of date, and Bateson had likewise found nothing at all to write home about in Muller. But Bateson was far more excited about the *Drosophila* work than he had expected to be. In fact, for Bateson his visit to Columbia had been a revelation. In particular, Bridges' demonstration that an aberrant distribution of the sex chromosomes during sex cell formation (a phenomenon Bridges had termed "non-disjunction") was invariably accompanied by the failure of a character (white eyes) to segregate according to Mendelian expectations had shocked Bateson out of his deep-set resistance to the chromosome theory.[26] In fact, already in 1907 Bateson had insisted that just such a parallelism would constitute the only valid demonstration of the chromosome theory, the definitive proof that genes were located on chromosomes.[27] His position was "untenable," Bateson confessed to his wife, and would soon have become "ridiculous."[28] Furthermore, the visit to the States had left him no doubt that "the people here have got ahead of us."[29]

Bateson's plenary address to the members of the AAAS in the grand Convocation Hall of the University of Toronto was the main event in the medley of biological meetings held in Toronto during Christmas week of 1921. Misrepresented by reporters, Bateson's address would be picked up by all the news media and used as a propaganda tool for William Jennings Bryan in his campaign to ban the teaching of evolution in the United States. The religious overtones in the title of his lecture, "Evolution Faith and Modern Doubts," and the lecture's being packed with biblical allusions, which were a staple of Bateson's lecture style, made him a ripe target for the sensationalist press. Sensing the danger, Bateson had included an unambiguous rejoinder to those who might be tempted to misinterpret his lecture: "Let us then proclaim in precise and unmistakable language that our faith in evolution is unshaken."

The talk itself, as Bateson recognized, had an antiquarian air about it, harkening back to arguments that Darwin himself had made sixty years earlier. Nostalgically Bateson recalled his first visit to America in 1888 when evolution had been the burning issue for any serious biologist. In that era, he recalled for the assembled crowd of scientists and journalists, "every aspiring zoologist was a embryologist" because embryology was then believed to hold the key to understanding evolution. By the turn of the century, Bateson continued, the old Haeckelian ideas had proved to be a dead end, and he and other evolution-minded biologists turned to the study of "variation and heredity," creating the discipline that Bateson himself had dubbed *genetics* in 1905. "We became geneticists in the conviction that there at least must evolutionary wisdom be found,"[30] he explained, but genetics, like embryology before it, had failed to deliver on the promise. Even with the advances of Mendelism and the further additions provided by the *Drosophila* workers at Columbia, biologists were no closer to understanding the "mode and processes of evolution" and to finally resolving the mystery of the origin of species.

The following day, Bateson gave another lecture, this one to the Society of Zoologists, which spelled out the nature of his conversion to the chromosome theory and elaborated on his various reservations. On the larger point—that genes resided in chromosomes—he did finally join the consensus, which included every major figure in genetics save Willem Johannsen, who was never to concede the point. But Bateson still refused to give his blessing to the theory of crossing over, which was the lynchpin of the whole argument, arguing that it was still too early to "contemplate so concrete and material a picture of the mechanism of crossing-over."[31]

On the same day that Bateson spoke before the Society of Zoologists, Muller addressed the American Society of Naturalists.[32] In his lecture, Muller touched briefly on the current state of the chromosome theory, including the finding that the two larger of the *Drosophila* chromosomes each contained hundreds of genes whose locations could be determined by crossing over, which he took as established facts. But the point of his lecture, which had an almost hallucinatory quality about it, concerned his bold and strikingly original discussion of the nature of the gene.

He began by observing that the most distinctive property of genes was that they were capable of reproducing themselves. More exactly, a gene had the ability to convert material from the protoplasm of the cell into an exact copy of itself. While there were many examples of complex biological molecules (enzymes) that catalyzed chemical reactions, the strange thing about genes was that of all the possible reactions genes might have catalyzed, they affected just those that resulted in the production of molecules with their own complex structure. But this "autocatalytic" property of genes was not even the most surprising aspect of the situation, Muller insisted. The truly astonishing thing was that when the structure of a gene changed due to an entirely random mutation, the altered gene was nonetheless still faithfully reproduced. On the face of it there was no obvious reason why a change in the gene's structure that resulted in a change in its function would simultaneously alter the gene's autocatalytic function in just the precise way that would result in the new variant of the gene being reproduced.

Without mentioning Bateson by name, Muller next turned to the subject of Bateson's plenary address of the night before. "It is commonly said that evolution rests upon two foundations—inheritance and variation; but there is a subtle and important error here," Muller declared, echoing

Bateson's famous formulation. Variation and heredity did not represent two separate components of the evolutionary path: Inheritance without variation could not result in evolutionary change, and, likewise, variation made no lasting impression unless it was heritable. It was not variation and heredity that mattered in evolution, but "inheritance *of* variation," as Muller rewrote Bateson's famous formulation, and it was the ability of genes to copy themselves and simultaneously retain the ability to copy their random mutations that made this possible.

Understanding the structure of genes was, Muller now proclaimed, the most fundamental problem of genetics. Even formulating the problem in these terms represented a paradigm shift, because it was not yet clear that genes must share a universal physical property. But the facts that genes possessed the peculiar property of being able to reproduce their variation and that a particular gene could undergo change after change without those changes affecting its ability to copy itself implied that all genes shared a common structural feature involved in replication. By the same logic, Muller argued, the specificity of an individual gene—the structural elements that accounted for its unique function in the cell and that were also capable of undergoing mutation—was an aspect of gene structure separate and distinct from the universal one. In a few years time, Muller would suggest that the two different structural properties lay in different dimensions.[33]

In the meantime, Muller presented a model for the gene as made up of different components whose various arrangements accounted for the differences between genes and also formed the basis of mutation.[34] In the process of duplication, each part of the gene attracted to itself a similar part from the protoplasm. In this way, the gene "molded" the newly recruited parts into an arrangement identical to its own, and those parts were then bound together to make an exact replica of the first gene.

At the heart of the theory was a mysterious attraction between the subunits of the gene and their corresponding parts from solution, and this same force of attraction was evidenced in the strong and highly specific attraction between genes of the same kind during the synapsis of chromosomes during sex cell formation. But the problem with the model was that there was no basis for believing that like matter attracted like matter. On the contrary, Muller pointed out, this seemed to be a property unique to genes, and the secret of the self-attraction of genes must depend on some fundamental feature of gene structure.[35]

In the concluding paragraph of his 1921 lecture, Muller made one further prophetic suggestion, which would eventually prove to be the starting point of molecular biology. Like genes, Muller pointed out, the microscopic viruses that grew on bacteria, which were initially called D'Herelle particles after their discoverer and later dubbed bacteriophages, were not only self-reproducing but appeared to be capable of copying their mutations. For all intents and purposes, he argued, they were genes, and as such they might offer huge advantages for the study of gene theory. It might soon be possible, he predicted, "to grind genes in a mortar and cook them in a beaker. Must we geneticists become bacteriologists, physiological chemists and physicists, simultaneously with being zoologists and botanists? Let us hope so."[36]

But the world was not yet ready to absorb Muller's message. As Muller was leaving the podium, the famous paleontologist Henry Fairfield Osborn pulled him aside and said, "I'm glad you have a sense of humor, Muller."[37] As was to be expected, Muller's prophetic lecture with its salient critique of Bateson's views on genetics and evolution was entirely ignored, and Bateson's address, which was widely misinterpreted, became a rallying call for the anti-evolution forces in America.

— ACCOMPANIED BY ALTENBURG, Muller made his first European trip in the summer of 1922. As it had been planned, Muller and Altenburg were to tour England, France, Switzerland, and Germany, and then, if they could gain admittance, their trip would culminate in a tour of Moscow and Leningrad. While Altenburg treated their trip as a holiday, Muller had a more serious goal. He planned to take advantage of the new airplane line connecting Moscow and Berlin to deliver a suitcase full of *Drosophila* cultures to the isolated geneticists in post-revolutionary Russia.

Cramped, noisy, overheated, and unventilated, their second-class cabin aboard HMS *Aquitania* was unacceptable to Altenburg. After the boat entered the Gulf Stream on the second day of the voyage, Altenburg refused to sleep another night in the sweltering room, and with detached amusement Muller observed as Altenburg stripped off his clothing, covered himself in a raincoat, and took his bedding up to the deck. The spectacle of the tall, half-naked, strikingly good-looking young American spreading his bedding under a lifeboat soon attracted a crowd, and Altenburg took advantage of the opportunity to deliver a spontaneous diatribe against the steamship company for taking advantage of the elimination of the German competition to exploit their customers. Forced back to the cabin by the nighttime cleaning crew, Altenburg assaulted the ship's purser the next day, threatening to bring the matter to the captain of the ship and to spread the word about the poor accommodations, claiming to be "in a position to reach a good many people." That afternoon after lunch Altenburg and Muller were transferred to a first-class cabin.[38]

Having solved their personal problem, Altenburg was content to let the matter drop, but Muller was more deeply affected by the shipboard inequalities, particularly after touring the third-class accommodations, where he was shocked at the crowded and unsanitary conditions in the employee cabins. Even more disturbing were the rumors he'd heard that the gates to the third class had been locked when the *Lusitania* went down, so that the passengers couldn't get out, the excuse being that they were excitable Italians and might create a panic. "Like caged animals," he wrote home to the woman, Jessie Jacobs, who would soon become his wife, "they all went down with the ship. And then they spoke in the papers of the 'frightfulness' and brutality of the Germans." At the end of their voyage, the sight of the British navy steaming out en mass to welcome home the Prince of Wales from a world tour disgusted Muller, one in a growing list of reminders of "the strength of the English caste system."[39]

During the crossing Altenburg had struck up an affair with a well-known, married American actress. A freethinker, the actress told Muller about a woman, the sister of a close friend of hers, who had had an illegitimate child with H. G. Wells for purely eugenic purposes and, immediately after the birth of the child, married a different man for love. Since then Wells had had several other illegitimate children, and both Muller and the actress agreed that this was all for the better, and that decisions about reproduction should be made for eugenic reasons and divorced from considerations of love and attraction. Meanwhile, however, Muller's romantic life was fraught with conflict and difficulties.

For nearly a year he had been seeing Jessie Jacobs, a Kansas-born mathematician, who had earned her doctorate from the University of Illinois at Urbana in 1919 and had begun as an instructor in mathematics at the University of Texas in Austin the same year as he arrived there. Like Hermann, Jessie was extremely slight of build, hardly five foot tall, with fine features and a penetrating intellect. A confirmed atheist, sympathetic to the communist cause, and interested in Hermann's ideas about eugenics, Jessie seemed like a nearly perfect match for Hermann, but from the beginning their connection had seemed somewhat fragile and each was prone to misunderstand the other's motives.[40] They had had a falling out on the day Muller was leaving for New York, and Jessie had not returned any of his many letters.

In London the whirl of activity temporarily distracted Muller from his romantic problems, but alone in Paris after being abandoned by Edgar, who had now resumed his affair with the actress from the boat, Muller began to grow increasingly despondent over Jessie's continued silence. Trying a softer approach, he wrote that he was too rough for tender nature. "I must be fashioned with sharp corners, and the only remedy is to shatter me," he wrote, an omen of events to come.[41] By the time they arrived in Berlin at the end of July, Muller had fallen into a dark mood about his personal life and the world at large. "It would not take much more to knock all idealism out of me," he wrote, "and for me that means death—virtually or actually."[42] Despite the fact that he had now received two letters from Jessie, and in the second one she'd written that she loved him, he wrote back that he could not see "how it can mean very much." Her having written only twice in eight weeks, he wrote, was clear evidence to him of her lack of feeling. "It was funny how we thought in Austin it was the other way around," he reflected with self-pity.

The state of Germany further fueled his dark mood. In his letter he described the terrible economic conditions, where professors barely earned enough to feed their families, and the punitive reparations exacted by the allies was only making a bad situation worse. The runaway inflation was "impoverishing the German people at a tremendous rate," he observed, and "the people [were] getting restless."[43] To top it off, the Russians refused to grant him entry to the country and it appeared that he would not be able to deliver the *Drosophila* cultures, which he had labored to keep alive for nearly two months. But in the middle of August, after he'd given up hope, he was granted entrance to Russia and his depression lifted. "Heaven Earth and Hell were moved," he now wrote exuberantly, and it had been arranged that he would have a place on an eighteen-hour flight from Berlin to Moscow that was ordinarily available only to government officials.

In the end, the visit to Russia was more than Muller had even hoped for. In the course of several weeks he visited many of the newly organized state "institutes" that had developed in Moscow and Leningrad since the revolution. He was particularly impressed with the department of genetics, which was located in Anikovo thirty miles outside Moscow. Overcoming the grave deprivation of the war years, the scientists at the Anikovo station had been able create a vital and productive research center,[44] and Muller chose A. I. Serebrovsky, who had independently discovered crossing over among the sex-linked genes in chickens, to be the recipient of his precious *Drosophila* strains.[45] By the time Muller returned to the Soviet Union in the 1930s, the seed he had planted in 1922 had blossomed into a full-grown *Drosophila* genetics program.

But the most significant event of the trip would turn out to be Muller's brief introduction to N. I. Vavilov during a brief excursion to the Bureau of Applied Botany in Leningrad, which had been the seat of agricultural research in pre-revolutionary Russia before the Bolsheviks destroyed it. The genetics work was done in the estates of the former nobility at Tsarskoye Selo on the outskirts of the city. Situated in derelict but once-beautiful parks, mansions had been converted to laboratories, and rooms that were still filled with gilding and Byzantine decorations were now outfitted with lab benches.[46] Vavilov himself lived in his office in the former country house that had been presented by the late Queen Victoria to her godson the ex–grand duke Boris Vladimirovitch.[47] Under Vavilov's direction, the genetics station was coordinating the work of plant breeders throughout Russia with the goal of developing a vast collection of cereals that were to be the basis of an ambitious breeding program aimed at creating improved agricultural varieties.[48] Fifteen years later, after Muller's and Vavilov's fates had become deeply entwined, they would see each other for the last time at Tsarskoye Selo, which had then been renamed Detskoye Selo (Children's Village).

— HOWEVER BLEAK their prospects for happiness together, Jessie and Hermann were married in June 1923, and on November 2, 1924, Jessie gave birth to a son, David Eugene, whose middle name expressed their shared belief that a child who was the product of two highly intelligent, socially progressive atheists was the best hope for the improvement of the gene pool. In high spirits, Hermann wrote Edgar about the birth of his new baby: "Rice beat Texas but Texas came back the next day by having David Eugene Muller born." But the birth of their first child was physically punishing for Jessie, who was already frail. Adding to her woes, the math department insisted she resign her position as instructor because she'd had a baby, at the same time praising her for her "loyal and efficient service."[49] Sleep-deprived from staying up nights with the new baby and unable to keep up with his fly work, Muller nevertheless entered a period of intense creativity. He was now more convinced than ever that the time was ripe to proceed with the mutation work.

Already in the spring of 1923, Altenburg and Muller had discussed possible new directions to take in the mutation work, and just before David was born Muller wrote to encourage Altenburg to begin treating flies in different stages of development with radium, and to test its effect on males and females to determine which sex tolerated the radiation best. After the birth of the baby, while Jessie recovered in the hospital, Muller began to think in earnest about the radium experiments. Several people had made similar attempts with varying degrees of success, beginning with Morgan, who had, following a suggestion of De Vries, tried to induce mutations with radium around 1910 and succeeded in isolating the beaded and truncate mutants. But with the flood of new visible mutants and results, Morgan hadn't felt the need to continue treating flies with radium.

Starting in 1917 a former Morgan student, Harold Plough, had also experimented with the effects of radium on flies and had found that radium exposure increased the rate of crossing over between chromosomes, but Muller thought Plough's experiments were poorly designed and the results in his recently published paper were internally inconsistent.[50] Plough's work had been followed up by Mavor, a radiologist at Union College in Schenectady, who found that machine-produced X-rays also worked to accelerate the rate of exchange between chromosomes,[51] but neither Plough nor Mavor was focused on the mutation problem. "The field here is still open," Muller wrote Altenburg, brimming over with confidence.[52]

After settling on an experimental plan, Muller invited Altenburg to visit Jessie, David, and himself for Thanksgiving, when they could complete the experiment. He also planned to do an independent study to test the effect of machine-produced X-rays at the same time, but "of course," he assured Altenburg, "if you want to do it with Ra. [radium] I'll do that with you first."[53]

Altenburg's visit had to be postponed after Jessie began to have difficulty with breast-feeding the new baby, who was being fed seven times a day on artificially pumped breast milk delivered by spoon in order that he not "mis-learn to suck human nipples thru the differently constructed bottle nipples."[54] Consumed by the demands of the new baby and his teaching, Muller was forced to abandon his fly crosses for several months, and in the spring of 1925 Altenburg fell ill with severe stiffness in his back, which would later be diagnosed as ankylosing spondylitis. With Altenburg's illness the radium experiments appear to have been put on hold, and in the meantime Muller began experimenting with machine-produced X-rays.

In March 1925 he sent off a paper reporting on the effect of X-rays on the frequency of crossing over,[55] but following the completion of his X-ray paper, the experiments were again interrupted while Muller relocated to the long-overdue new biology building, where he was given his own fly room.

The new fly lab was a scaled-up version of the original Morgan fly room, with one significant difference: the new labs were air-conditioned and had special temperature-controlled warm rooms.[56] With the new facilities, Muller was finally in a position to complete his long-planned experiment on the effect of temperature on mutation rate, and beginning in May 1925 with 106 pairs of flies, Muller launched into the mother of all mutation rate experiments. A year later he had definitively demonstrated the fact that there was a temperature effect on the rate of mutation.[57]

As he would later characterize the result, it was "the first demonstration of the effectiveness of any specified agent whatever in influencing the mutations of numerous genes."[58] In an uncharacteristically modest assessment of the impact of the nearly six-year study, he added, "The temperature effect on mutation is worth having, but by itself it stands as an isolated beam in the largely unseen structure of mutation and gene theory." However, Muller's modesty was entirely beside the point, for even as he was writing about the effect of temperature in the summer of 1927, he and most geneticists throughout the world already knew that the temperature experiments had led him to something far grander, a finding that would shake the foundations of genetics.

Beginning in November 1926 with the help of a local radiologist, Muller had resumed his X-ray experiment, this time to see if he could detect an effect on mutation rate. Although he used several different approaches, it was the tried-and-true C$\ell$B method, which had first come to him in a flash in the summer of 1920, that provided the most reliable estimate of the effect of X-rays on the rate of appearance of lethal X-linked mutations in male flies. The only significant difference between the early experiment and this one was that the parental flies were now exposed to X-rays before mating. Muller, who had been disappointed by the surprisingly low frequency of spontaneous lethals (0.1 percent) that the CℓB method yielded back in 1920, was not surprised by the fact that it did not yield a single lethal among the 198 cultures grown from the untreated males that served as controls for the X-ray experiment. However the X-raying had a nearly magical effect, and the higher the dose, the greater the effect. From 676 fertile cultures grown from flies X-rayed for 24 minutes there emerged 49 lethals, while 772 cultures grown from flies given the same radiation for 48 minutes gave 89 lethals, corresponding to frequencies of 7.2 percent and 11.5 percent, respectively.

Muller also studied the effect of X-rays on mutation following the older pre-CℓB protocol that Altenburg and he had used to compute their first lethal mutation rate in 1919, and, as in the earlier X-chromosome experiments, the recessive marked X-chromosomes could be used to identify and roughly map the newly formed lethals. Using this approach, it was possible to X-ray both males and females and to confirm that the mutations appeared only in the expected chromosomes.[59] When the results from both approaches were combined, Muller found a total of 5 lethals in 6,016 control chromosomes (0.083 percent), compared with 143 lethals in 1,177 chromosomes (12 percent) of flies subjected to the highest dose of radiation, giving a stunning 145-fold increase in mutation rate due to X-raying.

X-raying turned out not only to increase the rate of lethal mutations, but it also increased the rate of visible mutations. In the early days Bridges and others had picked up visible mutations in single individuals only, but because Muller's technique allowed an entire group of individuals bearing the same mutant gene to be observed, Muller could identify mutants that were previously too inconspicuous to notice.[60] In the course of only a few months, Muller would find more mutations that had been found in all *Drosophila* labs over the previous sixteen years of *Drosophila* research, and the X-ray-induced mutations included *all* of the early mutations that had been previously found.[61] According to legend, Muller, who was then working the night shift, was so excited by the appearance of the new mutants that he shouted the news of each one out the window down to another member of the department, named Buchholtz, who occupied the office directly below his. But Muller himself captured the excitement of those first months best in a lecture for the general public given in 1929:

> All types of mutations, large and small, ugly and beautiful, burst upon the gaze. Flies with bulging eyes or with flat or dented eyes; flies with white, purple, yellow or brown eyes; flies with curly hair, with ruffled hair, with parted hair, with fine and with coarse hair, and bald flies; flies with swollen antennae, or extra antennae, or legs in place of antennae; flies with broad wings, with outstretched wings, with truncate wings, with split wings, with spotted wing, with bloated wings and with virtually no wings at all. Big flies and little ones, dark ones and light ones, active and sluggish ones, fertile and sterile ones, long-lived and short-lived ones. Flies that preferred to stay on the ground, flies that did not care about the light, flies with a mixture of sex characters, flies that were especially sensitive to warm weather. The roots of life—the genes—had indeed been struck, and had yielded.[62]

Muller's first public announcement of the result appeared in *Science* in July 1927, and the article was unusual for its absence of primary data. Although the paper generated quite a stir, not all of the feedback was positive. As far as Morgan was concerned, the fact that Muller had been unable to repeat the early report of a 1 percent rate of lethal formation, and that later experiments had showed a frequency one-fifth to one-tenth that rate, was an embarrassment, and, with the *Science* article, he had finally "hung himself."[63] In addition to publicly doubting Muller's scientific integrity, Morgan spread the word that Muller was difficult to get along with. But Muller was not troubled by Morgan's attempts to undermine the confidence in his results. He freely acknowledged the mystery of the variable mutation rate, and considered getting to the bottom of it an interesting scientific challenge. Explaining the situation with Morgan to an old friend, Muller wrote, "All that will simply make my position stronger in time, as the work will stand."[64]

As it turned out, Morgan could not succeed in shaking confidence in Muller's science. Many of those scientists present at Muller's 1927 address to the Fifth International Congress of Genetics in Berlin, where Muller gave his first detailed account of the X-ray work, recognized that they had been witness to a major turning point in the history of science.[65] Man had for the first time willfully manipulated the genetic material. When Muller returned to the United States, he was an instant celebrity.

**Hermann Muller and Nikolai I. Vavilov in Muller's Moscow lab.**

CHAPTER 13

# Triumph of the Modern Gene

— WHILE THE SCIENTIFIC WORLD WAS STILL REELING from the news that X-rays generated mutations at undreamed of rates, Muller had already moved on. For he had quickly realized that in addition to generating localized changes in individual genes at unprecedented rates, X-rays caused large-scale rearrangements of blocks of genes composing whole sections of chromosomes.[1] In a matter of months he had found flies containing deletions and inversions, and most interesting of all, in December 1926 he found a case in which a group of genes formerly associated with the X chromosome had become linked to the third chromosome. Hoping to find physical confirmation that the genes had moved, Muller began to analyze the structure of the implicated chromosomes under the microscope. Although cytology was a tricky business, and the female *Drosophila* chromosomes were particularly difficult to work with, Muller plunged into the new line of work. By December 1927 he had clear pictures of the two chromosomes that showed the unusual linkage. The slides showed that one of the X chromosomes was broken and the missing fragment appeared to have become attached to the third chromosome in a process he dubbed "translocation."[2] The result, which Muller announced at the December meeting of the Society of American Naturalists, represented the first physical confirmation of the correctness of the abstract gene maps.[3] Later that winter he and Altenburg discovered that translocations were nearly as abundant as ordinary gene mutations, and by the spring of 1928 they had isolated over seventy of them.[4]

It wasn't long before Morgan, who had confidently predicted that Muller had dug his own grave by exaggerating his claims about X-raying, was scrambling to get a piece of the action for himself. That summer, while he was visiting Jessie's family in Kansas, Muller heard from a friend in Woods Hole that X-rays were all the rage there.[5] Sturtevant wanted to get translocations, and a recently arrived Russian cytologist, Theodosius Dobzhansky, was already busy at work on the Woods Hole X-ray machine. By the end of the summer, Dobzhansky had nine translocations, and he accompanied Morgan and "the boys" to Caltech, where Morgan had just been hired to start a new biology department.[6] That fall Dobzhansky began analysis of his translocations, correlating the cytological changes with the changes in the linkage maps, in exactly the way that Muller had first illustrated in his lecture to the Society of American Naturalists. For Muller the situation had a quality of déjà vu about it. Once again, it seemed, members of the Morgan lab were preparing to steal credit for his ideas.

This time, however, Muller had the manpower and resources to match the Morgan group. In addition to Altenburg at Rice, Muller recruited Theophilus Shickel Painter, a skilled and ambitious cytologist, to work on the translocations.[7] As Painter told the story years later, he had found Muller crawling on the floor trying to recover some fly ovaries with a pipette. "As skillful as he was in genetic analysis, he didn't have great skill in handling such small material," Painter wrote, and he offered Muller his services.[8] Whether Painter sought out Muller or vice versa, Painter was clearly the right man in the right place at the right time, and he was soon analyzing deletions and translocations full-time.

In the winter of 1928–29, Muller learned from a friend that

Dobzhansky was doing his own cytological analysis of genetically determined translocations, and that he planned to publish a preliminary report as early as the spring of 1929.[9] While Muller had mentioned the possibility of using X-rays to obtain physical evidence for the correctness of the linkage maps at the Sixth International Congress of Genetics in the fall of 1927, and he had sketched out the method in the analysis of his first X-ray-induced transposition at the Nashville meeting in December of the same year, he now realized that he had to publish a definitive account of the status of the translocation work or risk being scooped by the Morgan group.

Painter and Muller's complete analysis of the parallel genetic and cytological studies of dozens of deletions and translocations appeared in the *Journal of Heredity* in the summer of 1929. Their very first example, the "Star-curly transposition," demonstrated the power of the new method. In the chromosome plates it was possible to see that the left end of chromosome 3 had broken off and become attached to the right end of chromosome 2, while the genetic analysis of the same transposition revealed that the genes that formerly mapped to the left end of chromosome 3 were now linked to the right end of chromosome 2, and that they appeared there in the order predicted by the original genetic maps.[10]

While the cytological maps provided dramatic proof that genes that mapped together really did lie together physically, and also that the order of the genes on the linkage maps corresponded to the actual physical order in the chromosomes, they also made it perfectly clear that the linkage maps did not provide an accurate measure of distance.[11] Translocations that seemed to involve regions of 30 map units, for example, were hardly visible under the microscope at all, and conversely, a fragment that could be seen under the microscope to represent a sizable fraction of an entire chromosome might represent only a few map units genetically.[12] As both Sturtevant and Muller had realized in 1912, the accuracy of the crossover maps depended on the assumption that the rate of crossing over was constant across the entire length of the chromosome. If this was not the case, then in regions with a higher than average crossover rate, distances given by the percentage of recombination would be too large, and, conversely, in regions where crossing over was suppressed, the distances on the old gene

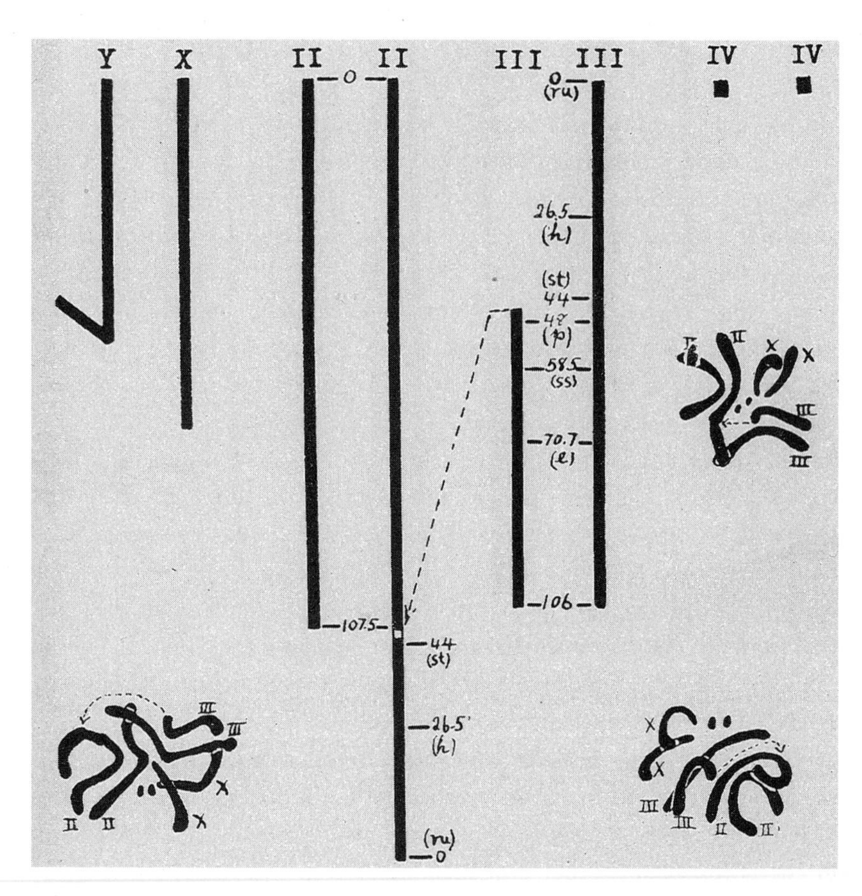


**The Star-Curly translocation.** From Theophilus Shickel Painter and Hermann J. Muller, "Parallel Cytology and Genetics of Induced Translocations and Deletions in Drosophila," *Journal of Heredity* 20 (1929): 289.

maps would be too short. "It now appears that through a combined genetic and cytological study of these translocations, deletions, etc., by the methods outlined in this paper," Muller and Painter wrote, "we shall eventually be able to construct truer chromosome maps, which will show not only the order of the genes but their actual spatial relations."[13]

— WHILE MULLER'S X-RAY WORK brought him and his department fame, it only seemed to exacerbate the problems in his already troubled marriage. In the crush of all the new work, Muller had been working in the lab until four or five in the morning, and Jessie was growing ever more estranged from him. In the summer of 1930, while Jessie was visiting her family in Kansas, he puzzled over what had gone wrong in their marriage, "You do not know how keenly I have felt, all along, having to be with you and yet separated from you—trying to get isolation. I think it is really that, not being together so much, but being together while we must be a source of irritation to each other of that kind and so little together essentially that has been trying."[14] Although he hoped that they might turn over a new leaf when she returned, it seemed that they had already drifted too far apart, and the arrival of three new lab members, two Russians and a German, marked the beginning of a new era for Muller both professionally and personally.

The two Russians, Israel Agol and Solomon Levit, had both trained in classical genetics under Serebrovsky, whom Muller had met in his 1922 trip to Leningrad. Levit, who had a medical degree as well as a graduate degree in genetics, was the head of his own institute for human genetics in Moscow and was a far more serious person than the flamboyant Agol, who had a particular passion for cars, wrecking two of them during his stay in Texas before the rental company refused to give him another.[15] The third new member of the lab was Carlos Offermann, who had been educated in Argentina after his family emigrated from Germany. While Muller treated Agol and Levit as equals, he took a more paternal attitude toward Offermann, who was younger, less self-confident, and suffered from a spinal deformity, the result of a massive dose of X-rays wrongly administered in a German hospital to treat a childhood spine injury.

Agol and Levit were obligated to return to the Soviet Union when their one-year Rockefeller fellowships ran out, but Offermann's plans, like Offermann himself, were much more fluid. When he arrived in Texas he was in his late twenties, but he had a youthfulness about him that made him appear far younger. In addition to his boyish good looks, Carlos had a charm that married women seemed to find nearly irresistible. Men seemed to like him too, and even husbands he'd cuckolded tolerated him, finding it difficult to conceive of the youthful Carlos, with his extremely slight physique and physical disability, as a serious threat.

By the summer of 1931, when Offermann left to take a summer course with the renowned British statistician R. A. Fisher in Ames, Iowa, it was clear that he had already made a deep impression on both Jessie and Hermann. Writing to him shortly after he'd departed, Muller struck an uncharacteristically sentimental note, confessing, "For Jessie and me, especially, you had become and are, like one of the family, and one is always rather at a loss when the family becomes reduced,"[16] and Jessie wrote separately about the vacuum created by his absence.[17]

Sometime in the fall of 1931 Jessie and Carlos began to have an affair. While Hermann and Jessie both adhered to the principles of free love and open marriage, as did many in their socialist circle, Hermann found the reality of Jessie's involvement with another man devastating. Despite their difficulties, he had never given up on their marriage and still hoped to mend the rift. As he would later explain, he found the contrast between her feelings for him and her feelings for Carlos impossible to bear, and his depression led Muller to attempt suicide.[18] "Until the time I attempted my final solution," he wrote several years after the fact, "I was only longing for you."[19]

Just after eight on an overcast January morning, Muller left the house and walked away from the university toward what was then open parkland surrounding Austin. After downing a bottle of barbiturates, he stopped to write a final note to Altenburg on the back of an invitation from "The Society of Sigma XI." As was his habit, the letter was carefully dated, "Jan 10, 1932 11 AM." Even in this moment of total despair, he was careful to protect Jessie, making no mention of her role in his decision to end his life.

"It seems I am too psychologically old to continue the struggle," he began.

> This agony throughout the past twenty years has been due in no small part to the direct and indirect effects of the predatory operations of T. H. Morgan (and his bootlickers, past & present) at least they ~~must~~ play the primary role. Doubtless there would not have been the necessary inducements and conditions for these were it not for the predatory spirit and structure of our capitalist system in general. As things are, my period of usefulness, if I had one, now seems about over, given the conditions that exist. Though without personal hope, I still have my pristine hope in the

> world and humanity in general and I wish that you would have my book "Out of the Night—the Development of Man and Science" published as soon as possible.[20]

The remainder of the letter contained detailed instructions for Altenburg regarding the publication of *Out of the Night,* Muller's book-length treatise on eugenics, and his unfinished history of *Drosophila* genetics, which had been commissioned by the German geneticist, Erwin Baur, as part of a series of monographs. He also asked Altenburg to convey various specific directions regarding joint work to Offermann, Painter, and a half dozen or so others, as well as to communicate his regret to Levit, with whom he had hoped to launch a Soviet eugenics program. He left a thousand dollars to the Communist Party and the remainder of his assets to Jessie. Although the last lines were badly water stained, enough remained to convey Muller's deep affection for Edgar: "The world is beautiful today—May there be many more such days. Anyway there . . . people like you . . . true friendship."

That evening when he did not return home, Jessie and some friends went out looking for him.[21] On Monday some of his colleagues and students joined the search along with six policemen. Early Tuesday morning a large search party was convened, consisting of more than a hundred students, faculty, and police as well as thirty-two members of the cavalry on horseback. He was found later that morning badly bruised and incoherent. It had rained hard Sunday into Monday morning, and Monday was another cold night. The sheriff who drove him to the hospital told the newspaper reporters that Muller would not have survived another day in the wilderness.

— AFTER HIS SUICIDE ATTEMPT, Muller sprang back to action almost instantly. The loss of hope about his marriage seemed to have been transformed into an intensification of his interest in politics and eugenics. In the spring he participated in the launching of a radical campus paper that was called *The Spark,* after the paper that Lenin had edited while in exile in Switzerland and smuggled into tsarist Russia. While Muller was careful not to implicate himself in the publication of *The Spark,* he nonetheless wrote many of the articles and arranged for its distribution on the campus in Austin and, through Altenburg, at Rice. Later that summer he attended the Third International Eugenics Conference in New York City, where he delivered a lecture entitled "The Dominance of Economics over Eugenics," a scathing attack on the practice of eugenics in America. When Charles Davenport, who had organized the meeting, saw the title of Muller's address, he threatened to cut Muller's talk down from an hour to fifteen minutes or to have it removed from the program altogether, but Muller, who insisted on his right to be heard, eventually prevailed.[22] In the talk, which received international attention, Muller for the first time publicly declared his communist sympathies and his deep disillusionment with the American system.[23]

The world was no closer to implementing a true eugenics program, Muller asserted, than it was when Galton first promulgated the doctrine of eugenics fifty years ago. The problem was that a true eugenics program was not possible under capitalism, and the most obvious conflict between eugenics and capitalism, Muller maintained, was that a system in which the primary motive of action was private profit left little place for children. A successful eugenics program required that the genetically better endowed had more than an average number of children, but children were a poor investment. For the rich, there was no incentive to have more than one child to inherit the estate, and for the already strained middle classes, who did not have estates, each additional child only intensified their economic hardship and further enslaved them. Under these circumstances, it was not reasonable to expect people of superior genetic worth, rich or poor, to have the four or more children that would begin to make a difference in the gene pool. Furthermore, given the vast sacrifices involved in bearing and raising children, no intelligent and self-respecting woman could be expected to have a large number of children, no matter how superior her genetic endowment. The lot of women was not better than that of a slave, Muller insisted, and "there is no profit in bettering the lot of a slave." What was needed was nothing short of "a revolution in our attitude toward women," and this would happen only when society was reorganized so that those who served the common good, those who had the children and took care of them, were justly rewarded for their service.

But the incompatibility between eugenics and capitalism went even deeper, Muller argued. Not only did capitalism take away the incentive to have greater numbers of children, but it also made it impossible to identify the genetically more fit. The effects of superior genes could not be separated from the effects of material and social advantages. Presenting recent studies showing the large differences in IQ between identical twins who were reared apart, and other studies on race and IQ, Muller demonstrated that the known differences in intellect between different social classes and races could be accounted for exclusively by the known effects of the environment, and the same was bound to be even more true of differences in temperament and moral qualities.

In order to justify social inequality, Muller continued, the dominant classes claimed that it was their genetic superiority that accounted for their success, but, Muller argued, the very opposite was more likely to be true. Many in the dominant classes were likely to have risen to the top "based on predatory rather than constructive behavior," and therefore they would be the repositories of the least desirable traits. The best genes were to be found in the "the high-minded, the scrupulous, the idealistic, the generous and those who are too intelligent to wish to confine their interests to their personal monetary success, [those who] were apt to be left behind in the present-day battle." It was only under socialism, Muller concluded, where the injustices of private ownership were leveled, that it might be possible to institute a functioning eugenics, one that cultivated genes for intelligence, cooperation, and health.

— IMMEDIATELY AFTER his lecture to the Eugenics Conference, Muller gave a major address to the 1932 International Congress of Genetics. "Oceans of words were spilled in formal and informal gatherings to discuss the vital question: what is the gene?" wrote R. C. Cook, the editor of the *Journal of Heredity*,[24] but once again Muller seemed to see farther than his peers. In fact, his suicide attempt appeared to have stimulated both his political and his scientific ideas. Forty years later, reflecting on Muller's address, Curt Stern wrote that it was obvious that "before them stood a man whose whole being was working to reach fundamental conclusions. Every muscle in his face seemed to be thinking."[25]

While Muller touched on a variety of problems, his most penetrating insight concerned the long-standing mystery of *Bar,* the eye-shape mutant that had played a key role in determining the frequency of new lethals. As first discovered by University of Illinois geneticist Charles Zeleny, around 1920, *Bar* mutants reverted to wild type at a strikingly high frequency (of 1 in 1,600 flies). Zeleny also noted that the high rate of reversion was found in only females, and he inferred that the reversion might involve crossing over, which was known to take place exclusively in females. Zeleny also observed that homozygous *Bar* females occasionally gave rise to mutants with extra-slitty eyes, which he called *ultra-Bar,* and these *ultra-Bar* mutants themselves could revert either to normal *Bar*-eyed mutants or to wild-type flies.

Following up on Zeleny's work in 1925, Sturtevant hypothesized that the reversion and production of *ultra-Bar* from *Bar* was due to an unusual type of recombination. As Sturtevant explained it, two homologous chromosomes underwent crossing over at a point lying to the right of the *Bar* locus in one chromosome and to the left of it in the other, producing one chromosome that contained two *Bar* genes and one that contained neither. According to Sturtevant, the chromosome with no *Bar* gene was the wild type. At the same time, he cautioned that such a crossover seemed to defy the normal mode of crossing over in which like genes were imagined to lie side by side, and he was at a loss to explain how this could occur.[26]

Muller had revisited Sturtevant's idea about unequal crossing over in 1930, after his discovery that chromosomes were broken and rejoined with a surprisingly high frequency and that, in the process, fragments were lost or reattached to new chromosomes. In particular, he hypothesized that if a chromosome broke into two (or more) fragments and, at the same time, a nearby chromosome underwent a similar breakage into fragments, the fragments might undergo a reattachment in a new configuration at their broken ends. "In adopting this scheme," Muller observed, "we accept the principle that the rearrangements occur by a process which is virtually

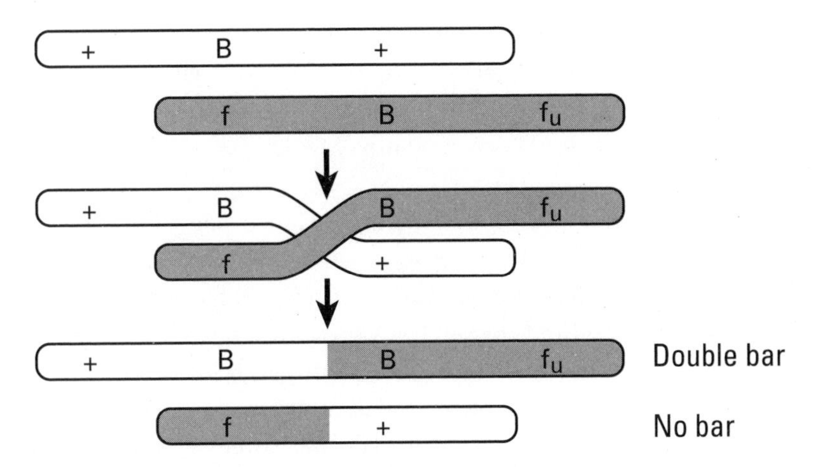


**Illustration of unequal crossing over, where B standing for *Bar*, and f and $f_u$ are two recessive mutations that are used to track the crossover types.** Illustration by Tamara L. Clark.

crossing over, except that it is between non-homologous regions and at a time other than the proper synaptic period."[27]

While he accepted Sturtevant's suggestion that unequal crossing could take place between nonhomologous regions, Muller categorically rejected Sturtevant's idea that normal flies lacked a *Bar* locus. Instead he proposed that *Bar* was caused by a duplication of a small section of the chromosome containing the normal *Bar* gene—a "duplication in situ," he called it—accompanied by a simultaneous mutation of the new copy.[28] The idea that a gene was duplicated and then underwent a simultaneous mutation was, Muller admitted, a weakness of his theory. Nevertheless, he went on to point out, if *Bar* were indeed caused by some kind of duplication, then the unequal crossing over that Sturtevant had suggested to explain the regeneration of the normal type from *Bar* would not generate a chromosome lacking the *Bar* locus but rather a chromosome containing a single unmutated copy.

Three years later, in 1936, Muller would revisit the *Bar* problem, and his ideas about chromosomes duplications would turn out to provide not only a complete resolution of the *Bar* case, but also the key to understanding the mechanism of gene evolution. But in the meantime he set off for Europe once again, this time to Germany, having accepted a Guggenheim fellowship that would support his research for a year at the Kaiser Wilhelm Institute for Brain Research in Buch, a suburb of Berlin.[29]

— MULLER SPENT THE TRIP across the Atlantic reading Lenin's "Materialism and Empirocritism" in order to get up to speed on Marxist thought, which, as he wrote Altenburg, he found quite compatible with his own beliefs. When he arrived in Germany, he visited his sister Ada and her husband in Munich, where he hoped to be able to finish his *Drosophila* monograph for Baur, but he found he was too distraught to be able to write, nor could he bring himself to return to the lab.[30] By November, filled with disgust for everything German—the stuffy sealed rooms, the heavy fried food with no fruits or vegetables and an excess of cakes, the feather comforters that were too hot when it was warm and not warm enough when it was cold, beds that were too short, even for him, the surveillance of one's every move by police, the slavery of women, the horrible parsimony, the backwardness of society using oxen as in Asia—Muller finally left Ada's and took up residence in the Institute.[31]

As a "curator," Muller was entitled to two meals a day and other special amenities, and life at the Institute seemed to have a salutary effect. By December Muller was beginning to get back on his feet, but on January 30, 1933, Hitler was appointed chancellor and the situation in Germany began to deteriorate rapidly. Communist meetings and newspapers were forbidden, and communists were murdered regularly. As the political climate in Germany became progressively more intolerable, Muller's resolve to move to Russia strengthened. It was nearly perfect timing that N. I. Vavilov, whom Muller had first met in 1922 in Petrograd and had seen again briefly in Austin in the winter of 1929–30,[32] stopped in Berlin in February on his return from a trip through Central and South America.

Since the time Muller and he first met, when he had been head of a small research institution inherited from the pre-revolutionary Ministry of Agriculture, Vavilov had been appointed first director of the All-Union Institute of Applied Botany and New Cultivated Crops, which was the fulfillment of Lenin's ambition to build a national Soviet academy of agricultural sciences modeled on the U.S. Department of Agriculture.[33] With his legendary capacity for work, remarkable organizational abilities, and working knowledge of twenty-two languages and dialects,[34] Vavilov had overseen a remarkable expansion of the government agriculture program that now constituted one of the greatest research organizations in the world.[35] From the All-Union Institute of Applied Botany, which itself employed 1,500 people, Vavilov had built a network of experimental stations extending the length and width of the country, from the extreme north region of Murmansk on the Arctic sea down to the southern Caucasus and from the western regions all the way across the vast expanse of Siberia. At the height of his reach in 1932–33, Vavilov oversaw the federation of more than 1,300 research institutes that employed over 26,000 people.[36] His plant collection consisted of over 150,000 plant varieties collected from every corner of the Soviet Union and more than sixty-five countries.[37]

Vavilov, who viewed Muller's participation as a key element in the successful modernization of Soviet biology, held nothing back. Muller would be put in charge of the twenty researchers and technicians who worked at an institute for pure genetics research in Leningrad that operated under the auspices of the Soviet Academy of Sciences.[38] The official invitation for a one-year stay in Leningrad, which would follow shortly after Vavilov's return to Russia, would grant Muller the highly prestigious title "Corresponding Member of the Academy of Science." In time, he would probably be given a Russian position as well, with the title "Chief Geneticist of U.S.S.R." In addition to his responsibilities as head of his own institute in Leningrad, Muller would make regular trips to Moscow, where he would consult with various other institutes, including those of Serebrovsky and Levit. Furthermore, Vavilov offered to arrange a one-month tour of the country in which he and Muller would travel by automobile and plane south over the Caucasus into Armenia and across the Caspian Sea, along the border of Persia into Central Asia, northward into Siberia, then west over the Urals and back to Moscow. In the process Muller would be able to view firsthand the progress that had been made in modernizing and collectivizing the peasant farms. Muller would speak at various places on the way, where scientific audiences could be found, in order to raise interest in genetics.

Wowed by Vavilov's dynamism and charisma, Muller accepted the offer on the spot, promising to take up his post in Leningrad in early September. "Despite the talking, late hours, inefficiency, the meetings, and the demand for general articles, the division of work and attention and bad living conditions," Muller wrote Jessie, it was impossible to pass up the opportunity "to get something worthwhile done." Though he was excited to be part of the Soviet experiment, his main interest was still genetics, and he hoped that working conditions might be different in the Soviet Union: "I am glad to give out my ideas if it doesn't merely mean a personal rivalry in which some one climbs up on my back to a higher position and pushes me down—as in the U.S.," he wrote Jessie.[39]

In March the Kaiser Wilhelm Institute was vandalized by the brownshirts. A few weeks later, under pressure from the new regime, the director of the Institute asked Muller to participate in a torchlight procession for the Nazis. "It is all very interesting," he wrote home, careful to strike a neutral tone for the benefit of the Nazi censors.[40] In May, Muller witnessed the official Burning of the Books in Opera Square. This time, he managed to provide a harrowing description of the scene while disguising his true feelings:

> It was a remarkable celebration, and the leaders spoke powerfully as they always do. The books were chiefly novels of Wassermann, Thomas Mann and Henry Mann, Upton Sinclair, etc. etc. gathered from the city and private libraries. It was all very inspiring. Physically it made a great heat, that we felt markedly even tho in the midst of the greatest crowd I have ever seen. Of course there are many other books too that should be destroyed, not novels; I guess that will come later.[41]

"When I see the regeneration and how well things are managed," Muller added with bitter irony, "I am proud of my German blood."

In April, Offermann, who had been fired from the University of Texas over his involvement with the communist groups on campus and for his role in the publication of *The Spark,* came to work with Muller in Buch. After some negotiation between Jessie and Muller, who preferred she remain in Texas and dreaded the prospect of resuming the triangle, it was decided that Jessie and David would join them in the summer and the four of them would make the move to Russia together.[42] In June, Jessie and David made the crossing to Europe, and that fall, as planned, the three Mullers accompanied by Offermann took a boat from Bremen across the North Sea to Leningrad. As a visiting dignitary of the Academy of Sciences, Muller was given a relatively large apartment with two bedrooms. Jessie and Carlos stayed in one room, and Hermann and David, who would turn 9 that fall, slept in another.[43]

In the summer of 1934, Muller and Offermann left with Vavilov for the tour of the country. Afraid that Hermann would try to use David as a weapon in their marital disputes and that the Russian authorities would take his side in a custody dispute, Jessie took advantage of his absence to leave the country with David and return to Texas. When Muller returned to Leningrad in the fall and found they were gone, he was devastated.[44] In December, after learning that she'd emptied their joint bank account, he filed for a Russian divorce, which required neither side to lodge a complaint.[45] A few months later in April, Jessie filed her own motion for divorce in Texas, threatening to take sole custody of David. Muller wrote in shock, "If people who have so much in common as you and I have to use force and compulsion on one another instead of reasoned agreement and mutual concessions when they differ on some important question, I think human nature is sadly failing."[46] More than ever, he now believed the only hope was in eugenics—"that we may some day make much better people than the best we know, but it is hard to sink so much to such a distant goal, never to be attained by ourselves and human beings are as yet hardly strong enough to do so except those too callous to care to,—and so all hangs in the balance."

Meanwhile, under the influence of the corrupt and ruthlessly ambitious Trofim Lysenko, the prospects for a genuine genetics program in the Soviet Union were grower dimmer with each passing month. Two disastrous wheat harvests in 1928[47] and 1929 had created an opening for Lysenko, who claimed to have discovered a method for treating winter wheat (sprouting seed in the fall and then freezing it for early spring planting) that supposedly increased its yield and quickened its course of development.[48] Although *vernalization,* as Lysenko dubbed it, had not actually been shown to be effective in increasing yields, and Lysenko was entirely lacking in the skills or inclination necessary to do a scientific analysis of the effect, he used his formidable skills as self-promoter and propagandist to sell the press and government on its benefits. By 1930 the "peasant scholar," as he had become known, had moved from a position of junior associate at a small experimental station in Azerbaijan to being the head of a large lab at the leading research institute in the Ukraine.[49] Under the direction of Y. A. Yakovlev, who was head of agriculture for the party, Lysenko was given his own journal, the *Bulletin of Vernalization,* and other lavish support. In the following four years vernalization first subsumed the entire field of plant physiology and then began to expand into plant breeding and genetics.[50] Built on a collection of meaningless but impressive scientific-sounding phrases and forged data designed to win over the press, the entire Lysenko program was a fraud.[51]

Lysenko's big break came in a speech given in February 1935 to an agricultural conference attended by Stalin, as well as Vavilov and other prominent scientists. As had become his standard practice, Lysenko began his lecture by invoking the treachery of kulaks, as the better-off peasants were called, who he claimed tried to poison the minds of less educated peasants by turning them against vernalization, but on this day Lysenko expanded the circle of deceivers and betrayers. "You see, comrades," he declared with rising hysteria, "saboteur-kulaks are found not only in your kolkhoz life . . . they are no less dangerous, no less accursed in science . . . A great deal of mortification has had to be endured in defending vernalization in all kinds of battles with so-called scientists . . . Both within the scientific world and outside, a class enemy is always an enemy, even if a scientist." At this point in his speech, Lysenko turned to the praise of the collective farms and the teaching of Comrade Stalin, and Stalin stood up and shouted "Bravo, comrade Lysenko, bravo!" The audience instantly sprung to its feet and broke out into rapturous applause. A detailed account of Lysenko's speech and Stalin's spontaneous interruption, accompanied by a portrait of Lysenko, appeared in *Pravda,* certifying the fact that the peasant scholar had overnight become one of the most powerful men in Soviet biology.[52]

For Muller's genetics department, which had been transplanted to Moscow in November 1934, the rise of Lysenkoism was deeply demoralizing. He and his staff were constantly watched by party informants and distracted by endless bureaucratic obligations that sapped their time and energy. One of the department's leading researchers, Y. Y. Kerkis, was on the verge of a nervous breakdown, and "the clique was after his blood," as Muller put it in a letter to Jessie.[53] In addition, funding had dried up for technicians and as a consequence the fly strains were "going to pieces."

But against all odds, Muller's own research continued to advance with the help of Dan Raffel, a recent Ph.D. from Johns Hopkins, Offermann, and most of all the highly gifted Russian cytologist Aleksandra Prokofyeva, who was doing excellent work, "better than anyone in the institute." Late in the fall of 1935, Prokofyeva made a discovery that finally led to the resolution of the mystery of the *Bar* mutations, the culmination of a two-year search for a duplication of a small region of the X chromosome that Muller had, in 1932, hypothesized was associated with the *Bar* phenotype. Making use of a powerful new technique that made it possible to identify each region of a chromosome based on a highly particular pattern of light and dark bands, Prokofyeva found that *Bar*-containing chromosomes contained two extra bands not seen in normal chromosomes.[54] In a letter of December 15, 1935, to Altenburg, Muller sketched the pattern of repeated bands that Prokofyeva had observed.[55]

Below the picture of the repeated section, Muller indicated how a duplication could be created by the breakage and rejoining between sister chromosomes at nearly, but not quite the same, point. As he had first described it in 1932,[56] when a chromosome, A B C D E F G H, broke between E and F, generating fragments A B C D E and F G H, and its homologue broke between C and D, generating fragments A B C and D E F G H, it was possible for the left fragment of the first chromosome to be joined to the right fragment of the second. In the end, two new recombinant chromosomes would be created, A B C D E D E F G H and A B C F G H, containing two and no copies of the D E region, respectively.

Three years after his initial idea, Muller now saw how the duplication could also explain the other unusual properties of the *Bar* mutants. Both the unusually high frequency of reversion of *Bar* and the creation of the

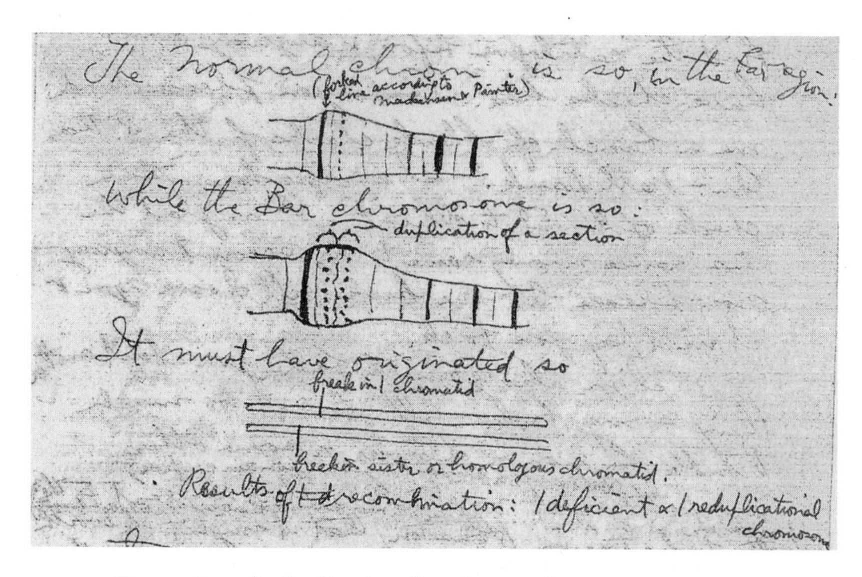


**Generation of a duplication, from letter of Hermann Muller to Edgar Altenburg dated December 15, 1936.**

Muller Manuscripts, Lilly Library.

more extreme forms like Zeleny's *ultra-Bar* could be accounted for by a crossing over between the left- and right-hand members of the twin sections of a pair of duplication chromosomes, giving two new chromosomes. The first recombinant chromosome, containing one copy of D E, would give a normal fly, and the other, containing the triple repeat, D E D E D E, would result in Zeleny's *ultra-Bar* mutant.

```
  A B C D E D E F G          A B C D E F G
         X              →
A B C D E D E F G            A B C D E D E D E F G
```

**Mechanism for reversion of *Bar*.**

Illustration by Tamara L. Clark.

The Muller and Prokofyeva paper appeared in the January 1936 issue of *Comptes Rendus de L'Académie des Sciences de L'URRS,* and in it Muller now used the term *unequal crossing over* to describe the generalized process of breakage and reunion of chromosomes, whether homologous or not, that resulted in the formation of new chromosomes.[57] The "unequal crossing over" that Sturtevant had observed in the case of *Bar* homozygotes, Muller now explained, was a "secondary unequal crossing over, resulting indirectly from the primary unequal crossing over, which established the duplication in the first place."[58]

However, the chief interest of the *Bar* case, Muller now suggested, was that it illustrated the manner in which new genes arose in evolution. Echoing Remak and Virchow, he now saw "no reason to doubt the application of the dictum *all life from pre-existing life* and *every cell from a pre-existing cell,* to the gene: *every gene from a pre-existing gene.*"[59] Genes that would begin as exact duplicates of others would tend to pick up different mutations in the course of generations and ultimately give rise to new functions.

The moment he saw the February 28, 1936, *Science* with an article by Bridges entitled "The Bar 'Gene': A Duplication," Muller knew that Bridges had done it again.[60] "He'd obviously sent it in because he'd heard of my work," Muller wrote Altenburg, explaining that he'd written about the result to an American friend who was also in communication with Bridges.[61] Having heard that *Bar* was actually a repeat, it was a trivial matter for Bridges to use his own highly refined salivary chromosome maps to find it and to compare the pattern of bands to the reverted type and to the *double Bar.* Bridges's manuscript was received on February 21 and published in record time one week later. By publishing in *Science* where the result would be widely read, Bridges would be credited with finally solving the decades-old mystery of *Bar,* and the fact that Muller's article had appeared first in a Russian journal would be ignored.

While it was nothing new for Bridges to use one of Muller's ideas without attribution, this time he had slipped up, failing to properly understand the full subtlety of Muller's idea. In particular, Bridges had mistakenly assumed that the duplication was caused by the "insertion" of a section of one chromosome into its homologue somewhere to the right or left of the original section, and that the exact point of insertion was of no particular consequence. However, the beauty of the primary unequal crossing over mechanism was that it created a very special kind of duplication in which the repeated fragments followed one after the other with no intervening region, and the absence of an intervening sequence was required for the regeneration of the wild type through a subsequent crossover event involving the duplication chromosomes.[62]

— THIS TIME, MULLER was determined to do his best to expose Bridges's shenanigans, and in a *Science* article published that May he laid out the facts: "Our results had been publicly announced at a meeting of geneticists in Moscow on December 26," he wrote, "and privately communicated to a number of American and English colleagues in December and January."[63] The article then went on to describe Bridges's faulty reasoning, making it perfectly clear that Bridges had simply rushed to repeat the result that he had heard through the grapevine without understanding its significance. "We should, of course, in our original article, have called the attention of the readers to Bridges' independent or confirmatory work, had we known of it," he added, a slap on the hand of the incorrigible Bridges, who was incapable of restraining himself when it came to borrowing other men's ideas.[64]

— IRONICALLY, JUST AS the closing jaws of Stalinism were casting doubt on his long-term prospects in the Soviet Union, Muller received a letter from Harry Yandell Benedict, the president of the University of Texas, informing him that the university intended to try him for breach of professional ethics and for breaking university rules for his role in the publication of *The Spark*. Based on a letter found on an alleged communist organizer arrested in San Antonio in 1932, the university was accusing Muller of defying the school policy by participating in an anonymous publication. Muller, who had not expected the paper to appear anonymously, denied the charges and threatened to come to Texas and create a public stir if they didn't drop them.[65] The university backed off, Muller officially resigned his professorship on April 17, 1936, and the newspapers reported that he had done so in "order to take a job with the Russian government."[66] There was no mention of his role in the publication of *The Spark* or a breach of professional ethics.

On November 14, 1936, the Politburo resolved to cancel the International Genetics Congress that had been scheduled to take place the following year in Moscow. At that meeting, Yakovlev, the head of Agriculture, proposed that the Communist Party take control of the proceedings and delay the conference until August 1938.[67] "Meanwhile it is said that the internal front must be cleaned," Muller explained in a letter to David from Moscow. Agol had been court-martialed and Levit "demoted and disgraced (for a beginning)."[68] On December 14, 1936, the *New York Times* ran an article that shocked the world with the announcement that the 1937 International Congress for Genetics had been delayed and that Vavilov and Agol had been arrested. A few days later, the *Times* retracted the story about Vavilov, but *Pravda* confirmed that Agol had been arrested.[69] On December 21, Levit and his Medico-Genetics Institute were attacked in the press. On December 22, Vavilov denied the rumors of his arrest to the *Times*[70] and further moved to reassure his Western supporters by reporting that he would be giving an address that evening at a meeting currently being held at the Lenin All-Union Academy of Agriculture Sciences. Lysenko, Serebrovsky, and Muller were scheduled to speak the next day.

Sensing that the fate of genetics in the Soviet Union hung in the balance, over 3,000 people—the largest crowd of the conference—gathered to hear Vavilov's lecture on the third evening of the meeting. Vavilov began by defending the accomplishments of the Lenin Academy under his leadership and gave an entirely objective critique of Lysenko's new program of "*intra*-varietal crossing,"[71] which had become Lysenko's new focus after it had become clear that vernalization was not increasing yields and in many cases actually lowered them.[72] While Vavilov defended the practice of ordinary *inter*varietal crossing, which was standard practice in America and elsewhere, he did not dismiss Lysenko's embrace of *intra*varietal breeding out of hand, instead dwelling on special cases in which intravarietal crossing might result in improvement of strains.[73] But it is impossible to know if Vavilov still believed Lysenko might be susceptible to persuasion and was hoping to win him over by rational discourse or whether he was holding back because he was an astute politician himself and knew just how far he might be allowed to go in his criticism of science that he knew to be entirely lacking in merit.[74] The closest he came to addressing his frustration with the antisocial and immoral tactics of Lysenko, who lied about his experimental results and slandered his opponents, was to make a plea for "greater attention to one another's work, more respect for one another."[75]

Many were curious to see how Lysenko, who was to be the opening speaker on December 23, would answer Vavilov and the many others who had made a seemingly irrefutable scientific case against him over the preceding days. The first clue appeared on the morning of his lecture when he entered the foyer on the first floor of the presidium of the Lenin Academy where Vavilov had asked Muller and Prokofyeva to arrange a display including microscope slides of stained chromosomes illustrating the various stages of mitosis and meiosis as well as pictures of Painter's salivary chromosomes.[76] Lysenko walked through the lobby, bending over one of the microscopes, glancing at the explanatory illustration and quickly moving on to the next microscope. All told, he spent less than five minutes examining the evidence for the modern chromosome theory.

As it had been the previous night, the auditorium was filled to capacity to hear Lysenko's address. Following his usual line, Lysenko presented himself as the true disciple of Darwin who was defending his "fundamental principles of evolution" against the "formalist" geneticists who were intent on misunderstanding and perverting them.[77] Those who were hoping that Lysenko might answer his critics were sorely disappointed. In answer to Vavilov, he simply asserted that the introduction of intravarietal crossing was a success, improving both yield and winter-hardiness, without providing any data to back up the claim. Moreover, he disregarded all the flatly contradictory evidence by declaring that "these facts nowadays are no longer needed."[78]

Later in the session Lysenko discussed the transformation of winter grain into spring grain, the problem that had first brought him to national attention. No longer mentioning vernalization of grains, he now insisted that his intention was to effect a complete hereditary change of them. He declared that "winter plants at certain moments in their life, under certain conditions, can be transformed, can change their hereditary nature into a spring type, and vice versa,—which we are quite successfully doing experimentally now."[79] At this point, even the deferential Vavilov could not contain himself, calling out from the floor, "You can refashion heredity?" Without a glimmer of self-doubt, Lysenko replied, "Yes, Heredity!" going on to acknowledge openly that his findings were incompatible with the concepts of modern genetics, and in particular with the idea that it was the gene, and not the whole cell, that was the physical basis of heredity.[80]

— SEREBROVSKY, WHO WOULD soon fall victim to the raging terror, gave a very brief presentation on the nature of genes, enough to show that he was not prepared to accept the new line, leaving it to Muller to deliver the knockout punch. It was clear from his opening lines that Muller, who was the last to take the podium December 23, was prepared to fight. "There is no such thing as 'formalistic' genetics, separable from real genetics for genetics today embraces one coherent body of principles on which virtually all geneticists agree."[81] Without qualification, he dismissed Lysenko's central premise, the Lamarckian idea that the development of the phenotype can influence the genotype. Such an idea, Muller asserted, "abolishes the most distinctive properties of the gene as a gene," the idea that the gene reproduced a likeness of itself and that the copying process was unaffected by the phenotype or protoplasm surrounding it, that it was highly stable but could, nonetheless, undergo rare mutations that left it still capable of reproducing its likeness and that these changes were entirely random and did not in any way reflect any particular circumstance in the growth and development of the plant.

Muller went on to give a remarkably lucid description of the development of the modern chromosome theory and the evidence supporting it, beginning with a defense of Mendel and the most basic facts of Mendelian segregation, which were being actively challenged by Lysenko. In his lecture, he traced the way that breeding results had led to the concepts of linkage groups, the linear arrangement of genes, and pointing out the perfect congruence between the breeding work and the cytology. He also described the modern view of mutation as a highly localized submicroscopic process that could hardly be affected by developmental and physiological changes of the individual. Although many people had claimed to be able to affect the direction of mutation, Muller asserted, no such claim had withstood the test of time, "The latest claims of this kind, those of Lysenko," he added, "give every evidence of suffering from the same errors as those of the past."

While the body of his speech was given in Russian by his friend N. K. Koltsov, as Muller looked on, he insisted on speaking the concluding lines himself in Russian: "If our outstanding practitioners are going to support theories and opinions that are obviously absurd to everyone who knows even a little about genetics—such views as those recently put forward by President Lysenko and those who think as he does—then the choice before us will resemble the choice between witchcraft and medicine, between astrology and astronomy or between alchemy and chemistry." Many of the key participants at the December 1936 meeting died soon after in the escalating terror, but those who survived and were later interviewed by the Russian writer Mark Popovsky in the 1960s and 1970s still remembered the thunderous applause that broke out as Muller spoke those words.[82]

After his lecture Muller was taken to a private room by Bauer, the chief of science of the Communist Party, and Yakovlev, who both expressed their vehement opposition to his views on the stability of genes. The idea that it was possible to "change the genes so as to make all races equal," Yakovlev said, represented the view in the highest official circles. "Just better the conditions and you better the genes." Furthermore, the idea that gene mutation was random was against state policy, Muller was informed, and he was forced to issue a public apology for his statements.[83] In a major policy speech delivered on January 5, 1937, Yakovlev declared that those who would defend the stability of the gene were "reactionaries and even saboteurs," signaling a change in official policy.[84] His speech, which was published in Lysenko's journal, also contained an article by Lysenko's top advisor, I. Prezent, which referred to Agol as a "Trotskyite bandit" and attacked Levit for supporting a fascist view of hereditary predisposition.

— IN THE EARLY SPRING Muller completed a bibliography of *Drosophila* literature dating back to 1905, a project he'd been asked to take over from Bridges. After submitting the manuscript to the Soviet Academy of Sciences for publication, he was summoned by the secretary of the Academy and told that his list would need to be revised because it contained unacceptable names, including those of Agol and the three Russian expatriates Boris Ephrussi, Theodosius Dobzhansky, and Timoféef-Ressovsky. When Muller objected that it would be impossible to remove the offending names from a bibliography that purported to be a complete list, the secretary suggested changing the title to *List of More Important Works on Drosophila.* After much debate, Muller succeeded in keeping all names except that of Agol.[85]

Ever since the December meeting, the situation had gone precipitously downhill, and intimidated by Lysenko and Prezent, many geneticists had ceased working. Although Muller himself was permitted to continue his fly work, the party was interfering to such a degree that it had become impossible to get research done. There was also increasingly pressure on him to publicly proclaim his support for Stalin and the executions. But he had been holding out, still hoping that the authorities could be persuaded to institute a national eugenics program. At Levit's urging, Muller had sent a personal letter to Stalin along with a copy of *Out of the Night,* and his hopes had been raised by Stalin's personal secretary, who had translated the book for him and had been extremely enthusiastic about it. But in early March, Muller learned that Stalin himself had been extremely displeased and had ordered that an attack be prepared against it in the Soviet press.

With no more reason to stay in Russia, Muller immediately began to look for a way to leave that would not harm Vavilov and his allies, who might be punished for his defection. Also, he was acutely aware that if he simply withdrew without explanation, socialists outside the Soviet Union would misconstrue his actions and he'd do harm to the cause. As he explained in a letter to Huxley, "I'd be classed as just another 'bourgeois idealist' who failed to accept the realities of the revolution, and expected a paradise and then, in reaction became *anti-soviet,*" and his eugenic ideas would also be tainted, passed off as "bourgeois" and "counterrevolutionary." When, a few days later, he was offered a chance to help in the fight against Franco by joining a group in Madrid that was developing new blood transfusion techniques, he jumped at the chance. He would spend a short period in Spain, and thereby demonstrate his continuing commitment to the cause. To avoid the perception that he was using the trip to Spain as a way to "sneak out," he would then return to Russia, and, only after he'd been back for a reasonable period, then announce that he intended to leave permanently.

— WHEN HE ARRIVED in Madrid in the middle of March, the city was under siege, subject to daily bombing raids from freshly minted German fighter planes. Drawing on his study of physiology, Muller assisted in the blood transfusion studies designed to extract blood from the recently killed. After one air attack, Muller helped a fellow scientists remove his equipment from a bombed building, and, as he'd hoped, a Moscow paper ran a story about his exploits.[86] After eight weeks on the front lines, Muller stopped briefly in England where he lobbied his influential British friends, including Huxley, Haldane, and Cyril Darlington, who was head of the John Innes Horticultural Institution, to accept the Soviet invitation to hold the genetics conference in Moscow in 1938.[87] In May he took a boat to the United States and arrived unannounced in Texas, where he immediately began proceedings to regain partial custody of David in the hopes that he would be able to take him to England the following summer. By early July he was back again in England awaiting a reentry visa from the U.S.S.R., where he planned to spend the summer before making his final exit.[88] However, word leaked of his intention to leave the country, and his reentry visa was delayed for two months. Finally after protracted negotiation he was granted a visa in early September and arrived in Moscow on September 10.[89] With only two weeks left before he was due in Paris to chair a meeting on mutation and radiation, he rushed to make subcultures of some 250 *Drosophila* stocks, many of which were irreplaceable.[90]

Muller spent his last night in Leningrad, where Vavilov had arranged a farewell banquet in his honor. After the dinner, the two friends went to the newly released movie *Peter the First,* which chronicled the life of Peter's son, Aleksei, who had died after being subject to interrogation and torture for aiding the anti-Petrine conspirators. After the movie, the two men wandered around the city until after midnight, when Vavilov dropped Muller off at his hotel. Vavilov returned at 5 a.m. the next morning looking completely rested and drove Muller to Detskoye Selo, where they had met for the first time fifteen years earlier. The labs, which would soon be destroyed by Lysenko, were still operational, and Vavilov proudly showed off the gardens where samples of cultivated plants were sown. The Abyssinian flax with its magnificent large blue flowers was in full bloom. After the tour, they breakfasted on smoked fish, chocolate bars, and fresh bread prepared in one of the labs used to test the qualities of different cereals. Vavilov's driver ate with them. After breakfast, Vavilov dropped Muller off at the dock where he boarded a boat for England.[91]

The attack on genetics had triumphed, Muller wrote Jessie from the boat. "Many people are being unjustly accused nowadays. The country is now in a bad state of hysteria. About 10% of the high officials have vanished. A significant sign is that the best communists now no longer seem to consider arrest a disgrace."[92] Levit was now completely out of science and his Institute of Medico-Genetics destroyed, and Muller had not dared to visit him for fear of causing him even more trouble. Anti-genetics was now compulsory in all higher educational institutions and the attack on genetics by Yakovlev had been distributed in pamphlet form by the millions. Except for a few who were deemed essential, foreigners were being forced out of the country. To help Vavilov, Muller had asked the Academy for a twelve-month leave of absence, but he had no intention of returning.[93]

At the Paris meeting, Muller was offered three jobs, one by the Soviet-born developmental geneticist Boris Ephrussi at the Radium Institute in Paris, another by Frederick Crew at the Institute of Animal Genetics in Edinburgh, and a third at the leading Swedish genetics institute—the State Institute for Race Biology, in Uppsala near Stockholm— "so I suddenly feel like Cinderella," he wrote Altenburg.[94] After a short period of deliberation, Muller chose Crew's Institute, where Crew had promised to "see to it that while he was here he should lack nothing" and there were already nearly a dozen *Drosophila* workers for him to guide. Having barely escaped the "closing jaws of Stalinism," Muller wrote Edgar from Scotland in December, he felt more vigorous than he had in years, excited to have the freedom to do his own research and oversee the thriving *Drosophila* group at the Institute without interference from the authorities.[95] In addition, he was consumed with the planning for the Seventh International Congress of Genetics, which the planning committee had moved from Moscow to Edinburgh while at the same time appointing Vavilov president, as a sign of support.

Not only was Muller's professional life looking up, but his relationship with Jessie and David seemed to be normalizing a bit, and for the first time in years David and Jessie were writing him.[96] In the summer of 1938, he returned to the United States and was finally allowed to bring David to the East Coast, where they spent the summer together in Woods Hole. When he returned to Edinburgh in the fall, Muller was convinced that England's entry into the war was imminent, but he was happier than he'd been in years, writing to Edgar, "Might as well make hay, till Hitler sweeps it all away."[97] In January, Muller met Thea Kantorowicz, a 29-year-old German emigrant with an M.D. degree whom Crew had hired to work in the Pregnancy Diagnosis Department of the Institute, and it was love at first sight. She was "exactly of our crowd," he wrote Edgar in a swoon, and absolutely devoted to him. The only dark cloud was that she had just been diagnosed with tuberculosis and needed to recuperate in a hospital, but they would not let this stand in the way of their plans to marry.[98] In the same letter, he urged Edgar to attend the congress that was scheduled for August, as it would probably be his last chance to see this part of the world "before it disappears in the worst cataclysm in human history."[99]

From Russia, Vavilov had written of a steadily deteriorating situation where, in the latest meeting, "doubts [were] centered on the reality of the 3:1."[100] On July 26, a week before the congress was scheduled to begin, Vavilov cabled Muller that none of the Russians would be permitted to attend. The official letter from Vavilov, which was obviously written for him, arrived a few days later. In it he refused the presidency and withdrew all the Russian participants on the grounds that the congress should have not been take away from Moscow.[101] Forty-four papers had to be deleted from the page proofs of the conference proceedings.[102]

The conference took place in the last ten days of August, and throughout the meeting small groups of scientists huddled around radios that had been set up in the meeting halls to monitor the latest dispatches on the impending war. On August 23, Germany and Russia signed a nonaggression pact and the German geneticists withdrew from the conference. A week later, on September 1, Hitler invaded Poland, and two days after that Britain and France declared war on Germany. Crew, who expected to be called away any day and didn't expect real research to resume in England until after the war, advised Muller to relocate to America and took it upon himself to write to President Conant at Harvard, where he'd heard there might be a job opening, as well as to Rollins Emerson, a leading American geneticist at Cornell. Meanwhile German reconnaissance planes had been spotted flying high over the Institute on their way to more valuable targets.[103] In the spring, all "alien" males between the ages of 16 and 60 were required to report in to the Edinburgh police once a day and to observe an 8 p.m. curfew. As he had in Russia, Muller managed to continue to generate interesting results in the lab even in difficult times.[104]

In April Muller heard from Huxley, who had been in touch with a number of American geneticists, that he didn't have a chance for even a temporary job in the United States. As the threat of German invasion mounted, and his fears for his half-Jewish wife intensified, Muller grew more frantic about leaving. Muller's sister Ada petitioned the Rockefeller Foundation, the president of the National Academy of Sciences, as well as the National Council of Jewish women, and finally the State Department granted Thea's application for a visa. In late August, the Mullers took a plane from London to Portugal, where they were fifty-fifth on the list for a flight on the transatlantic Pan American clipper, a huge seaplane that made a daily trip from Lisbon to New York.[105] Due to stormy seas, the Clipper flights could not get off, and the Mullers anxiously waited their turn. Meanwhile, Muller and Thea went out in search of yeast and bananas for the flies, and Muller washed out the vials and supplied them with fresh food in the hotel bathroom. When it was finally time to go, they collected their luggage and made their way to the pier where the giant seaplane was docked. Half an hour out of Lisbon, Muller jumped out of his seat frantically asking after his flies. They had been in a white box, he explained, with the word "BREAD" stenciled on the side, and it had been sitting with their luggage on the pier before they left. A search was made of the entire plane, through the baggage compartment and the washrooms, all in vain. Fearing the flies would die of the heat if they were left sitting on the pier, Muller begged the pilot to return. Instead he radioed Lisbon, but the white box was not there. When all hope was lost, the steward suddenly came running in with the white enamel box. It had been brought to the pantry, of course, where the bread was kept.[106]

— IN NEW YORK MULLER and Thea moved into a run-down hotel near Columbus Circle and Muller began looking for a position, but just as Huxley had warned, no offers were forthcoming. The problem, Muller believed, was that Morgan and Sturtevant with the support of one or two of his former Texas colleagues had successfully spread the impression that he was "bad in his personal relationships, clamoring for credit, unable to make good anywhere, suffering from a persecution complex, unstable as well as red and underhanded and over rated."[107] But still there were some people who appreciated the breadth and depth of his contributions, including Harold Plough, who had done some of the pioneering X-ray studies on *Drosophila.* When Plough heard of Muller's plight, he prevailed upon his colleagues at Amherst to give Muller a one-year research appointment in the biology department, which Muller immediately accepted.[108] When the months continued to pass without further offers, Muller suspected that he had been blacklisted by the major universities for his involvement in the *Spark* affair, but he also recognized that he had many other vulnerabilities, including "Russia, my book, my 'disappearance,' my divorce and my having aroused the opposition of Morgan—a factor the effects of which I see growing." Still he felt "by no means beaten yet," and he was determined to continue searching for a job in genetics.[109]

In the meantime, Altenburg was convinced that the best hope was to take on Morgan and Sturtevant, whom they both believed to be the driving force behind calumny of Muller, and to write another textbook for Holt in which he laid out the truth in regards to the early history of the *Drosophila* room. "As I see it," he wrote Muller in the early spring, "I am the only person in a position to tell the truth, since the Baur paper [Muller's never-finished history of *Drosophila*] has not been published and you yourself could not very well give the facts in a text since a text is supposed to be impersonal."[110] In a letter a few days later, Altenburg wrote out a list of Muller's major accomplishments of the past thirty years:

> As I see it they are: 1. the discovery of linear linkage (including interference), 2. the theory of genic interaction and the proper interpretation of the relationship of genes and traits, 3. the synthesis of compound stocks and the use of markers for genetic analysis, 4. the theory of genic balance (together with a tough argument to convince Morgan and Sturtevant), 5. the theory of balanced lethals, 6. the use of balanced lethal stocks in the attack on the mutation problem and in the partial solution of the Oenothera case, 7. the first quantitative determinations of the mutation rate, 8. the artificial production of mutations, 9. the first cytogenetic maps, 10. the use of x-rays in dissecting the chromosomes for the attack on various genetic problems (as the determination of sex, the production of the deletions, inversions, and translocations which furnished the basis for Painter's cytogenetic maps), 11. the generalized theory of the position effect. These represent the main lines of advance since 1911 (I haven't mentioned such things as your insistence on the stability of the gene, the proper interpretation of Mendel's principle, etc.). All of this should certainly put you in the hall of fame, alongside of Darwin and Mendel.[111]

Although there was some debate about the relative contribution of Muller and Sturtevant in the development of the theory of linear mapping, and Bridges had failed to credit him for the theory of genic balance,[112] the other nine items on the list were incontestably Muller's.

Despite his remarkable record of achievement, Muller's applications for permanent positions continued to be rejected, but that fall, when Plough enlisted in the army, Muller was recruited to take his place and appointed interim professor of biology without a fixed termination date.[113] In the spring of 1944 Thea discovered that she was pregnant, and Muller's son, David, who was then 19 and a student at Caltech, announced his engagement and marriage to a 17-year-old art student, Mimi Held. The following year, in February 1945, Muller was offered a tenured professorship by Indiana University in Bloomington and the long period of wandering was finally over. Thea gave birth to a baby girl, Helen Juliet Muller, in August 1945, and eleven months later David and Mimi had their first child. Suddenly the 55-year-old Muller had an infant daughter and a grandson.[114]

In August 1945 Muller heard from Huxley that Vavilov had died in Siberia around 1941, after his bread card was taken away from him.[115] But the true story of Vavilov's death would remain sealed in state archives until 1964, when, just after Khrushchev's fall from power, there was a brief thaw that made it possible for an intrepid Russian writer, Mark Popovsky, to piece together the facts.[116]

Vavilov had been arrested during a collecting trip in a remote mountainous region of Putilya in August 1940. After being flown back to Moscow, he was subjected to eleven months of brutal interrogation in which he was often woken up in the middle of the night and questioned for twelve or more hours. As Vavilov would have immediately gleaned, the elaborately falsified case against him, which drew on years of data, had been long planned and approved in the highest circles, including both Stalin's foreign minister, Vyacheslav Molotov, and his infamous head of internal security, Lavrenti Beria. Vavilov's many scientific supporters rallied to his defense, appealing to Beria and writing directly to Stalin, but their efforts were to no avail, and in July 1941 a commission of three generals condemned Vavilov to be shot. In August he was spared by direct intervention of Beria, who had intended to allow him to resume his botanical research. Before Beria's orders were carried out, though, the Nazis began to close in on Moscow, and in October Vavilov and many other prisoners was transferred to a prison in Saratov, where Vavilov was put in solitary confinement. By January 1942 he had contracted scurvy and his health began to seriously decline. In response to Vavilov's appeals, Beria once again intervened on his behalf, petitioning to have his death sentence commuted and to allow him to resume his work. On June 23, 1942, after barely a minute's deliberation, the Presidium of the Supreme Soviet commuted his death sentence, and Vavilov was transferred out of the death cell and moved to a regular cell where he was allowed to exercise and bathe, but he was never transferred to a camp where he could have worked. Instead, he appears to have spent his final six months sick with dysentery resulting from lack of food. He died in the camp hospital on January 26, 1943, showing all the classic signs of starvation.[117]

— IN JANUARY 1946, a few months after word of Vavilov's death had begun to leak to the world, Morgan also died, and, in the span of a few months Muller had lost the two most powerful influences in his professional life. Fittingly, Muller was asked to write the eulogy of Morgan for *Science,* and Altenburg promptly warned him against getting carried away in the "emotional intoxication of a eulogy" and forever forfeiting his claim to credit for the early *Drosophila* work.[118] "Undoubtedly Morgan had streaks of occasional brilliance and went right to the point," Altenburg admitted, "but of course this was not a sustained brilliance; he went from the sublime to the ridiculous, and most of the time he stayed with the ridiculous." Altenburg then proceeded to enumerate several examples of his most "ridiculous" misconceptions, and included references and page numbers to facilitate Muller's writing. When he spoke about Morgan's profound misconceptions about Mendelism, Altenburg suggested he refer to his "broadmindedness," rather than his confusion, "thus damning with faint praise." But however he chose to handle the task, Altenburg begged Muller, "don't get soft all of a sudden; remember that Morgan did you incalculable damage, not only by his steals, but also by keeping you from getting a job."[119]

When six days later Altenburg read Muller's first draft of the eulogy, his only reservation was that it failed to communicate the "utter confusion which existed in Morgan's mind in the early days on such fundamental matters as Mendelism, mutation, natural selection and sex determination."[120] Taking Altenburg's comments under advisement, Muller wrote another draft, but he was now worried that he had erred too far in the other direction and that he had better tone it down. In any case, Muller assured Altenburg that "there should be no confusion on the main points: (1) that Morgan was badly confused on fundamentals and (2) that Wilson and the three younger men had a very important influence on Morgan, the first (Wilson) in imparting to him, in large measure, the chromosome theory, and the second (younger men) in the viewpoint which he later came to accept on fundamentals and in the further development of the chromosome theory."[121] Altenburg replied immediately: "The only thing he did, is give the genetic proof of crossing over."[122]

The final version of the eulogy appeared in *Science* in May, the most prominent of the many eulogies of the great man.[123] For the purposes of the eulogy Muller put aside his bitterness over Morgan's stealing and, borrowing a phrase from Altenburg, he instead celebrated Morgan for his "democratic spirit." Morgan's two great contributions, Muller wrote, were in the discovery of the meaning of linkage, which laid the foundation for all that followed, and in fostering the conditions under which his students could develop the full-blown chromosome theory. It was to his "enduring credit," Muller wrote, that despite his own "very active opposition" to the theory of natural selection, Morgan allowed himself to be won over by the younger men. In his recognition of the significance of Morgan's greatest discovery, Muller was unstinting, proclaiming that the recognition of the evidence for crossing over among sex-linked genes and the insight that genes farther apart crossed over more frequently was a "thunderclap hardly second to the discovery of Mendelism."

Later that year, on an October morning, a reporter called the Muller residence with the news that Muller had been awarded the Nobel Prize. The support for him had been overwhelming, with over fifty people separately writing in to propose his name, and the committee had been unanimous in their choice.[124] After the traditional reindeer dinner served at the rectory of the Caroline Institute, Muller was asked to take the presidency of the Eighth Genetics Congress in 1948. He was the successor to Vavilov, who had been chosen as the president of the seventh in 1939 (though not permitted to serve) and Morgan as the president of the sixth in 1932.

# EPILOGUE

— IN FEBRUARY 1944, Oswald Avery and two colleagues, Colin MacLeod and Maclyn McCarty, published a brilliant and penetrating paper showing that it was possible to transform a nonvirulent strain of the pneumococci bacterium into a virulent form by treating it with a purified extract of nucleic acid extracted from virulent cells. Avery was reluctant to declare that DNA, as the deoxyribose nucleic acid would soon be known, was the transformative molecule, fearing that his highly purified extract might still contain trace amounts of protein, and it would be the protein that turned out to be the active agent. Although he was not yet ready to take the leap in public, shortly before his epic paper was published Avery wrote his brother that the transformative agent "sounds like a virus, may be a gene."

It was left for Hermann Muller to push the analysis to its logical conclusion. During a 1945 address to the Pilgrim Trust, Muller conjectured that Avery's transformations of pneumococci were due to the incorporation of the injected genes into the genome of the receiving bacteria by a process of crossing over entirely analogous to ordinary crossing over.[1] Furthermore, he suggested that a similar mechanism explained the observation that the properties of one bacterial virus could be acquired by another, the process that he had first called attention to in his 1921 address at the Toronto meeting of the American Society of Naturalists. Like so many of his ideas, Muller's suggestion that these viruses, later called bacteriophage, could be the model organism of choice in which to study the nature of genes was prophetic, and by the late 1940s their study had become the driving force in genetics.

In 1947 the 18-year-old James Watson left the University of Chicago and enrolled as a graduate student at Indiana University, where he had been drawn by the presence of Muller. In Muller's course on advanced genetics for first-year graduate students, Watson was exposed for the first time to gene theory, much of which Muller himself had developed. In particular, Muller had long argued that genes must contain two distinct structural elements. One feature, which would allow for the copying of a pattern to take place and whose function would not be influenced by the particular pattern being copied, would have existed since the dawn of life. The second feature, somehow embedded in the first, would vary from gene to gene, determining each gene's particular function, and at the same time it had to be capable of mutational change. Figuratively, at least, "and perhaps . . . even literally," Muller wrote, the two kinds of arrangements lie in different "dimensions" from one another.[2]

But Muller was even more explicit in his description of the properties of the molecule, insisting that the embedded feature "must be arranged in only one or two dimensions."[3] This was the only way to explain how a daughter gene copied the pattern of the parent gene. The gene acted as a template, it was "a modeller, and forms an image, a copy of itself, next to itself," he explained. "A duplicate chain is produced next to each original chain, and no doubt lying in contact with a certain face of the latter."[4] Furthermore, he speculated, genes must be made up of smaller units, and he believed these smaller units would come in relatively few different types (it would turn out that there are four). It was the unique order in which these blocks were laid down that accounted for the differences between genes. A sudden change in the pattern, due to a "micro-chemical accident," he brilliantly deduced, was what biologists called a "mutation." The gene of the new type now reproduced the new pattern. The most fundamental outstanding enigma in genetics, Muller asserted, was the structural property of genes that allowed them to copy themselves and to copy their changes.

In almost every detail, Muller was spectacularly correct, going wrong only in the suggestion that the subunits in the gene to be duplicated were paired with identical subunits from the surrounding solution. In 1940 this idea was challenged by the prodigiously talented American physical chemist Linus Pauling, who published a paper with the young physicist-turned-phage-geneticist Max Delbrück arguing that "complementary structures in juxtaposition" accounted for the attraction of biologically active molecules. In a way they could not fully explain, Pauling and Delbrück believed that complementary structures could be identical. The problem for the structural chemist was "to analyze the conditions under which complementariness and identity might coincide."[5] Eight years later, in 1948, Pauling spelled out a solution to the paradox that was fully borne out when the structure of DNA was solved three years later.

> If the structure that serves as a template (the gene or virus molecule) consists of, say, two parts, which are themselves complementary in structure, then each of these parts can serve as the mould for the production of a replica of the other part, and the complex of two complementary parts thus can serve as the mould for the production of duplicates itself.[6]

— IN THE FALL of 1950, Watson left for Denmark on a postdoctoral fellowship in order to study the biochemistry of nucleic acids. The following spring he visited Naples, where he attended a lecture by Maurice Wilkins from King's College London on the three-dimensional structure of DNA. From Wilkins, Watson gleaned that DNA formed crystals, which meant that the structure was repeating and could in principle be solved by doing X-ray crystallography. At the end of his lecture, Wilkins put up a slide showing the X-ray diffraction picture that stuck in Watson's mind. Convinced that X-ray crystallography would provide the key to the gene, Watson managed to transfer to the Cavendish Laboratory in Cambridge, where X-rays were used to study the three-dimensional structures of proteins.

Almost immediately after arriving in England, Watson met Francis Crick, with whom he felt an instant rapport. Crick was quickly won over by Watson's enthusiasm for genes, and the two teamed up to solve the structure of DNA. Watson brought Crick his knowledge of the gene and of bacteriophage, which had replaced the fruit fly as the experimental organism of choice for geneticists, and Crick in turn provided a keen mathematical mind combined with a working knowledge of the basic tools of X-ray crystallography, many of which had been developed at the Cavendish Laboratory.

Meanwhile in March 1951 at Caltech, Pauling announced that he had solved the structure of the alpha helix, a common structural feature of proteins, largely by means of model building. Convinced that Pauling would next turn his attention to DNA, Watson and Crick decided to try to beat him at his own game by building a model of DNA. Providentially, Maurice Wilkins, whose X-ray pictures of DNA had so impressed Watson, was in close contact with Crick and had kept him abreast of the latest developments on the X-ray structure of DNA. Based in part on a new series of X-ray photos collected by Rosalind Franklin, who had recently come to work at King's College, Wilkins was now firmly convinced that DNA was a helix consisting of at least two or more central chains.[7] While Pauling's model was constructed from amino acid subunits, Watson and Crick's would be made from the four nucleotide building blocks—each consisting of a sugar molecule, a phosphate, and one of four differing bases, adenine, thymine, cytosine, or guanine—that were the constituents of DNA. From the work of Cambridge chemist Alexander Todd, they knew that the nucleotides were linked together in a chain by their sugar-phosphate groups. Watson and Crick's operating assumption was that this sugar-phosphate backbone was very regular, which accounted for the regularity seen in the X-ray pictures, but that the order of the bases would be irregular, the differences in the order of the bases explaining the differences between genes.[8] This principle, that the differences among genes would be accounted for by a change in the pattern of the subunits, was first enunciated by Muller in 1937.

In November 1951, Watson was invited to attend a colloquium at King's College where Rosalind Franklin presented her latest data. The following week Watson and Crick began to construct a model, making use of what they had gleaned from Wilkins over the previous months, the few facts that Watson recalled, it turned out wrongly, from Franklin's seminar, and some theoretical ideas about helix diffraction recently derived by Crick. Within the week, they had a model consisting of three sugar-phosphate chains spiraling up a central core with the nucleotide bases pointing outward, and Wilkins and Franklin were called up to Cambridge to view the final product. In a matter of minutes, Franklin found a fatal flaw. In order to account for the data, the sugar-phosphate chains had to be on the outside of the structure, not buried within it.[9]

More than a year after the fiasco of their first model, Watson and Crick received word via Linus Pauling's son, Peter, that Pauling had solved the DNA structure. The preprint of the paper, however, revealed that Pauling's model, like Watson and Crick's first model, was a three-chain structure with the bases pointing out. Frantic to find the correct structure before Pauling realized his mistake, Watson and Crick immediately resumed their model building. This time, however, they did not have to rely on Watson's memory for the crystallographic facts, but instead were in possession of Franklin's actual data thanks to an internal Medical Research Council report, which she had not intended to be made available to them.[10] After briefly reconsidering models with the chains on the inside and bases facing out, Watson began building a two-chain model with the helix spiraling around a central core containing some later-to-be-determined configuration of bases. At first Watson considered models in which the two chains were held together by interactions between like bases. This model was pleasing from a genetic point of view because it suggested that the one chain had served as a template for the other, the scheme favored by Muller. But the fact that the bases came in two distinct size classes—adenine and guanine being two-ring structures while cytosine and thymine had only one ring—presented a problem. In particular, the fact that a pair of two-ringed adenines or guanines would occupy more room within the central core of the helix than a pair of single-ringed thymines or cytosines was difficult to reconcile with the perfect regularity of the crystal structure.

After some timely advice from his crystallographer office mate, Jerry Donahue, about the location of the hydrogen atoms in thymine and guanine, on the morning of February 28, 1953, Watson saw that adenine (A) could be paired with thymine (T), and guanine (G) with cytosine (C). Further support for this scheme was provided by the recent work of the Austrian biochemist Erwin Chargaff, who had shown that the amount of A and T in DNA samples was roughly equal, and that the same was true for C and G.[11] Using cardboard cutouts, Watson showed that the two pairs had almost identical shapes. In this new scheme, a pair of complementary strands reproduced itself, each strand serving as a template for its complement. That is, during the process of DNA replication, the helix was peeled apart and complementary bases were laid down on each of the two complementary strands (later called "Watson" and "Crick"), resulting in two new DNA molecules. When he arrived at the lab later that morning, the ordinarily skeptical Crick was quickly won over.[12]

Watson and Crick announced the double helical structure of DNA in the April 25, 1953, issue of *Nature*.[13] As the structure famously made clear, the specificity of the gene was determined by the order of the nucleotide bases, and mutations amounted to changes in their order. The two-layered structure also illustrated both how the molecule could be self-replicating and how variations in particular genes could be inherited, the two properties that Muller had long insisted were fundamental to the hereditary particles. The gene's ability to reproduce its own variation was simply a reflection of the fact that the particular order of the bases did not affect the copying mechanism.

— THE DISCOVERY of the structure of DNA catalyzed the development of molecular biology and led to spectacular advances in the knowledge of genetics and cell biology. Nevertheless, more than two decades would pass before this knowledge was brought to bear on the human genome. Fifteen years after Watson and Crick's discovery, the only genes that had been mapped in humans were X-linked, revealed by their characteristic inheritance patterns—affected fathers producing normal daughters half of whose sons, on average, again showed the trait. Not a single gene had been mapped to the 22 pairs of other (non-sex) chromosomes called autosomes.

The discovery in 1968 of new methods for staining chromosomes (analogous to the methods developed by Painter for Drosophila chromosomes in the 1930s), and the use of human-mouse hybrid cells grown in culture, made possible the unique identification of each human chromosome. By the late 1970s roughly 220 human genes had been located by various methods, about 110 of them on the X chromosome and the remaining ones on the autosomes, but progress was extremely slow, hampered by a dearth of good genetic markers.[14]

What was needed to make progress in locating human disease genes was a set of marker genes of known location spread across the genome. Furthermore, to be useful in linkage studies, these loci had to be heterozygous in a large fraction of the population. Only then would it be possible to map unknown disease genes. In fruit flies, where it was possible to arrange crosses between flies that were heterozygous at each of two loci, mapping was fairly straightforward, but the problem was more complicated in humans, where matings could not be designed to extract genetic information. Instead extensive pedigrees spanning several generations of a disease-carrying family were required. But even the most complete pedigrees often did not contain enough information to determine the configuration of the two pairs of alleles, making it impossible to distinguish parental types from recombinants.[15] Using sophisticated statistics, it was possible, however, to estimate the degree of linkage between two pairs of allelic factors even if the configuration of the parental chromosomes was unknown, but in that case the analysis of linkage was actually a statistical statement based on the most likely configuration of the parental genes.

Meanwhile, the development of recombinant DNA technology continued apace during the 1970s. One of the principal discoveries was a new class of enzymes, so-called *restriction* enzymes. By cutting DNA at sequence-specific sites, restriction enzymes created a characteristic set of DNA fragments of different lengths. Mutations in the restriction enzyme recognition sites resulted in changes in the pattern of fragments generated when the DNA was cut, as did insertions or deletions of blocks of DNA within particular fragments. Such alterations in the DNA resulting in distinct cutting patterns were referred to as RFLPs (restriction fragment length polymorphisms). The recognition in 1974 that that RFLPs could serve as genetic markers was a major conceptual breakthrough.[16] In 1978 an RFLP was found on a segment of DNA containing the human beta globin gene. Although one of the two RFLP variants was much more frequent in the population at large, among blacks who suffered from sickle-cell anemia, which was caused by the possession of defective beta globin protein, 87 percent were found to contain the less common form.

Earlier that spring, at a small meeting in Alta, MIT molecular biologists David Botstein and colleagues realized that, in principle, RFLPs could be used to map human disease genes even in the absence of any prior knowledge about their locations or the biochemistry of their actions.[17] The idea was first to identify a large set of human RFLPs that was spread out across the genome, just as had been done for visible traits in Drosophila. Then by collecting small blood samples from disease-carrying families, it would be possible by track the inheritance of RFLPs in affected individuals and look for ones that segregated with the disease. But such a program was predicated on the existence of large set of high-frequency that is, heterozygous in a large fraction of the population) RFLPs spread evenly across all 24 distinct chromosomes.[18]

Several years after the Alta meeting, James Gusella at Massachusetts General Hospital set out to look for an RFLP that was linked to the inheritance of Huntington's disease. In a remarkable stroke of good fortune, after screening only twelve known polymorphic DNA regions, Gusella found one that appeared to be associated with the inheritance of Huntington's disease in an American family consisting of 40 individuals. The same polymorphic region was found to be associated with the inheritance of the disease in a much more extensive pedigree consisting of over 3,000 individuals living on the shore of a remote lakeside community in Venezuela. By the end of 1983, Gusella had narrowed the location of the Huntington's gene to a region on chromosome 4 encompassing about ten million base pairs.

Over the next several years, scientists following the program Botstein and colleagues had laid out were able to locate the defective genes for cystic fibrosis, Duchenne muscular dystrophy, retinoblastoma, familial colon cancer, neurofibromatosis, and others.[19] In each case, the successful localization of the gene led eventually to the cloning of the gene and a biochemical characterization of the disease. In 1986 the invention of PCR (polymerase chain reaction) by Kary Mullis, a biochemist at Cetus Corporation, made it possible to make millions of copies of DNA fragments in a few hours, radically simplifying the process of DNA sequencing. As a result of PCR and the development of automated sequencing technology, the idea of sequencing the human genome in its entirety began to take off. By the end of 1986, human genome sequencing had been the focus of five major biological conferences throughout the country and an independent initiative by scientists at the Department of Energy. At an estimated cost of one dollar per base pair, the total cost of sequencing the entire human genome was projected to exceed three billion dollars.

Initially many highly respected scientists, including David Botstein, questioned the value of getting a full genome sequence, arguing that money was better spent on mapping and other smaller-scale initiatives. In 1987 the National Academy of Science convened a committee on the genome project to study the problem. In their report, issued in the spring of 1988, the committee advised focusing first on genetic and physical maps (made from DNA markers). The result would be a collection of well-ordered, overlapping DNA fragments, each about 40,000 base pairs in length, spanning the complete set of chromosomes. James Watson, a strong advocate of genome sequencing from the start, argued that the project ought to be run by a prominent scientist who could advocate effectively for the project and make sure that the money was well spent, and that fall the director of the National Institutes of Health (NIH) created the Office of Human Genome Research and appointed Watson himself as its first director.[20] Still associated with the glory of the double helix, Watson would prove to be the perfect advocate for the Human Genome Project.

Over the next decade, the project unfolded very much according to plan, turning out increasingly detailed maps and sequences of smaller genomes ahead of schedule. In 1998 former NIH researcher-turned-biotech-entrepreneur Craig Venter announced the formation of a privately funded company to sequence the human genome four years ahead of the NIH-led effort at a fraction of the cost. Venter's idea was to sequence fragments selected at random from a pool of sheared whole genomic DNA and to assemble them into a coherent sequence after the fact, with the aid of massive computing power. In the end, this led to a much-publicized contest between Celera Genomics, Venter's company, and a frantic public consortium motivated by the fear that Venter would try to patent or in other ways slow the dissemination of valuable genomic information. In February 2001 the public consortium and Celera simultaneously published rough drafts of the entire genome in *Nature* and *Science*, respectively. While there was great bitterness on both sides, each group had clearly benefited from the other, the public consortium by making use of Venter's "shotgun" technique to get out a first draft of a sequence two years ahead of schedule, and Venter by taking advantage of the free public sequence data to reduce his company's work.

— THE FIRST PUBLIC human genome sequences were in fact consensus sequences determined from many different individuals (Celera's sequence was largely that of Venter). These sequences captured 99.9 percent of the bases that were expected to be identical across all races and continents. From the beginning, a separate effort had been planned to characterize the remaining 0.1 percent of the sequence that varied among people. These bases would hold the key to inherited differences among people: differences in susceptibility to disease and drugs and perhaps even differences in temperament, intelligence, and other complex psychological traits.

The problem was complicated, though, by the fact that most human traits were not caused by single, clearly segregating Mendelian factors. As Muller had first argued in the early part of the twentieth century, most traits were likely to be influenced by a large number of genes, each of which exerted only a small effect, and this was likely to be particularly true of the genetic basis for complex human characters. Even in Drosophila, as he and Altenburg had so cleverly demonstrated, the simplest morphological feature, wing shape, was influenced by several genes, none of which resulted in a predictable phenotype. As we saw in Chapter 10, Muller invented an entirely new method—which he called the "method of linked factors"—to prove that there were at least three different genes on three different chromosomes that influenced the development of truncated wings, including a master gene on chromosome II that was absolutely required for the expression of the trait (but did not always result in truncation) and two factors on chromosomes I and III, respectively, that increased both the rate and the severity of truncation but could not, by themselves, cause the mutant phenotype. The method turned on being able to demonstrate the presence of genes for truncation, not only in affected flies, but also in flies that did not overtly manifest the trait. To do this, Muller and Altenburg had taken advantage of the fact that male flies did not undergo crossing over, which made it possible to track the inheritance of whole chromosomes (and all the genes on them) by single heterozygous genes (specifying eye or body color) that had nothing to do with the trait under study.

As Muller pointed out, his method would require modification when applied in humans, where crossing over takes place in both sexes (making it impossible to follow an entire chromosome with a single marker). Whereas it might be possible to get away with a limited number of well-defined markers to map traditional single-gene diseases, mapping complex human traits would require a huge number of identifying factors. Given a large enough collection of markers spread evenly across the entire genome, Muller reasoned, these markers could be employed to track small adjacent sections of chromosomes that would be highly unlikely to be separated from a neighboring marker by crossing over. But to find these markers, geneticists would need to collect a different class of mutant genes. "Study of factors [markers] which are inconspicuous, or unimportant in actual life, has been largely avoided," Muller pointed out, but "any two parents of a human family would probably differ nearly always in a very large number of factors." With a detailed map of linked factors on every chromosome, one could begin the search for genes behind the more complicated and important human traits.

— EIGHTY YEARS WOULD PASS before Muller's dream of using a set of linked identifying factors to identify the genes involved in complex human traits became possible. By the late 1990s it had become clear that there were far more DNA polymorphisms in the form of single base pair changes than any one had previously anticipated. Not only did these single-nucleotide polymorphisms (SNPs) blanket the genome, but many of them existed at a high enough frequency in the population to be useful in mapping complex traits. In a 1996 study, Eric Lander, the head of the human genome sequencing center at MIT, estimated that there were on the order of ten million SNPs where the less frequent variant (allele) appeared in a least 5 percent of the population.[21] On average, then, each individual contained such a SNP once in every 300 base pairs.

Because each new mutation giving rise to a SNP took place on a chromosome that already contained other SNPs, each new allele became associated with a particular set of nearby alleles. These groups of highly correlated alleles were called "haplotypes." At first the assumption was that the correlation among SNPs, which was highest at close distances, would fall off in a regular way at increasing distances, but by 2001 it had became clear that this simplified picture wasn't correct. When a particular chromosome was sequenced in a small group of ethnically diverse individuals, it was found that the genome fell naturally into blocks of varying sizes, each block consisting of only four or five prominent haplotypes that were representative of the vast majority of individuals.[22]

— THE DISCOVERY that the genome can be partitioned into blocks, and that within each block there is far less diversity than might have been expected, significantly reduced the labor involved in testing for genes of small effect. In each block, a small subset of SNPs can be used to uniquely identify each of the four or five common haplotypes. Thus a fairly complete picture of the state of the more than ten million SNPs could in principle be gleaned by looking only at a tenth or fewer of them. Armed with a complete set of indicator SNPs, the search for genes behind complex traits could be undertaken in earnest.

With this in mind, the International HapMap Project was initiated in 2002. DNA was collected from a group of 269 individuals of diverse ethnic and geographic backgrounds—90 Nigerians, 90 Americans, 45 Han Chinese, and 44 Japanese.[23] By 2005, each individual had been genotyped at over 600,000 common SNPs spread out across the entire genome, giving at least one SNP for every 5,000 base pairs. As had been observed in earlier studies, the genome could be subdivided into discrete blocks. Nearly all the blocks consisted of more than four common SNPs, with nearly all haplotypes in each block matching one of a few common types.

With the completion of the first phase of the HapMap Project in 2005, the systematic hunt for genes involved in complex diseases has now begun in earnest. To look for genes involved in adult-onset diabetes, for example, one would first collect a large sample of patients and controls, and determine their haplotypes in each block. As in the truncate case, the method will consist in searching for a correlation between the possession of a particular chromosome (or haplotype) and the expression of a complex trait. For most blocks, the frequency of each haplotype would be expected to be the same in the trait group and the control group, but if a haplotype contains a gene that influences the expression of the disease, it would be expected to appear at a higher frequency among patients compared to controls. To date, researchers have uncovered genes implicated in Alzheimer's disease, types 1 and 2 diabetes, rheumatoid arthritis, macular degeneration, inflammatory bowel syndrome, and many other diseases with complex inheritance patterns, but it remains to be seen whether the completion of the HapMap will usher in a new era of understanding of the basis of complex human traits.

— "THE ROAD OF EVOLUTION is controlled by man himself," Muller wrote late in life. "Even as he challenges the great constellations of stars without, [he] challenges the even more amazing constellations of genes within." Intelligence and cooperation had brought us to our ascendant place in the hierarchy of organisms, he believed, but the process of our development had only just begun, and it was our business to take it as far as we could: "We can, in securing and advancing our position, increasingly avoid the missteps of blind nature, circumvent its cruelties, reform our own natures, and enhance our own values."[24] Even after the Nazi abuses of eugenics, Muller remained convinced that the best hope for humankind lay in reforming our genetic nature. Coming at it from a very different background, Francis Galton had arrived at exactly the same conclusion a hundred years earlier.

While the HapMap may shed new light on the genetic basis of complex diseases like schizophrenia, which is known to be 70 to 80 percent heritable, it is also possible that the interactions of genes and their relationship with the surrounding environment will prove to be too complex and elusive to dissect in man. Association studies, for example, will only succeed to the extent that investigators can divide the population into affected individuals and controls, and the judgment about who is affected with altruism or intelligence is a daunting problem at best. For this reason alone, the discovery of genes influencing complex psychological characters must be viewed with great caution. While few doubt that genes play a significant part in the mix of genetic, cultural, and environmental factors that are involved in the development of complex human attributes, the challenge is to use what we know about them to preserve ourselves and the life around us. Whether we wish it to be so or not, humans are increasingly in control of their destiny, genetic and otherwise.

NOTES

ACKNOWLEDGMENTS

INDEX

# NOTES

## 1. VIVA PANGENESIS

1. Adrian Desmond and James Moore, *Darwin: The Life of a Tormented Evolutionist* (New York: W. W. Norton, 1994), 549.

2. Janet Browne, *Charles Darwin,* vol. 2, *The Power of Place* (New York: Knopf, 2002), 263.

3. Desmond and Moore, *Darwin,* 531.

4. Ibid., 532.

5. Charles Darwin, *The Variation of Animals and Plants under Domestication,* vol. 2 (London: John Murray, 1868), 374.

6. Letter to Hooker, Jan. 11, 1844, in Charles Darwin, *More Letters of Charles Darwin,* ed. Francis Darwin and A. C. Seward (London: John Murray, 1903), 1:41.

7. Charles Darwin, *The Variation of Animals and Plants under Domestication,* vol. 2 (New York: Appleton, 1894), 389.

8. Ibid., 35.

9. Even as a teenager, Galton had looked up to his older cousin, enlisting his

help in persuading his father to allow him to take time off from his medical training to study mathematics at Cambridge. In a letter to Darwin written in 1869 Galton wrote, "Your Origin of Species formed a real crisis in my life; your book drove away the constraint of my old superstition as if it had been a nightmare and was the first to give me freedom of thought," and nearly fifty years later he wrote in his autobiography that reading Darwin's book had "marked an epoch" in his mental development. The first letter is quoted in Karl Pearson, *The Life, Letters and Labours of Francis Galton,* 3 vols. (Cambridge: Cambridge University Press, 1914–1930), 1:6. The second appears in Francis Galton, *Memories of My Life* (London: Methuen and Co., 1908), 287.

10. Francis Galton, "Hereditary Talent and Character," *Macmillan's Magazine* 12 (1865), pt. 1: 157–166 , pt. 2: 318–327.

11. Ibid., 165.

12. For the history of Galton's use of the term, see Pearson, *Galton,* 2:249 n. 1.

13. Ibid., 161.

14. Ibid., 160.

15. D. W. Forrest, *Francis Galton: The Life and Work of a Victorian Genius* (New York: Taplinger, 1974), 88.

16. Ibid., 15.

17. Ibid., 16.

18. Francis Galton, *Hereditary Genius: An Inquiry into Its Laws and Consequences* (London: Macmillan and Co., 1869), 16.

19. Letter from Francis to his father, Nov. 3, 1840, in Pearson, *Galton,* 1:144.

20. Ibid.

21. Letter, Nov. 26, 1840, in Pearson, *Galton,* 1:145

22. Letters, Dec. 9, 1840, and Feb. 3, 1941, in Pearson, *Galton,* 1:148, 150.

23. Letter, Nov. 11, 1841, in Pearson, *Galton,* 1:163

24. Arthur Cayley would become a Senior Wrangler in 1842 and go on to become the Sadleirian Professor of Pure Mathematics at Cambridge, where he would make major contributions in the fields of algebra and geometry.

25. Letter of Jan. 21, 1842, transcribed in Pearson, *Galton,* 1:164.

26. Pearson unearthed the list of finishers; *Galton,* 1:164.

27. Letter of Mar. 22, 1842, in Pearson, *Galton,* 1:166.

28. Letter, Aug. 1, 1842, in Pearson, *Galton,* 1:169.

29. Letter, Nov. 2, 1842, in Pearson, *Galton,* 1:170.

30. Letter, ca. Nov. 28, 1842, in Pearson, *Galton,* 1:172–173.

31. Galton, *Memories of My Life,* 78–79.

32. Letters from his sister and mother are transcribed in Forrest, *Francis Galton,* 86.

33. Galton, *Memories of My Life,* 155.

34. Galton, *Hereditary Genius,* 19, 20

35. Ibid., 30.

36. Ibid., 31–32.

37. Ibid., 32.

38. Ibid., 33.

39. Ibid., 11.

40. Ibid., 364.

41. Ibid., 35–36.

42. Ibid., 373.

43. Letter from Galton to Darwin, Dec. 11, 1869, quoted in Pearson, *Galton,* 2:157.

44. Pearson, *Galton,* 1:6.

45. Letter from Galton to Darwin, Dec. 24, 1869, reproduced in Pearson, *Galton,* vol. 1, plate 2, after p. 6.

46. Pearson, *Galton,* 2:201 n. 1.

47. Letter from Galton to Darwin, Mar. 15, 1870, in Pearson, *Galton,* 1:157.

48. Letter from Mrs. Darwin to Henrietta dated Mar. 19, 1870; letter from Galton to Darwin, Mar. 31, 1870; letter from Galton to Darwin, May 12, 1870; all in Pearson, *Galton,* 1:158, 160.

49. Francis Galton, "Experiments in Pangenesis, by Breeding from Rabbits of a Pure Variety, into Whose Circulation Blood Taken from other Varieties Had Previously Been Largely Transfused," *Proceedings of the Royal Society of London,* 19 (1871): 402, 404.

50. Charles Darwin, "Pangenesis," *Nature* 3 (1871): 502–503.

51. Darwin, *Variation,* vol. 2 (1894), 350 n. 1.

52. Letter from Galton to Darwin, Apr. 25, 1871, transcribed in Pearson, *Galton,* 2:162.

53. Francis Galton, "Pangenesis," *Nature* 4 (1871): 5.

54. Lionel Beale, "Pangenesis," *Nature* 4 (1871): 25–26.

55. Letter from Galton to Darwin, May 12, 1871, in Pearson, *Galton,* 2:162.

56. Letters from Galton to Darwin, Sept. 13 and Nov. 21, 1871, in Pearson, *Galton,* 2:166, 167.

57. Letters, from Darwin to Galton, Jan. 23 [1872]; from Galton to Darwin, Feb. 1, 1872; from Galton to Darwin, May 26, 1872; all in Pearson, *Galton,* 2:167, 168.

58. Letter from Darwin to Galton, May 27 [1872], in Pearson, *Galton,* 2:168.

59. Letter from Galton to Darwin, May 28, 1872, in Pearson, *Galton,* 2:168–169.

60. Francis Galton, "On Blood-Relationship," *Proceedings of the Royal Society of London,* 20 (1872): 394.

61. Ibid., 400.

62. August Weismann posited the existence of special inviolable cells containing the hereditary material in specific nuclear bodies (chromosomes), but he later acknowledged that Galton's theory of heredity was nearly identical to the "main idea contained in my theory of the continuity of germ-plasm." Letter from Weismann to Galton, Feb. 23, 1889, in Pearson, *Galton,* 3:340n. Weismann had first introduced the term *germ-plasm* in his 1883 essay "On Heredity," which can be found in August Weismann, *Essays upon Heredity and Kindred Biological Problems, 1883,* ed. Edward B. Poulton, Selmar Schonland, and Arthur E. Shipley (Oxford: Clarendon Press, 1889), 104.

63. Letter from Darwin to Galton, Nov. 8, 1872, in Pearson, *Galton,* 2:175.

## 2. REVERSION TO THE MEAN

1. Francis Galton, "Hereditary Improvement," *Fraser's Magazine* 13 (1873): 116–130.

2. Ibid., 117.

3. Ibid., 125, 128.

4. Ibid., 123, 128–129.

5. Ibid., 129, 130.

6. Letter from Darwin to Galton, Jan. 4, 1873, in Karl Pearson, *The Life, Letters and Labours of Francis Galton,* 3 vols. (Cambridge: Cambridge University Press, 1914–1930), 2:176.

7. Galton, "Hereditary Improvement," 119.

8. Letter from Darwin to Galton, Jan. 4., 1873, in Pearson, *Galton,* 2:176.

9. Alphonse de Candolle, *Histoire des sciences et des savants depuis deux siècles* (Geneva: H. Georg, 1873), 181.

10. Ibid., 93–94. Translated in Raymond E. Francher, "Alphonse De Candolle, Francis Galton, and the Early History of the Nature-Nurture Controversy," *Journal of the History of the Behavioral Sciences* 19 (1983): 345.

11. Candolle, *Histoire,* data from tables 9 and 10 on pp. 176–177, 181.

12. Ibid., 196–197

13. Ibid., 104–105.

14. Ibid.

15. Pearson, *Galton,* 2:135.

16. Letter from Galton to De Candolle, Dec. 27, 1872, in Pearson, *Galton,* 2:135–136.

17. Letter from Darwin to Candolle, Dec. 11, 1872, in Francher, "Nature-Nurture Controversy," 346; letter from Darwin to Galton, Dec. 23 [1859], quoted in Pearson, *Galton,* 1:6.

18. Letter from Galton to De Candolle, Dec. 27, 1872, in Pearson, *Galton,* 2:135–136.

19. Francher, "Nature-Nurture Controversy," 345.

20. Francis Galton, "On the Causes Which Operate to Create Scientific Men," *Fortnightly Review* 13 (1873): 346.

21. Francis Galton, *English Men of Science: Their Nature and Nurture* (London: Macmillan and Co., 1874).

22. Galton, "On the Causes," 346.

23. Pearson, *Galton,* 2:177.

24. Galton, *English Men of Science,* 2,

25. Ibid., vi.

26. Steven M. Stigler, *The History of Statistics: The Measurement of Uncertainty before 1900* (Cambridge: Harvard University Press, 1986), 278

27. For a general discussion of the history and significance of DeMoivre's theorem, see ibid., 71–77.

28. Galton's report appears in Charles Darwin, *The Effects of Cross and Self Fertilisation in the Vegetable Kingdom* (London: John Murray, 1876), 16–18.

29. Letter from Galton to Pearson, Aug. 30, 1907, in Pearson, *Galton,* 3:325. See also D. W. Forrest, *Francis Galton: The Life and Work of a Victorian Genius* (New York: Taplinger, 1974), 187.

30. Francis Galton, "Opening Address to Anthropology Section for the British Association of Science at Aberdeen," *Nature* 32 (1885): 507.

31. In a letter of Apr. 14, 1875, Galton asks Darwin to plant sweet peas for him and reports that he had lost the previous year's crop that he had planted at Kew Gardens. See Pearson, *Galton,* 2:180.

32. Galton discusses the experiment in detail in his *Natural Inheritance* (London: Macmillan and Co., 1889), 81–82.

33. The graph, which Galton never published, shows that the means of parents and offspring differ, which was presumably a consequence of the fact that the two generations were grown in different environments.

34. Galton, "Opening Address," 507.

35. Letter from Galton to George Howard Darwin, May 17, 1876, in Ruth Schwartz Cowan, "Francis Galton's Statistical Ideas: The Influence of Eugenics," *Isis,* 63 (1972): 518.

36. Letter from Galton to George Darwin, Jan. 5, 1877, in Pearson, *Galton,* 3B:465.

37. Galton introduced a clever extension of the normal theory that required a revamped quincunx to illustrate. See Stigler, *History of Statistics,* 275–281, for a detailed explanation of the theory.

38. Cowan, "Galton's Statistical Idea," 521 n. 44.

39. Galton, "Opening Address," 507.

40. For example, the "1" in the upper right corner of the graph indicates that there was 1 adult child in the range between 4 and 5 inches above average height born to mid-parents who were between 3 and 4 inches above average.

41. Francis Galton, *Memories of My Life* (London: Methuen and Co., 1908), 302.

42. Letter from Galton to Tertius, Nov. 3, 1840, in Pearson, *Galton,* 1:144

43. Pearson, *Galton,* 1:302–303.

44. Francis Galton, "Regression toward Mediocrity in Stature," *Miscellanea of the Journal of the Anthropological Institute of Great Britain* 15 (1886): 255; and again in Galton, *Natural Inheritance,* 102

45. Galton, "Opening Address," 507

46. Galton, "Regression toward Mediocrity," 253

47. Ibid., 256.

48. Ibid, 253.

49. Galton's error had been to group together in each height class a variety of people from different ancestries. Under this procedure, a significant number of exceptionally tall (or short) people who came from families of unexceptional height and whose exceptional height was due to nutrition or other environmental factors that were not heritable were grouped together with people who carried tall genes. It was simply not true that exceptional parent types would necessarily have less exceptional children, or that the plants with large seeds would give rise to plants with seeds of smaller size. Contrary to Galton's belief, the offspring of a marriage of tall families would, on average, be as tall as their parents, and likewise, large seeds derived from a single ancestral line would have given rise to progeny peas that were, on average, as large as their parental seeds. Furthermore, if tall men and women were selected out over a period of generations (until they were homozygous for the genes coding for increased height), there would ultimately be no regression at all in height. Theoretically, selective breeding can create a race of giants.

The correct biological interpretation of Galton's classic regression experiments was only widely appreciated in the middle of the twentieth century when it was understood that the regression coefficient measured the degree to which a trait was inherited. This quantity was called the "heritability" of a trait. If the variability of a trait in the population is due largely to differences in the (additive) effects of genes, the slope of the regression line will be nearly 1, and parental traits will be passed on fully to their adult children. In other words, such traits are highly heritable. At the other extreme, if the variation of a trait is entirely due to the random effects of the environment, the regression coefficient will be nearly flat, and there will be no correlation between children and their parents. Stature in man, for example, is 65 percent heritable, as is body weight in cattle. Tail length in mice is only 40 percent heritable.

50. Francis Galton, "President's Address," *Journal of the Anthropological Institute* 15 (1886): 497, 496.

51. Francis Galton, *Hereditary Genius: An Inquiry into Its Laws and Consequences* (London: Macmillan and Co., 1869), 369.

### 3. GALTON'S DISCIPLES

1. Karl Pearson, "Walter Frank Raphael Weldon, 1860–1906," *Biometrika* 5 (1906): 13.

2. W. F. R. Weldon, "The Variations Occurring in Certain Decapod Crustacea.—I. *Crangon vulgaris*," *Proceedings of the Royal Society of London* 47 (1890): 445, 446.

3. Theodore M. Porter, *Karl Pearson: The Scientific Life in a Statistical Age* (Princeton: Princeton University Press, 2004), 236.

4. Weldon's influence on Pearson is discussed in ibid., 216, 237.

5. Karl Pearson, "Contributions to the Mathematical Theory of Evolution," *Philosophical Transactions of the Royal Society of London,* ser. A, vol. 185 (1894): 71–110.

6. An anonymous obituary in the St. John's College magazine, "In Memoriam: William Bateson, 1861–1926," *Eagle* 44 (1926): 3.

7. See for example, chapter 22 in Charles Darwin, *The Variation of Animals and Plants under Domestication,* vol. 2 (New York: Appleton, 1894), esp. 243.

8. Quoted in William Coleman, "Bateson and Chromosomes: Conservative Thought in Science," *Centaurus* 15 (1970): 248.

9. W. K. Brooks, *The Law of Heredity: A Study of the Cause of Variation and the Origin of Living Organisms* (Baltimore: John Murphy and Co., 1883), 301.

10. Written in his application for the Linacre Professorship of Comparative Anatomy at Oxford, which he did not win. Quoted in Beatrice Bateson, *William Bateson,*

*F.R.S. Naturalist: His Essays and Addresses Together with a Short Account of His Life* (Cambridge: Cambridge University Press, 1928), 34.

11. William Bateson, *Scientific Papers of William Bateson,* vol. 1, ed. R. C. Punnett (Cambridge: Cambridge University Press, 1928), 159.

12. Quoted by William Provine, *The Origins of Theoretical Population Genetics* (Chicago: University of Chicago Press, 2001), 41–42.

13. B. Bateson, *William Bateson,* 28.

14. At the conclusion of his long letter of Sept. 28, 1888, in which Weldon strenuously defends his recent scientific pursuits, Weldon asks Bateson: "And when are you coming to crush me???" William Bateson to W. F. R. Weldon, Sept. 28, 1888, William Bateson Manuscripts, microfilm, American Philosophical Society Library, Philadelphia (hereafter Bateson MSS), reel B, no. 13.

15. W. F. R. Weldon to William Bateson, Feb. 15, 1894, Bateson MSS, reel B, no. 13.

16. W: F. R. Weldon to William Bateson, Mar. 1, 1894, Bateson MSS, reel B, no. 13.

17. William Bateson to W. F. R. Weldon, Mar. 4, 1894, Bateson MSS, reel B, no. 13.

18. Ibid.

19. Following his return from the steppe in 1887, Bateson spent seven years working on his book (B. Bateson, *William Bateson,* 27).

20. W. F. R. Weldon, "The Study of Animal Variation," *Nature* 50 (1894): 25.

21. Francis Galton, "Discontinuity in Evolution," *Mind* 3 (1894): 369.

22. William Turner Thiselton-Dyer, "Variation and Specific Stability," *Nature* 51 (1895): 460.

23. William Turner Thiselton-Dyer, "Origin of the Cultivated Cineraria," *Nature* 52 (1895): 3.

24. William Bateson, "The Origin of the Cultivated Cineraria," *Nature* 52 (1895): 29.

25. W. F. R. Weldon, "The Origin of the Cultivated Cineraria," *Nature* 52 (1895): 54.

26. William Bateson to W. F. R. Weldon, May 24, 1895, Bateson MSS, reel B, no. 10.

27. William Bateson to Francis Galton, Oct. 15, 1896, Bateson MSS, reel C, no. 15.

28. Francis Galton to William Bateson, Oct. 17, 1896, Bateson MSS, reel C, no. 15.

29. William Bateson to Francis Galton, Oct. 18, 1896, Bateson MSS, reel C, no. 15.

30. W. F. R. Weldon to Francis Galton, Oct. 22, 1896, Bateson MSS, reel C, no. 15.

31. Francis Galton to William Bateson, Oct. 30, 1896, Bateson MSS, reel C, no. 15.

32. William Bateson to Francis Galton, Nov. 2, 1896, Bateson MSS, reel C, no. 15.

33. William Bateson to Francis Galton, Nov. 15, 1896, Bateson MSS, reel C, no. 15.

34. Galton, "Discontinuity," 372.

35. Francis Galton to W. F. R. Weldon, Nov. 17, 1896, in Karl Pearson, *The Life, Letters and Labours of Francis Galton,* 3 vols. (Cambridge: Cambridge University Press, 1914–1930), 3A:127.

36. William Bateson to W. F. R. Weldon, Dec. 6, 1896, Bateson MSS, reel B, no. 10.

37. W. F. R. Weldon to William Bateson, Dec. 8, 1896, Bateson MSS, reel B, no. 10.

38. Francis Galton to William Bateson, Jan. 1, 1897, Bateson MSS, reel C, no. 15.

39. William Bateson to Francis Galton, Jan. 3, 1897, Bateson MSS, reel C, no. 15.

40. Weldon reports on his private communication with Galton in his letter to Bateson dated Jan. 16, 1897, Bateson MSS, reel C, no. 15.

41. W. F. R. Weldon to William Bateson, Feb. 16, 1897, Bateson MSS, reel C, no. 15.

42. William Bateson to W. F. R. Weldon, Jan. 18, 1897, Bateson MSS, reel C, no. 15.

43. Karl Pearson to Francis Galton, Feb. 12, 1897, in Pearson, *Galton,* 3A:128.

44. Darwin, *Variation* (1894), 2:46, 47. Unknown to Darwin, the phenomenon he called "prepotency" had four years earlier been dubbed *dominance* by Mendel.

45. The contributions of the ancestors sum up to 1 because the sum of the geometric series approaches unity as more terms are added. Already with five generations it was getting close: $1/2 + 1/4 + 1/8 + 1/16 + 1/32 + 1/64 = .96$

46. Francis Galton, "The Average Contribution of Each Several Ancestor to the Total Heritage of the Offspring," *Proceedings of the Royal Society of London* 61 (1897): 401–413.

47. Pearson, *Galton,* 3A:40.

48. Karl Pearson, "Mathematical Contributions to the Theory of Evolution: On the Law of Ancestral Heredity," *Proceedings of the Royal Society of London* 62 (1897–1898): 412.

49. Pearson, *Galton,* 3B:504.

50. Bateson, "Hybridisation and Cross-Breeding as a Method of Scientific Investigation," *Journal of the Royal Horticultural Society,* 24 (1899): 63.

51. Letter from William Bateson to B. Bateson, Aug. 7, 1899, Bateson MSS, reel A, no. 1.

52. Ibid.

## 4. PANGENES

1. Peter W. van der Pas, "The Correspondence of Hugo de Vries and Charles Darwin," *Janus* 57 (1970): 179.

2. Ibid., 182.

3. De Vries wrote his grandmother that meeting Darwin had been "the actual goal of the trip" (ibid., 188).

4. Ibid., 187.

5. Ibid., 199.

6. Ibid., 200–201.

7. Ibid., 192.

8. Hugo de Vries, *Intracellular Pangenesis,* trans. C. Stuart Gager (Chicago: Open Court, 1890), 13.

9. This theme is the subject of a penetrating essay by Ida Stamhuis, "The Reactions on Hugo de Vries's Intracellular Pangenesis: The Discussion with August Weismann," *Journal of the History of Biology* 36 (2003): 131.

10. August Weismann, *Essays upon Heredity and Kindred Biological Problems,* vol. 1, ed. Edward B. Poulton, Selmar Schonland, Arthur E. Shipley (Oxford: Clarendon Press, 1889), 69.

11. Quoted in Stamhuis, "Reactions," 131.

12. It had been nearly twenty years since Haeckel had first suggested that the hereditary material was located in the nucleus and that the cytoplasm was responsible for the adaptive functions of a cell, but Haeckel had provided no evidence. In 1875 Haeckel's protégé, Oscar Hertwig, presented evidence that fertilization consisted of the fusion of a sperm nucleus with the nucleus of the egg cell in the sea urchin. Because the male nucleus was (essentially) the only contribution of the male to the newly formed individual, Hertwig argued, it must contain the hereditary material.

13. De Vries, *Intracellular Pangenesis,* 75.

14. Onno G. Meijer, "Hugo de Vries No Mendelian," *Annals of Science* 42 (1985): 203.

15. Stamhuis, "Reactions," 121

16. In the same year he conducted a similar experiment in *Oenothera,* showing that it was possibly to transfer a mutation of the flower (causing a withering of the female style, the columnar structure that receives pollen) in the so-called *brevistylis* form to the normal *O. lamarckiana.*

17. Erik Zevenhuizen, "The Hereditary Statistics of Hugo de Vries," *Acta Botanica Neerlandica* 47 (1998): 432.

18. Ibid., 433.

19. The 1:2:1 ratio of different outcomes resulting from two blind draws follows from the observation that there is only one way to get two white marbles (by get-

ting white on the first and second draw) and likewise only one way to get two blacks, but there are two ways to get one black and one white. The probabilities of the various combinations for a sample of any size are given by the rows of Pascal's triangle, which can be viewed as a tool to count the number of ways of getting all possible outcomes. In general, the relative frequencies of getting 0, 1, . . . , $k$ whites in a sample of $k$ randomly chosen marbles is given by the $(k+1)^{th}$ row. For example, the chances of getting 0, 1, 2, 3, or 4 whites in a sample of 4 marbles is 1:4:6:4:1, which is the 5th row of Pascal's triangle.

1<br>
1 1<br>
1 2 1<br>
1 3 3 1<br>
1 4 6 4 1

**First five rows of Pascal's triangle.**

20. Zevenhuizen, "Hereditary Statistics," 444.

21. In the 1895 edition of *Textbook of Plant Physiology*, De Vries discusses Newton's binomial curve, which he shows can be generated by $(b + n)^k$. The rows of Pascal's triangle correspond to the integer coefficients of the expansion of the $(b + n)^k$. For example the coefficients of the expansion of $(b + w)^2$ are 1, 2, 1, as seen by writing out $(b + w)^2 = \underline{1}b^2 + \underline{2}bw + \underline{1}w^2$.

22. The early determination of flower color had actually been made before the flowers bloomed, based the color of a plant's stem, which was a usually reliable indicator. By August, the grandchildren had bloomed and could be scored on the basis of flower color.

23. Zevenhuizen, "Hereditary Statistics," 449.

24. Ibid., 463.

25. Had De Vries considered a four-pangene instead of a two-pangene model to explain the *Lychnis* results, as he had in his 1896 flower color experiments, the grandchildren would then have been expected to follow a 1:4:6:4:1 law. If the expression of hairiness was assumed to require the inheritance of two or more "hairy" pangenes, the expected fraction of hairless types would have been as 5:11 (or .31), which was in good agreement with the data (.35 hairless).

26. Meijer, "No Mendelian," 211.

27. Ibid., 215.

28. T. J. Stomps, "On the Rediscovery of Mendel's Work by Hugo de Vries," *Journal of Heredity* 45 (1954): 294.

## 5. MENDEL

1. Today Heinzendorf lies in the eastern corner of the Czech Republic.

2. Mendel reports that he was forced to support himself in his third-person autobiography, which appears in translation in Robert C. Olby, *Origins of Mendelism* (New York: Schocken Books, 1966), 175–178. The nature of the accident is described in Hugo Iltis, *Life of Mendel* (London: George Allen and Unwin, 1932), 38. Iltis, who salvaged the remnants of Mendel's papers, which had been largely destroyed after his death, wrote the first and definitive biography of Mendel. See Olby, *Origins of Mendelism,* 103 for the history of Mendel's papers.

3. Iltis, *Life of Mendel,* 35–36. See also Vitezslav Orel, *Gregor Mendel: The First Geneticist,* trans. S. Finn (Oxford: Oxford University Press, 1996), 41.

4. Iltis, *Life of Mendel,* 38, 39.

5. When he became Abbot, Mendel provided all three of his sister's nephews with room and board at the monastery while they attended Gymnasium in Brunn, and later he covered the cost for their medical educations at Vienna University (Olby, *Origins of Mendelism,* 120–121).

6. Iltis, *Life of Mendel,* 39.

7. The autobiographical statement in Olby, *Origins of Mendelism,* 177.

8. Orel, *Gregor Mendel,* 52.

9. Iltis, *Life of Mendel,* 47–49, 54, 56.

10. In his autobiography, Mendel wrote that in the monastery "he had been relieved of that anxiety about the physical basis of existence which is so detrimental to study, the respectful undersigned acquired fresh courage and energy . . . His fondness for natural science grew with every fresh opportunity for making himself acquainted with it . . . he has ever since been so much addicted to the study of nature that he would shrink from no exertions which might help him, by further diligence on his own part and by the advice of men who have had practical experience, to fill the gaps in his information" (Olby, *Origins of Mendelism,* 177).

11. Iltis, *Life of Mendel,* 62.

12. Orel, *Gregor Mendel,* 52.

13. Iltis, *Life of Mendel,* 57.

14. Orel, *Gregor Mendel,* 57.

15. Iltis, *Life of Mendel,* 57, 63, 65, 66.

16. Orel, *Gregor Mendel,* 67, 68.

17. Gaertner's book was published in German as *Versuche und Beobachtungen über die Bastarderzeugung im Pflanzenreich* (Olby, *Origins of Mendelism,* 40).

18. H. F. Roberts, *Plant Hybridization before Mendel* (Princeton: Princeton University Press, 1929), 168.

19. Orel, *Gregor Mendel,* 77.

20. In 1820, Goss fertilized a true-breeding blue-seeded parent with a white-yellow-seeded pollen parent and got all white-yellow peas in the pod, a perfectly clear example of dominance. When he grew up these plants, he found some pods with all blue, some with all white, and others with a mixture of white and blue peas. When he separated out the blues, he further observed, he found that they bred true, while the white seeds (the dominants) "yielded some pods with all white and some with both blue and white seeds intermixed." But he was not interested in the uncovering the mechanism of heredity, only in the possibility of making improved varieties by crossings (Roberts, *Plant Hybridization,* 103).

21. Orel, *Gregor Mendel,* 77.

22. Unger rejected the Lamarckian idea that the evolution of species was driven by changing external conditions. He also did not believe that hybridization could account for it.

23. Orel, *Gregor Mendel,* 87. Fenzl was the grandfather of E. von Tschermak, who in 1900 became the third rediscoverer of Mendel's by-then long-forgotten paper.

24. Years later (in 1870), De Vries would be so inspired by this quote that he would include it on the title page of his Ph.D. thesis.

25. R. A. Fisher, "Has Mendel's Work Been Rediscovered?" *Annals of Science* 1 (1936): 133.

26. On average, *A* eggs would be fertilized by *A*- as often as *a*- pollen, and likewise for *a*- eggs.

27. It was critical that the starting strains be constant in order to distinguish between the strain's natural variability and the effect of crossing the strain with another related variety.

28. The seven traits selected were (1) shape of ripe seeds; (2) color of seeds, either yellow or green; (3) seed coat color, which corresponds to flower color; (4) shape of ripe pod; (5) color of unripe pod; (6) difference in position of flower; and (7) length of stem (Curt Stern and Eva R. Sherwood, *The Origin of Genetics: A Mendel Source Book* [San Francisco: W. H. Freeman, 1966], 6–7).

29. The advantage of experimenting with the characteristics of seeds was that one didn't have to actually grow an adult plant to score the phenotype of the offspring.

30. Stern and Sherwood, *The Origin of Genetics,* 9.

31. In modern terminology, the parental generation is called $P_1$, the hybrids offspring are $F_1$ (standing for the first filial generation), and the progeny of the hybrids are $F_2$. This nomenclature was introduced by Bateson in 1902 in order to simplify the discussion of crosses (W. Bateson and E. R. Saunders, *Royal Society Report to the Evolution Committee, Report* [London: Harrison and Sons, 1902], 159–160 n.).

32. Of 565 $F_2$-plants raised from round seeds, 193 were true-breeding for round seeds and 372 yielded a mix of wrinkled and rounds, giving a ratio of constant to hybrid of 1 to 1.93. Of 519 plants grown from yellow seeds, 166 yielded only yellow peas and 353 yielded a mixture of greens and yellows, giving a ratio of 1 to 2.13.

33. In modern textbooks the 3:1 segregation of types is called Mendel's first law.

34. Fisher, "Rediscovered?" 155.

35. In order to "see" the composition of the germ cells, Mendel devised a new genetic technique—the *backcross*—that would become a staple of early twentieth-century genetics. To determine the relative frequencies of the four types of germinal cells, Mendel fertilized the double hybrid *A/a B/b,* which produced egg cells of types *AB, Ab, aB,* and *ab,* with pollen from a pure-breeding *ab* plant. The resulting progeny seed were of genotypes *AaBb, Aabb, aaBb,* and *aabb,* which are *round-yellow, round-green, wrinkled-yellow,* and *wrinkled-green,* respectively, and thus easily distinguished. Likewise, pollen from the double hybrid was dusted on a true-breeding *ab* plant to determine the relative types and frequencies of pollen cells.

36. Fisher, "Rediscovered?" 115–137.

37. To test the purity of a dominant plant (as opposed to seed) character, Mendel collected self-fertilized seeds from it and grew them up. Because space was limited, Mendel sowed ten seeds from each of 100 ($F_2$ dominant) plants. If one or more of the ten progeny plants showed the recessive trait, he assumed the parent plant was a hybrid, and the absence of even a single recessive type among the ten progeny he took as proof that the plant was a true-breeding dominant. Remarkably, Mendel had failed to realize that there was a small but significant likelihood that ten randomly chosen seeds of a hybrid plant would contain no pure recessives among them. In fact, the chances were $(3/4)^{10}$ = .0563 (5.65 percent), or almost 6 in 100. In other words, Mendel's test misclassified 5 to 6 percent of the hybrids as pure-breeding dominants. The correct expectation of heterozygotes to recessives was therefore 1.8874 to 1.1126 rather than 2 to 1.

38. Fisher, "Rediscovered?" 129.

39. Daniel J. Fairbanks and Bryce Rytting, "Mendelian Controversies: A Botanical and Historical Review," *American Journal of Botany* 88 (2001): 741.

40. The reason is that not all of the seven traits would be expected to map to separate chromosomes (there being only seven chromosomes in *Pisum*), and traits mapping to the same chromosome would likely show some degree of linkage.

41. The argument that Mendel pursued an experimental strategy that allowed him to identify seven unlinked traits is presented in Frederico Di Trocchio, "Mendel's Experiments: A Reinterpretation," *Journal of the History of Biology* 24 (1991): 485–519.

42. Letter from Mendel to Naegeli, Dec. 31, 1866, in Stern and Sherwood, *The Origin of Genetics,* 56.

43. Iltis, *Life of Mendel,* 191.

44. Mendel quotes this phrase in his letter to Naegeli of Apr. 18, 1867, in Stern and Sherwood, *The Origin of Genetics,* 63.

45. Letter from Mendel to Naegeli, Apr. 18, 1867, in Stern and Sherwood, *The Origin of Genetics,* 60–71.

46. Stern and Sherwood, *The Origin of Genetics,* 61.

47. Ibid., 63.

48. Ibid., 71

49. Ibid., 79, 81, 89, 90.

50. Long after Mendel's death, the true explanation for the strange behavior of constant-breeding hybrids of hawkweed came to light: Pure species of *Hieracium* were capable of both sexual and asexual reproduction, possessing some eggs that were capable of giving rise to offspring in the normal way by pollination and others that gave rise to genetic clones by simple division. In the process of formation of hybrids, it appeared, plants lost the capacity for sexual reproduction altogether.

51. Orel, *Gregor Mendel,* 260.

52. His associates egged him on (Iltis, *Life of Mendel,* 255).

53. Ibid., 267–268.

54. Ibid., 36–37.

## 6. REDISCOVERY

1. Hugo de Vries, "Das Spaltungsgesetz der Bastarde," *Berichte der deutschen botanischen Gesellschaft* 18 (1900): 83–90; De Vries, "Sur les unités des charactères spécifiques et leur application a l'étude des hybrides," *Review général de botanique* 12

(1900): 257–271; De Vries, "Sur la loi de disjonction des hybrids: Note de M. Hugo de Vries, présentée par M. Gaston Bonnier," *Comptes Rendues Academy of Sciences* (Paris) 130 (1900): 845–847.

2. Robin Marantz Henig, *The Monk in the Garden* (New York: Houghton Mifflin, 2000), 179.

3. C. G. Correns, "G. Mendel's Law concerning the Behavior of Progeny of Varietal Hybrids," trans. Leonie Kellen Piternick, in Curt Stern and Eva R. Sherwood, *The Origin of Genetics: A Mendel Source Book* (San Francisco: W. H. Freeman, 1966), 119.

4. A. H. Sturtevant, *A History of Genetics* (Cold Spring Harbor, N.Y.: Cold Spring Harbor Laboratory Press, 2001), 25; Wilhelm Olbers Focke, *Die pflanzen-mischlinge: Ein beitrag zur biologie der gewächse* (Berlin: Gebrüder Borntrager, 1881). Focke's book was the acknowledged reference on hybridization since its publication in 1881.

5. Hugo de Vries, "Sur les unités des charactères spécifiques et leur application à l'étude des hybrides," *Review général de botanique* 12 (1900): 271.

6. Letter from Correns to Roberts, Jan. 23, 1925, in H. F. Roberts, *Plant Hybridization before Mendel* (Princeton: Princeton University Press 1929), 335.

7. H.-J. Rheinberger, "When Did Carl Correns Read Gregor Mendel's Paper? A Research Note," *Isis* 86 (1995): 612–616.

8. Beatrice Bateson, *William Bateson, F.R.S. Naturalist: His Essays and Addresses Together with a Short Account of His Life* (Cambridge: Cambridge University Press, 1928), 73.

9. Robert Olby points out that, strictly speaking, it was possible that it was Mendel's, not De Vries's, paper that Bateson read on the train. Bateson would have had to have received De Vries's *Berichte* paper no later than Monday, May 7, then read it, noted the reference to Mendel, gone to the Cambridge University library to borrow the Brunn *Verhandlungen* for 1865, and taken it on the train (Robert Olby, "William Bateson's Introduction of Mendelism to England: A Reassessment," *British Journal of the History of Science* 20 [1987]: 399–420).

10. William Bateson, "Societies: Royal Horticultural Lecture," *Gardeners' Chronicle* 3 (1900): 303, quoted in Olby, "William Bateson's Introduction," 401.

11. Hugo de Vries to William Bateson, Oct. 18, 1900, William Bateson Manuscripts, microfilm, American Philosophical Society Library, Philadelphia (hereafter Bateson MSS), reel C, no. 15.

12. Hugo de Vries to William Bateson, Oct. 25, 1900, Bateson MSS, reel C, no. 15.

13. Hugo de Vries, "On Crosses with Dissimilar Heredity," *Journal of the Royal Horticultural Society* 25 (1901): 251 (translated from *Berichte der Deutschen Botanischen Gesellschaft* 18 [1900]: 435).

14. Ibid., 254.

15. As it happened, the *Oenothera* varieties not only failed to obey the standard Mendelian splitting of hybrids, but they also violated the fundamental Mendelian principle of the equivalence of the role of the two sexes in crosses. For example, when *O. lamarckiana* was used as the seed plant and *O. biennis* as the pollen plant, the hybrids were constant, but when the roles of the male and female were reversed (that is, *O. biennis* was pollinated by *O. lamarckiana*) the hybrids split into the twin hybrids. Though it would turn out that De Vries had indeed stumbled on a new principle, neither *Hieracium* nor *Oenothera* operated according to entirely novel laws of heredity. In fact, the explanation for the strange behavior of the *Oenothera* mutants would ultimately be found in an ingenious application of orthodox Mendelism, but it would take another decade before the true nature of the mutants would be revealed by the American geneticist H. J. Muller.

16. De Vries, "On Crosses," 249.

17. William Bateson, "Problems of Heredity as a Subject of Horticultural Investigation," *Journal of the Royal Horticultural Society* 25 (1901): 55–61.

18. B. Bateson, *William Bateson,* 179–180.

19. A. G. Cock, "William Bateson, Mendelism and Biometry," *Journal of the History of Biology* 6 (1973): 4.

20. On Oct. 31, 1901, De Vries wrote Bateson, "I am very anxious to know your doubt as to the purity of Lamarckiana" (Bateson MSS, reel C, no. 15).

21. Bert Theunissen, "Closing the Door on Hugo de Vries' Mendelism," *Annals of Science* 51 (1994): 247.

22. De Vries laid out his hereditary theory in the final hundred pages of the second volume of *Die Mutationstheorie: Versuche und Beobachtungen über die Entstehung von Arten im Pflanzenreich* (Leipzig: Veit, 1901–1903), as well as in a short article in the *Revue genérale de botanique,* the journal in which he'd published one of his three rediscovery papers. Regressive and degressive mutations, which arose from normal and regressive pangenes, respectively, were always paired in the nucleus with their normal counterparts, and as a result they were capable of undergoing normal Mendelian segregation during sex cell formation, but the all-important progressive mutations had no partner with which to pair. It was the species-forming pangene's inability to find a matching partner that disrupted the normal process of segregation. Instead the species-forming pangene underwent a division into two identical copies, which were in turn distributed to the two daughter cells. This, De Vries believed, was the explanation for the constancy of crosses between new mutant races.

23. William Bateson and E. R. Saunders, *Royal Society, Reports to the Evolution Committee, Report 1* (London: Harrison and Sons, 1902), 126.

24. In the second half of the twentieth century, the second syllable, which emphasized the tie with a physical character, is most often dropped, and the variants of a single gene are now simply referred to as *alleles.*

25. Bateson and Saunders, *Report to the Evolution Committee,* 126.

26. Ibid., 81.

27. William Bateson, E. R. Saunders, and M. A. Punnett, *Experimental Studies in the Physiology of Heredity: Reports to the Evolution Committee, Report 2* (London: Harrison and Sons, 1904), 89. Bateson's coupling-repulsion theory was the first formal attempt to explain the failure of certain traits to behave independently. With the acceptance of the chromosome theory of inheritance, Bateson's theory was abandoned and it was recognized that the association between alleles was a function of their distance apart on the same chromosome—the closer together they were located, the more likely they were to be inherited together.

28. W. F. R. Weldon, "Mendel's Laws of Alternative Inheritance in Peas," *Biometrika* 1 (1902): 240, 252.

29. Letter from William Bateson to Karl Pearson, Feb. 10, 1902, Bateson MSS, reel C, no. 18.

30. Karl Pearson to William Bateson, Feb. 15, 1902, Bateson MSS, reel C, no. 18.

31. Attached to letter from Wright to William Bateson, Apr. 28, 1902, Bateson MSS, reel D, no. 26.

32. Karl Pearson, "Mathematical Contributions to the Theory of Evolution: On the Law of Ancestral Heredity," *Proceedings of the Royal Society of London* 62 (1898): 396.

33. Karl Pearson, "Mathematical Contributions to the Theory of Evolution—On the Law of Reversion," *Proceedings of the Royal Society of London* 66 (1900): 140–164.

34. Later that year, Pearson's disciple, Yule, soon showed that in cases of complete dominance Mendel's laws did in fact lead directly to a special case of the law of ancestral inheritance (G. Udny Yule, "Mendel's Laws and Their Probable Relations to Intra-Racial Heredity," *New Phytologist* 1 [1902]: 227).

35. Bateson and Saunders, *Report to the Evolution Committee,* 1, 158.

36. William Bateson, *Mendel's Principles of Heredity: A Defence* (Cambridge: Cambridge University Press, 1903).

37. Yule, "Mendel's Laws," 206.

## 7. MENDEL WARS

1. Walter Frank Raphael Weldon, "On the Ambiguity of Mendel's Categories," *Biometrika* 2 (1902): 45–55.

2. In a binomial distribution with $p = 3/4$ and a sample size of $N = 500$, sigma equals 9.68. Thus the observed result is $|333 - 375| / 9.68 = 4.30$ standard deviations from the mean. For a sample size $N = 153$, sigma equals 5.36 and the observed result is therefore $|99 - 114.75| / 5.36 = 2.94$ standard deviations from the mean.

3. Weldon, "Ambiguity," 53.

4. William Bateson to A. D. Darbishire, Dec. 31, 1902, William Bateson Manuscripts, microfilm, American Philosophical Society Library, Philadelphia (hereafter Bateson MSS), reel D, no. 27.

5. A. D. Darbishire, "Second Report on the Result of Crossing Japanese Waltzing Mice," *Biometrika* 2 (1903): 173.

6. William Bateson, "Mendel's Principles of Heredity in Mice," *Nature* 67 (1903): 462.

7. W. F. R. Weldon, "Mendel's Principles of Heredity in Mice," *Nature* 67 (1903): 512.

8. Bateson, "Mendel's Principles," 585.

9. Weldon, "Mendel's Principles," 610. Bateson's idea that there might be a heterogeneity in gametes caused by secondary segregating factors (later dubbed "modifier genes") would prove correct. See Chapter 10.

10. Ibid., 34.

11. A. D. Darbishire, "Third Report on Hybrids between Waltzing Mice and Albino Races," *Biometrika* 2 (1903): 285.

12. W. E. Castle and G. M. Allen "The Heredity of Albinism," *Proceedings of the American Academy of Arts Sciences* 38 (1903): 613.

13. This was the same point that Weldon had made in regard to the inheritance of shape and color among Mendel's peas in his first critique of Mendelism published in *Biometrika* in early 1902. Again and again Weldon would return to this same point. Bateson, in his *Mendel: A Defence* (Cambridge: Cambridge University Press, 1902), had argued that the failure of hybrids to show absolute dominance of one trait over another had never been an integral part of Mendel's theory, and pointed out that Mendel himself had freely acknowledged that he had focused on traits showing clear dominance only to make his case clearer (see 117). Discussing the nature of dominance, Mendel had observed: "Although the intermediate form of some of the more striking traits, such as those relating to shape and size of leaves, pubescence of individual parts, and so forth, is indeed nearly always seen, in other cases one of the two parental traits is so preponderant that it is difficult, or quite impossible, to detect the other in the hybrid" (8).

14. Letter from A. D. Darbishire to William Bateson, received by Bateson on Oct. 26, 1903, Bateson MSS, reel D, no. 27.

15. A. D. Darbishire, "On the Result of Crossing Japanese Waltzing with Albino Mice," *Biometrika* 3 (1904): 20, 14.

16. Although Bateson had never doubted it, Darbishire's subservient role in the interpretation and analysis of the experiments was confirmed by Pearson, who wrote after Weldon's death, "The work on these mice was for two years 'entrusted' to Mr A. D. Darbishire, but the whole plan for the experiments, the preparation of the correlation tables and the elaborate calculations were Weldon's" (Karl Pearson, "Walter Frank Raphael Weldon, 1860–1906," *Biometrika* 5 [1906]: 41).

17. Darbishire would later reveal that he had wrongly assumed that the pigmented parents were all true hybrids, failing to properly distinguish between homozygous dominants and heterozygotes.

18. William Bateson to C. C. Hurst, Mar. 24, 1904, Bateson MSS, reel D, no. 21.

19. William Bateson to A. D. Darbishire, Mar. 27, 1904, Bateson MSS, reel D, no. 27.

20. A. D. Darbishire to William Bateson, Apr. 1, 1904, Bateson MSS, reel D, no. 27.

21. William Bateson to A. D. Darbishire, Apr. 3, 1904, Bateson MSS, reel D, no. 27.

22. William Bateson to A. D. Darbishire, Apr. 23, 1904, Bateson MSS, reel D, no. 27.

23. This is the only letter in the entire series that Bateson failed to save; however, it is possible to glean its content from Bateson's response.

24. William Bateson to A. D. Darbishire, May 22, 1904, Bateson MSS, reel D, no. 27.

25. A. D. Darbishire to William Bateson, May 27, 1904, Bateson MSS, reel D, no. 27.

26. William Bateson to A. D. Darbishire, May 30, 1904, Bateson MSS, reel D, no. 27.

27. Beatrice Bateson, *William Bateson, F.R.S. Naturalist: His Essays and Addresses Together with a Short Account of His Life* (Cambridge: Cambridge University Press, 1928), 237, 241, 238.

28. Ibid., 240.

29. Pearson, "Weldon," 44.

30. Rona Hurst, "Evolution of Genetics" (unpublished manuscript), 769. For a description, see Rona Hurst, "The Hurst Collection of Genetical Letters," *Mendel Newsletter* 11 (1975): 1.

31. Ibid.

32. Pearson, "Weldon," 44.

33. Stebbings was a natural historian, specializing in crustacea, and was an early follower of Darwin. See *Dictionary of Scientific Biography* 12:8–9.

34. "Zoology at the British Association," *Nature* 70 (1904): 538.

35. R. C. Punnett, "Early Days of Genetics," *Heredity* 4 (1950): 8.

36. K. Pearson and A. Lee, "Mathematical Contributions to the Theory of Evolu-

tion. VIII. On the Inheritance of Characters not Capable of Exact Quantitative Measurement. Part I. Introductory. Part II. On the Inheritance of Coat-Colour in Horses. Part III. On the Inheritance of Eye-Colour in Man," *Philosophical Transactions A* 195 (1900): 79–150.

37. Ibid., 214.

38. Racehorses were divided into six groups—bays, browns, chestnuts, greys, roans, and blacks—by coat color.

39. Karl Pearson, G. U. Yule, Norman Blanchard, and Alice Lee, "The Law of Ancestral Heredity," *Biometrika* 2 (1903): 214–215.

40. Ibid., 215.

41. Hurst, "Genetical Letters," 1.

42. Following his father, in 1891 he joined the Royal Horticulture Society, where he had the opportunity to compare notes with other orchid breeders. Hurst's paper, which followed the inheritance of twenty independent characters in orchids, was precisely the kind of study that Bateson, in particular, had called for in his lecture earlier that afternoon. In addition to observing the fact that some characters were "predominant" over others, Hurst noted that some of his results seemed to defy Galton's ancestral law. Hurst learned of Mendel the following year from De Vries, who wrote asking Hurst to confirm that one of his results was in accord with Mendel's predictions.

43. William Bateson implies this in his letter to C. C. Hurst dated Oct. 22, 1904, and Mrs. Bateson confirms that this was the case in a note on her typescript of the letter from Bateson to Hurst dated Oct. 25, 1904,

44. Geikie to William Bateson, Oct. 21, 1904, Bateson MSS, reel D, no. 21.

45. William Bateson to C. C. Hurst, Oct. 22, 1904, Bateson MSS, reel D, no. 21.

46. C. C. Hurst to William Bateson, Oct. 25, 1904, in Hurst, "Evolution of Genetics," 6.

47. Hurst, "Genetical Letters," 3.

48. William Bateson to C. C. Hurst, Feb. 5, 1904, Bateson MSS, reel D, no. 21.

49. William Bateson to C. C. Hurst, Apr. 4, 1905, Bateson MSS, reel D, no. 21.

50. William Bateson to C. C. Hurst, Apr. 22, 1905, Bateson MSS, reel D, no. 21.

51. William Bateson to C. C. Hurst, Apr. 25, 1905, Bateson MSS, reel D, no. 21.

52. C. C. Hurst to William Bateson, Apr. 25, 1905, Bateson MSS, reel D, no. 21.

53. William Bateson to C. C. Hurst, Sept. 26, 1905, Bateson MSS, reel D, no. 21.

54. C. C. Hurst, "On the Inheritance of Coat Colour in Horses," *Proceedings of the Royal Society of London* 77 (1904): 388–394.

55. William Bateson to C. C. Hurst, Nov. 1, 1904, Bateson MSS, reel D, no. 21.

56. These events are described by Rona Hurst, who married Hurst in June 1906. In her account, Hurst suggested champagne as a "celebration-plus-pickmeup" ("Evolution of Genetics," 868).

57. Ibid., 869.

58. Punnett, "Early Days of Genetics," 8.

59. Hurst, "Evolution of Genetics," 870.

60. Ibid., 868.

61. William Bateson to C. C. Hurst, Dec. 7, 1905, Bateson MSS, reel D, no. 21.

62. C. C. Hurst to William Bateson, Dec. 9, 1905, Bateson MSS, reel D, no. 21.

63. William Bateson to The Secretaries of the Royal Society, Dec. 13, 1905, Bateson MSS, reel D, no. 21.

64. William Bateson to C. C. Hurst, Nov. 1, 1905, Bateson MSS, reel D, no. 21.

65. C. C. Hurst to William Bateson, Dec. 21, 1905, Bateson MSS, reel D, no. 21.

66. Hurst, "Evolution of Genetics," 875.

67. Pearson, "Weldon," 44, 46.

68. Ibid., 45.

69. W. F. R. Weldon, "Notes on the Offspring of Thoroughbred Chestnut Mares," *Proceedings of the Royal Society of London* 77 (1904): 394–398.

70. Pearson, "Weldon," 48.

71. Pearson's comment is quoted in a letter from C. C. Hurst to William Bateson, May 9, 1906, Bateson MSS, reel D, no. 21.

72. William Bateson to C. C. Hurst, May 10, 1906, Bateson MSS, reel D, no. 21.

73. B. Bateson, *William Bateson,* 102.

74. Ibid., 103.

## 8. CELL BIOLOGY

1. Schwann believed that new cells precipitated in the fluid surrounding the cells like salt crystal from a concentrate salt solution, while Schleiden maintained that they were formed inside the mother cell like an embryo gestating in a womb (Bruno Kirch, "Forgotten Leaders in Modern Medicine: Part III. Robert Remak [1815–1865]," *Transactions of the American Philosophical Society* 44 [1954]: 253).

2. J. R. Baker, "The Cell-Theory: A Restatement, History and Critique. Part IV. The Multiplication of Cells," *Quarterly Journal of Microscopical Science* 94 (1953): 435–436.

3. Kirch, "Forgotten Leaders," 260–261, 280–281.

4. Virchow's views on the subject are quoted in Baker, "The Cell Theory," 437.

5. Stephen Jay Gould cites Erik Nordenskioeld's view that Ernst Haeckel was more influential than Darwin in spreading the doctrine of evolution (Stephen Jay Gould, *Ontogeny and Phylogeny* [Cambridge: Harvard University Press, 2002]: 77). For Haeckel's views on race, see Daniel Gasman, "The Scientific Origins of National Socialism" (New York: American Elsevier, 1971), xxv.

6. Ernst Haeckel, *Generelle Morphologie der Organismen* (Berlin: G. Reimer, 1866), 2:287–288.

7. Leopold Auerbach's discoveries were reviewed by the American zoologist E. L. Mark in 1881; see Kirch, "Forgotten Leaders," 311. Mark's pictures inspired Wilson to study chromosomes (T. H. Morgan, "Edmund Beecher Wilson: 1856–1939," *National Academy of Sciences Biographical Memoirs* 22 [1940]: 318).

8. Bruno Kirch, "Forgotten Leaders in Modern Medicine: Part IV. Leopold Auerbach (1828–1897)," *Transactions of the American Philosophical Society* 44 (1954): 311.

9. Ibid., 310. Kirch quotes Oscar Hertwig on the effect Auerbach's paper had on him. He had been on a collecting trip in Corsica with Haeckel when he read the paper. From Corsica he went to Villefranche, on the French Riveria, where he studied the sea urchin *Toxopneustes lividus* (William Coleman, "Cell, Nucleus, and Inheritance: An Historical Study," *Proceedings of the American Philosophical Society* 109 [1965]: 137, 146).

10. Coleman, "Cell, Nucleus, and Inheritance," 137. Several years later, the mysterious rays were understood to be part of the cellular machinery known as the spindle, which distributed the chromosomes to the two daughter cells.

11. A. Schnieder, "Untersuchungen über Plathelminthen, Vierzehnter Bericht der Oberhess," *Gesell. für Nature- u. Heilkunde* (1873): 114 (quoted in German in Coleman, "Cell, Nucleus, and Inheritance," 131), translation by Gerhard Hochschild. See also W. Waldeyer, "Karyokinesis and Its Relation to the Process of Fertilisation," *Quarterly Journal of Microscopical Science* 30 (1889): 159; Waldeyer credits Schnieder with discovering mitosis.

12. Flemming refers to "picric acid" preparations. It was William Perkins who discovered the purple-staining dye (photoscience.la.asu.edu/photosyn/courses/BIO_343/lecture/history.html).

13. J. T. Cunningham, "Review of Recent Researches on Karyokinesis and Cell Division," *Quarterly Journal of Microscopical Research* 22 (1882): 37.

14. In the same year Walther Flemming also introduced the term *chromatin,* for the chromatic dye-absorbing material in the nucleus. Waldeyer coined the term *chromosome* in 1888 (Coleman, "Cell, Nucleus, and Inheritance," 131).

15. Edouard Van Beneden, "Recherches sur la maturation de l'oeuf et la fécondation," *Archive de Biologie* 4 (1883): 265–640.

16. August Weismann, *Essays upon Heredity and Kindred Biological Problems,* ed. Edward B. Poulton, Selmar Schonland, and Arthur E. Shipley (1883; Oxford: Clarendon Press, 1889), 69.

17. August Weismann, "The Continuity of the Germ-plasm as the Foundation of a Theory of Heredity," translated in *Essays upon Heredity,* 161–249.

18. Ibid., 212.

19. August Weismann, "On the Number of Polar Bodies and Their Significance in Heredity," translated in *Essays upon Heredity,* 362.

20. Not only had Galton insisted on the necessity of halving the number of hereditary determinants in the sex cells, but he had insisted that the hereditary material, like Weismann's germ-plasm, passed unaltered from one generation to the next. Weismann's letter acknowledging Galton's priority is reprinted in Karl Pearson, *The Life, Letters and Labours of Francis Galton,* 3 vols. (Cambridge: Cambridge University Press, 1930), 3A:340.

21. Weismann, "On the Number of Polar Bodies," 363.

22. Frederick B. Churchill, "August Weismann and a Break from Tradition," *Journal of the History of Biology* 1 (1968): 112.

23. E. G. Conklin, "Obituary Notice of Members Deceased: August Weismann," *Proceedings of the American Philosophical Society* 54 (1915): viii.

24. J. Bretland Farmer and J. E. S. Moore, "On the Maiotic Phase (Reduction Divisions) in Animals and Plants," *Quarterly Journal of Microscopical Sciences* 48 (1905): 489–557.

25. Frederick B. Churchill, "Hertwig, Weismann, and the Meaning of Reduction Division circa 1890," *Isis* 61 (1970): 432.

26. The characterization of their dinner conversations comes from Fritz Baltzer, who was Boveri's student. Fritz Baltzer, *Theodor Boveri: Life and Work of a Great Biologist, 1862–1915,* trans. Dorothea Rudnick (Berkeley: University of California Press, 1967), 53.

27. Ibid., 72.

28. The idea that the male and female contributed equally to heredity had long been accepted among botanists. Already in 1764 the German plant hybridizer Joseph Köelreuter, who was the first to take a scientific approach to plant crossing, had noted the equivalence of reciprocal crosses and recognized that this implied an equivalence of the fertilizing elements of males and females. But the botanists were way ahead of the zoologists, some of whom still doubted these basic facts as late as 1918 (H. F. Roberts, *Plant Hybridization before Mendel* [Princeton: Princeton University Press, 1929]: 57). Sturtevant discussed the confusion of the zoologists in the lecture

on the history of genetics he gave at Caltech on Apr. 12, 1962 (lectures by A. H. Sturtevant on audiotape, Archives, California Institute of Technology).

29. Baltzer, *Theodor Boveri,* 66.

30. Observation made in 1877 (Thomas Hunt Morgan, *The Development of the Frog's Egg: An Introduction to Experimental Embryology* (New York: Macmillan Co., 1897), 30.

31. We now know that the centrosome is continually shooting out rays (microtubules) in an exploratory fashion in different directions and retracting them in a process that can be compared to a fisherman casting a line. If there is no bite at the end of the line, the line is quickly withdrawn and a new cast is made (Bruce Alberts, Dennis Bray, Alexander Johnson, Julian Lewis, Martin Raff, Keith Roberts, and Peter Walter, *Essential Cell Biology: An Introduction to the Molecular Biology of the Cell* [New York: Garland, 1998], 523).

32. Theodor Boveri, "On Multipolar Mitoses as a Means for the Analysis of the Cell Nucleus," *Verhandlungen der Physikalische-medizinischen Gesellschaft zu Würzburg* 35 (1902): 67–90, translated in B. R. Voeller, ed., *The Chromosome Theory of Inheritance* (New York: Appleton-Century-Crofts, 1968), 93. See also Baltzer, *Theodor Boveri,* 86–95. In "On Multipolar Mitoses" Boveri gives the probability that each of the poles in a quadripolar egg inherits a complete set of chromosomes as $(.4)^9$, which is .026%, not .0026% as he wrote in the text.

33. Boveri, "Multipolar Mitoses," 93.

34. From descriptions of wheat harvests collected by the Kansas State Historical Society.

35. *Walter Stanborough Sutton, April 5, 1877–November 10, 1916* (published by his family, 1917), 57, 66.

36. Elof Axel Carlson, *Mendel's Legacy: The Origin of Classical Genetics* (Cold Spring Harbor, N.Y.: Cold Spring Harbor Laboratory Press, 2004), 80; A. Baxter, "Edmund Beecher Wilson and the Problem of Development: From the Germ Layer Theory to the Chromosome Theory of Inheritance" (Ph.D. diss., Yale University, 1975), 280. Baxter also agrees that it was Henking who named the X. Crow says the X got its name because of McClung's habit of labeling it that way in drawings (Ernest W. Crow and James F. Crow, "100 Years Ago: Walter Sutton and the Chromosome Theory of Heredity," *Genetics* 160 [2002]: 1). McKusick says both McClung and Sutton labeled it X in their drawings, and Wilson coined the name (Victor A. McKusick, "Walter S. Sutton and the Physical Basis of Mendelism," *Bulletin of the History of Medicine* 34 [1960]: 488 n. 1).

37. C. E. McClung, "A Peculiar Nuclear Element in the Male Reproductive Cells of Insects," *Zoölogical Bulletin* 2 (1899): 187.

38. McClung gives a thorough review of the literature on accessory-chromosome-like structures in C. E. McClung, "The Accessory Chromosome: Sex Determinant?" *Biological Bulletin* 3 (1903): 43–84.

39. Clarence McClung, "The Spermatocyte Divisions of the Acrididae," *Kansas University Quarterly* 9 (1900): 95.

40. Walter S. Sutton, "Spermatogonial Divisions in Brachysola Magna," *Kansas University Quarterly* 9 (1900): 135–160.

41. Spermatogonia arise from undifferentiated germ cells. After eight or more generations of division, spermatogonia give rise to *primary spermatocytes,* each of which divides twice to give four *spermatids,* the immediate precursors of the sperm cells (also called *spermatozoa*).

42. The paper was rejected by the *Journal of Morphology.* More than two years after its first submission, the paper was published in *Biological Bulletin.* McClung refers to the long delay in publication in C. E. McClung, "Notes on the Accessory Chromosome," *Anatomischer Anzeiger* 20 (1901–1902): 220–221.

43. C. E. McClung, "The Accessory Chromosome, Sex-Determinant," *Biological Bulletin* 3 (1902): 74–75.

44. Ibid., 73–74.

45. Sutton first reported that the female lubber had 22 chromosomes, and the male 23, in 1902. See Walter S. Sutton, "On the Morphology of the Chromosome Group in Brachystola Magna," *Biological Bulletin* 4 (1902): 35.

46. McClung, "Accessory Chromosome," 82. During the year and a half that had elapsed since the submission of the original manuscript, McClung had also witnessed the splitting of the accessory chromosome into two chromatids, which further strengthened the view that it was present in exactly half the spermatozoa.

47. *Walter Stanborough Sutton,* 61–62.

48. There were two opinions concerning the mechanism of the reduction in number of the chromosomes. Sutton assumed that the homologous chromosomes were joined end to end, which was known as telosynapsis, and that each pair was then cut in half in the next division. The other view was that the homologous chromosomes lay side by side, which was known as parasynthesis, and that the pairs were split lengthwise. It wasn't until 1912 that Wilson, who was by then the leading exponent of the chromosome theory, weighed in favor of the side-by-side conjugation (Wilson, "Chromosomes VIII," *Journal of Experimental Zoology* 13 [1912]: 348–349).

49. Sutton mistakenly believed that the reduction in chromosome number took place in the second meiotic division. Confusion over this point persisted in the lit-

erature until the early 1930s. See Matthew Hegreness and Matthew Meselson, "What Did Sutton See? Thirty Years of Confusion over the Chromosomal Basis of Mendelism," *Genetics* 176 (2007): 1939–1944.

50. *Walter Stanborough Sutton,* 69.

51. Boveri, "On Multipolar Mitoses," 67–90, translated in Voeller, *Chromosome Theory of Inheritance,* 85–94.

52. The deduction that the chromosome pairs that formed just before the reductive divisions were each composed of one chromosome contributed by the female and one from the male was first suggested by Thomas Montgomery in a lecture he gave to the American Philosophical Society in January 1901, and later appeared in print in Montgomery's "A Study of the Chromosomes of the Germ Cells of the Metazoa," *Transactions of the American Philosophical Society* 20 (1902): 222.

53. Walter S. Sutton, "On the Morphology of the Chromosome Group in *Brachystola magna,*" *Biological Bulletin* 4 (1902): 39.

54. C. E. McClung, "A Plain Tale from the Hill," transcript of an unpublished 1927 lecture found in the University of Kansas Medical Center Archives, p. 9.

55. Letter from William Bateson to Beatrice Bateson, Oct. 2, 1902, in Beatrice Bateson, *William Bateson, F.R.S. Naturalist: His Essays and Addresses Together with a Short Account of His Life* (Cambridge: Cambridge University Press, 1928), 81.

56. Walter S. Sutton, "The Chromosomes in Heredity," *Biological Bulletin* 4 (1903): 231–251.

57. Ibid., 240.

## 9. SEX CHROMOSOMES

1. T. H. Morgan, "The Determination of Sex," *Popular Science Monthly* 64 (1903): 107.

2. In 1905 Stevens reported that three generations were involved in the production of sexual females. An apterous parthenogenetic female gave rise to a winged migrant, which was still parthenogenetic, which in turn gave rise to the sexual females (N. M. Stevens, "The Germ Cells of Aphids," *Journal for Experimental Zoology* 2 [1905]: 328).

3. From a letter of recommendation Morgan wrote for Stevens (S. G. Brush, "Nettie M. Stevens," *Isis* 69 [1974]: 171).

4. This is from a copy of Morgan's grant application to the Carnegie Institution dated Nov. 19, 1903 (Brush, "Nettie M. Stevens," 171).

5. T. H. Morgan, "Sex-Determination in Phylloxerans and Aphids," *Journal of Experimental Zoology* 7 (1909): 239–352, here 241.

6. Ibid., 315, 316, 317.

7. Ibid., 317.

8. T. H. Morgan to Charles Zeleny, Jan. 28, 1905, Charles Zeleny Papers, University of Illinois Archives, Urbana (hereafter Zeleny Papers).

9. T. H. Morgan to Hans Driesch, July 25, 1905, Morgan-Driesch Correspondence, Universitäts Bibliothek, Leipzig (microfilm, American Philosophical Society Library; hereafter Morgan-Driesch Correspondence).

10. Ibid.

11. Edmund B. Wilson, "Studies on Chromosomes: 1. The Behavior of the Idiochromosomes in Hemiptera," *Journal for Experimental Zoology* 2 (1905): 371–405.

12. He proposes calling them "sex chromosomes" (Edmund B. Wilson "Studies on Chromosomes: 3. The Sexual Differences of the Chromosome Groups in Hemiptera, with Some Considerations of the Determination and Inheritance of Sex with Six Figures," *Journal for Experimental Zoology* 3 [1906]: 28). The "Y" terminology was introduced in 1909 (Edmund B. Wilson, "The Sex Chromosomes," *Archiv für Mikroskopische* 77 [1909]: 253).

13. Wilson, "Studies on Chromosomes: 1," 402.

14. Ibid., 398.

15. A study of *Anasa* seemed to demonstrate that the spermatogonia cells contained two copies of the accessory chromosome, and that these two chromosomes were fused in a late stage of sperm development to form a single accessory chromosome.

16. Wilson, "Studies on Chromosomes: 1," 398–399.

17. Ibid., 399 n.

18. N. M. Stevens, "Studies in Spermatogenesis with Especial Reference to the Accessory Chromosome," Carnegie Institution, Publication No. 36 (1905), 18.

19. The fact that Wilson read Stevens's unpublished manuscript and then hastened to reexamine the *Anasa* chromosomes was discovered by S. G. Brush by reading the Carnegie archives. In his essay, Brush argued that without the timely push her paper provided, Wilson would have taken longer to move forward in his thinking, and Stevens, not Wilson, would be remembered for the chromosome theory of sex (S. G. Brush, "Nettie M. Stevens," *Isis* 69 [1974]: 163–172).

20. Tragically, Nettie Stevens would die from breast cancer in 1912 at the age of 41. Her contribution to the chromosome theory was effectively minimized, mostly by Morgan in his obituary of Wilson (T. H. Morgan, "Edmund Beecher Wilson: 1856–1939," *National Academy of Sciences Biographical Memoirs* 22 [1940]: 318). But even Morgan,

who would write after her death that she lacked "inspiration" and "a wider vision," grudgingly admitted that Stevens's discoveries represented a "remarkable achievement" (T. H. Morgan "The Scientific Work of Miss N. M. Stevens," *Science* 36 [1912]: 468–470).

21. T. H. Morgan to Driesch, Oct. 23, 1905, Morgan-Driesch Correspondence.

22. The assumption leads to a contradiction as follows: If one assumed that an egg fertilized by an X-containing sperm produced a female because the X contained the female qualities, then half of this female's eggs would inherit the same X. According to Wilson's theory, when those eggs were fertilized by sperm lacking an X, they had to produce males. However, at the same time, it was also the case that these eggs possessed an X chromosome containing female qualities, which should have resulted in them developing as females. The apparent contradiction was later resolved by the suggestion that it was X dosage that determined sex, not the particular determinants carried on the X.

23. T. H. Morgan to Driesch, Oct. 25, 1905, Morgan-Driesch Correspondence.

24. T. H. Morgan, "The Male and Female Eggs of Phylloxerans of the Hickories," *Biological Bulletin* 10 (1906): 201.

25. He described it this way in a letter to Driesch dated Jan. 30, 1909, Morgan-Driesch Correspondence.

26. Morgan, "Male and Female Eggs," 206.

27. In fact, we now know that early development is determined by chromosomes in conjunction with maternally derived cytoplasmic factors.

28. T. H. Morgan, "Zielger's Theory of Sex Determination, and an Alternative Point of View," *Science* 22 (1905): 841.

29. While spermists insisted that the sperm contained the embryo and ovists argued that it was in the ovum, both sides agreed that procreation involved the growth of a preformed miniature human.

30. In 1910, W. E. Castle and C. C. Little would suggest that the yellow gene was a recessive lethal and that this explained why it was not possible to find true-breeding yellows (Castle and Little, "On a Modified Mendelian Ratio among Yellow Mice," *Science* 32 [1910]: 868–870).

31. "Once crossed always mixed" is in Morgan to Davenport, Dec. 12, 1905, and the longer explanation is in Morgan to Davenport, Aug. 20, 1906, Charles B. Davenport Papers, American Philosophical Society.

32. T. H. Morgan, "Sex-Determining Factors in Animals," *Science* 25 (1907): 382–384.

33. At a 1909 meeting of the American Breeders Association in St. Louis, Morgan delivered a paper entitled "What Are 'Factors' in Mendelian Explanations?" in which he challenged the entire Mendelian theory, objecting to the idea that "the tall-factor and the dwarf-factor retire into separate cells after having lived together through countless generations of cells without having produced any influence on each other" (T. H. Morgan, "What Are 'Factors' in Mendelian Explanations?" *Report of the American Breeders' Association,* no. 5 [1909]: 366).

34. T. H. Morgan to Hans Driesch, Sept. 15, 1907, Morgan-Driesch Correspondence.

35. William Bateson to Beatrice Bateson, Aug. 8, 1907, William Bateson Manuscripts, microfilm, American Philosophical Society Library, Philadelphia (hereafter Bateson MSS), reel A, no. 3.

36. Even after Morgan was widely perceived to be the prophet of the new genetics, Bateson was still baffled by Morgan's success, writing to Beatrice, "He is of no considerable account. His range is so dreadfully small. Off the edge of a very narrow track, he is not merely puzzled, but lost utterly . . . he is totally free from pretense—he is almost without shame in his ignorance—I mean of things scientific" (William Bateson to Beatrice Bateson, Dec. 20, 1921, Bateson MSS, reel F, no. 35). Bateson insisted that, intellectually, Morgan was "no size above the ordinary," and he couldn't see "any quality of greatness in him" (William Bateson to Beatrice Bateson, Dec. 24, 1921, Bateson MSS, reel F, no. 35).

37. William Bateson to Beatrice Bateson, Dec. 20, 1921, Bateson MSS, reel F, no. 35.

38. T. H. Morgan to Hans Driesch, Nov. 27, 1908, Morgan-Driesch Correspondence.

39. T. H. Morgan, "The Production of Two Kinds of Spermatozoa in Phylloxerans—Functional 'Female Producing' and Rudimentary Spermatozoa," *Proceedings of the Society for Experimental Biology and Medicine* 3 (1908): 56.

40. Morgan, "Sex-Determination."

41. Ibid., 274–275.

42. Ibid., 275.

43. A. H. Sturtevant characterized Morgan's work on sex determination as "one of his most brilliant achievements" ("Thomas Hunt Morgan," *Biographical Memoirs of the National Academy of Sciences* 33 [195]: 288).

44. Ibid., 290. Morgan also refers to his plan to visit Amsterdam in a letter to Zeleny, Aug. 14, 1902, Zeleny Papers.

45. When Morgan asked his Ph.D. advisor, Brooks, who also doubted the efficacy of selection of small variations, for permission to dedicate the book to him, Brooks

replied with decided indifference, "Do just as you like, Morgan." Morgan was quite hurt but went ahead with the dedication anyway. T. H. Morgan to Ernest Brown Babcock, Dec. 29, 1917, Morgan Letters, American Philosophical Society.

46. In his account of the meeting, De Vries recalled with pleasure how the two had come across a specimen of *Oenothera cheirantifolia* during their hike (Cornelius van Bavel, *Hugo de Vries: Travels of a Dutch Botanist in America, 1904–1912* [Center Point, Tex.: Pecan Valley Press, 2000], 11).

47. T. H. Morgan to Hans Driesch, Dec. 13, 1904, Morgan-Driesch Correspondence.

48. T. H. Morgan to Charles Zeleny, Jan. 28, 1905, Zeleny Papers.

49. T. H. Morgan to Hans Driesch, Apr. 17, 1906, Morgan-Driesch Correspondence.

50. Hugo de Vries, "The Aim of Experimental Evolution," *Carnegie Institution of Washington Yearbook* (1904): 5.

51. T. H. Morgan to Hans Driesch, Jan. 20, 1908, Morgan-Driesch Correspondence.

52. De Vries, "Aim of Experimental Evolution," 48.

53. Sturtevant, "Thomas Hunt Morgan," 293.

54. Ibid., 47.

55. Ross G. Harrison, "Embryology and Its Relations," *Science* 85 (1937): 370.

56. T. H. Morgan, "Hybridization in a Mutating Period in Drosophila," *Proceedings of the Society for Experimental Biology and Medicine* 7 (1910): 160–161.

57. Among 1,237 red-eyed offspring, 3 were now white-eyed males.

58. T. H. Morgan to H. D. Goodale, June 15, 1910, Goodale Papers, American Philosophical Society.

59. T. H. Morgan, "Sex Limited Inheritance in Drosophila," *Science* 32 (1910): 120.

60. Elof Axel Carlson, *Genes, Radiation, and Society: The Life and Work of H. J. Muller* (Ithaca: Cornell University Press, 1981), 52.

61. Nettie Stevens had recently published a paper on *Drosophila* showing that half of the male sperm inherited an X and half inherited no X at all, and it was this difference that determined sex.

62. Morgan, "Sex Limited Inheritance in Drosophila," 120.

63. T. H. Morgan, "The Origin of Nine Wing Mutations in Drosophila," *Science* 33 (1911): 496.

64. A decade later, Hermann Muller, Morgan's most brilliant graduate student, would finally explain the behavior of beaded mutants and in the process show the way to the resolution of the mystery of De Vries's "new species" in *Oenothera.*

65. He refers to it as "S" for "short" in T. H. Morgan, "The Method of Inheritance

of Two Sex-Limited Characters in the Same Animal," *Proceedings for the Society for Experimental Biology and Medicine* 8 (1910): 17–19. Later, "short" was renamed "rudimentary," probably to distinguish it from "miniature"—another wing mutant with very short wings.

66. Ibid., 18. Morgan's confusion resulted from his mistaken assumption that males, like females, must have two copies of each factor. This second copy, he hypothesized, might be found on a chromosome that contained the male determining factors.

67. T. H. Morgan, "The Application of the Conception of Pure Lines to Sex-Limited Inheritance and to Sexual Dimorphism," *American Naturalist* 45 (1911): 76. He mentioned the result in a symposium of the American Society of Naturalists on Dec. 29, 1910, and the paper was published in Feb. 1911.

68. Translated in C. Stern, "Boveri and the Early Days of Genetics," *Nature* 166 (1950): 445.

69. T. H. Morgan, "An Attempt to Analyze the Constitution of the Chromosomes on the Basis of Sex Inheritance in Drosophila," *Journal of Experimental Zoology* 2 (1911): 404. Morgan first presented the data in a public lecture given at Woods Hole on July 7, 1911.

70. Frans Alfons Janssens, "La théorie de la Chiasmatypie: Nouvelle interpretation des cinèses de maturation," *La Céllule* 25 (1909): 389–400.

71. T. H. Morgan, "Random Segregation versus Coupling in Mendelian Inheritance," *Science* 34 (1911): 384.

## 10. THE FLY ROOM

1. C. D. Darlington, "Morgan's Crisis," *Nature,* 278 (1979): 786–787. Theodosius Dobzhansky also quotes Sturtevant in his oral history interview at Columbia ("Reminiscences of Theodosius Dobzhansky," 1962–1963, p. 271, interview, Columbia University Oral History Research Office Collection). In the Dobzhansky-Sturtevant version, Sturtevant alone is said to be Morgan's greatest discovery.

2. A. H. Sturtevant, *A History of Genetics* (Cold Spring Harbor, N.Y.: Cold Spring Harbor Laboratory Press, 2001), 49–50.

3. Muller, in Harriet A. Zuckerman interview, "Reminiscences of H. J. Muller: Oral History, 1964," Nobel Laureates on Scientific Research project, Columbia University Libraries, 10–11.

4. See, for example, Morgan's presentation of the development of the gene map

and the idea of interference. He does not mention either Sturtevant or Muller (Thomas Hunt Morgan, *A Critique of the Theory of Evolution* [Princeton: Princeton University Press, 1916], 137).

5. Zuckerman, "Reminiscences of H. J. Muller," 21.

6. "Reminiscences of Theodosius Dobzhansky," 271.

7. Robert Kohler describes the demoralization of Sturtevant when he was at the California Institute of Technology (Kohler, *Lords of the Fly:* Drosophila *Genetics and the Experimental Life* [Chicago: Chicago University Press, 1994], 122–126).

8. A. H. Sturtevant, "Thomas Hunt Morgan," *Biographical Memoirs of the National Academy of Sciences* 33 (1959): 295.

9. H. J. Muller to A. H. Sturtevant, Apr. 30, 1959, in Muller Manuscripts, Lilly Library, Indiana University (hereafter Muller MSS).

10. H. J. Muller, "Autobiographical Notes," prepared in 1936–37 at N. I. Vavilov's request, p. 1, Muller MSS.

11. Edgar converted Hermann over from tepid Unitarian to confirmed atheist (Elof Axel Carlson, *Genes, Radiation, and Society: The Life and Work of H. J. Muller* [Ithaca: Cornell University Press, 1981], 22).

12. Muller and Altenburg's relationship was described to me by Altenburg's younger son, Lewis, and Helen Muller, Hermann's daughter.

13. From an interview conducted by Elof Carlson in Houston, Nov. 1966, in the Hermann J. Muller Collection, Cold Spring Harbor Laboratory Archives (hereafter Muller Collection). There are now two major Muller collections. The one at the Cold Spring Harbor Laboratory Archives is new. The other is at the Lilly Library at Indiana University.

14. In the preface, Lock explained that the book was meant to be a thorough treatment of the subject of "genetics," the term that had been coined by Bateson, his mentor, the previous year, but that he had settled on the more descriptive "Variation, Heredity and Evolution" for fear that nonspecialists wouldn't know the new word (R. H. Lock, *Recent Progress in the Study of Variation, Heredity and Evolution* [New York: E. P. Dutton and Co., 1907], vii).

15. Muller describes his reaction to reading Lock in autobiographical data he supplied in response to a request from the National Academy of Sciences (Hermann J. Muller, "Autobiographical Data Requested of Hermann Joseph Muller by National Academy," early 1940s, Muller MSS). In his reminiscences, transcribed by Charlotte Auerbach, Muller says that Lock, even more than Wilson, realized the connection between chromosomes and the laws of heredity, even down to the consequence of

multifactor inheritance ("Some Reminiscences of H. J. Muller," autumn 1965, Madison, Wisconsin, notes taken by Charlotte Auerbach, Muller Collection). When Muller finally met William Bateson in New York in 1921, he was dismissive of Bateson and described him as "old-school" and "unjustifiably famous" (H. J. Muller to Miss Jessie Jacobs, Dec. 26, 1921, and June 28, 1922, Muller Collection). But Muller had failed to appreciate that it was Bateson, via his disciple and colleague Robert Lock, who had won him over to the study of genetics in the first place.

16. H. J. Muller, "Obituary Notices: Dr. Calvin B. Bridges," *Nature* 143 (1939): 191–192.

17. R. C. Punnett, *Mendelism* (Cambridge: Macmillan and Bowes, 1905). The story of Sturtevant's early life is given in "Biographical Notes" written by Sturtevant himself (Box 7, Sturtevant Papers, Millikan Library Collection, Archives, California Institute of Technology). See also E. B. Lewis "Remembering Sturtevant," *Genetics* 141 (1995): 1227–30.

18. In the first place, Sturtevant proposed that coat color is determined by a dominant factor for black pigment, which he called the Hurst factor and denoted as *H*. The absence of black (the recessive form), he denoted as *h*. In addition, he introduced a new gene, a second factor, which he called *B* for bay, and he denoted its recessive form (which he believed, following Bateson, to be an absence) as *b*. Horses homozygous for the absence of black—of genotype *hh*—were always chestnut regardless of the bay gene. Bays, on the other hand, contained at least one copy of the dominant Hurst factor (*H*) and one copy of the dominant bay gene (*B*). Browns likewise contained the dominant Hurst gene, but might or might not contain the dominant bay factor, and pure blacks were always of the form *HHbb*. With the introduction of two further gene pairs, *G*/*g* and *R*/*r*, Sturtevant accounted for the complete range of coat colors—chestnut, bay, brown, black, grey, and roan. (A. H. Sturtevant, "On the Inheritance of Coat Color in the American Harness Horse," *Biological Bulletin* 19 [1910]: 204–216.)

19. Muller describes T. H. Morgan's reaction in a lecture entitled "An Episode in Science" he gave at the Biological Laboratory of the Brooklyn Institute, Cold Spring Harbor, on July 25, 1921 (Muller MSS).

20. T. H. Morgan, "Personal Recollections of Calvin B. Bridges," *Journal of Heredity* 30 (1939): 355.

21. T. H. Morgan, "Random Segregation versus Coupling in Mendelian Inheritance," *Science* 34 (1911): 384.

22. A. H. Sturtevant says the idea came to him in the latter half of 1911 (Sturtevant, *A History of Genetics*, 47).

23. Ibid.

24. This memory is Sturtevant's as told to Elof Carlson at California Institute of Technology in the spring of 1967 (Muller Collection).

25. Muller, "An Episode in Science."

26. Muller laid out the early history of the development of gene mapping in a letter to Hugo Iltis (H. J. Muller to Hugo Iltis, Jan. 11, 1951, Muller MSS). He also reviewed the early history of the *Drosophila* lab in a monograph he wrote for Erwin Baur in the late 1920s, which he never completed. This handwritten Baur manuscript is stored at the Lilly Library (Hermann J. Muller, "Baur Ms.," 8 folders, Muller MSS). This history is also discussed by Edgar Altenburg in two letters (Edgar Altenburg to Jack Schultz, June 12, 1961, and Edgar Altenburg to A. H. Sturtevant, June 19, 1940, Muller Collection). Muller reviews the history in a letter to Altenburg (H. J. Muller to Edgar Altenburg, Jan. 6, 1933, Muller MSS).

27. In the effort to explain the coupling and repulsion of alleles, William Bateson and his longtime associate R. C. Punnett formulated the rococo theory of "reduplication" in which the ratios of parental to recombination types were thought to occur in a regular series (Sturtevant, *History of Genetics,* 40; also see Hermann J. Muller, "The Mechanism of Crossing-Over. II: The Manner of Occurrence of Crossing-Over," *American Naturalist* 50 [1916]: 287).

28. Using Muller's distance formula, the distance between two genes was given by the fraction of single crossovers added to twice the fraction of double crossovers.

29. Two markers are defined to be 1 centimorgan apart if they undergo recombination in 1 percent of meioses. Muller's new unit of distance also made it clear that the frequency of double crossovers was suppressed over shorter distances, as if the chromosomes had some rigidity that hindered the occurrence of two breaks. To account for the unexpectedly low number of double crossovers, Muller introduced the concept of "interference," which was a measure of the extent to which the first crossover inhibited a double crossover. He found that there was complete interference (that is, a complete absence of double crossovers) at distances of 10 centimorgans or less, and interference gradually decreased up to 40 map units, where it was zero. By restricting himself to only closely spaced triples (in which the three genes were less than 10 centimorgans apart), Muller showed that "AB + BC = AC" held exactly across the entire chromosome and thus provided the first rigorous proof that the genes were arranged in a line.

30. A. H. Sturtevant, "The Linear Arrangement of Six Sex-Linked Factors in Drosophila, as Shown by Their Mode of Association," *Journal of Experimental Biology* 14 (1913): 45.

31. Muller describes the attitude in the early fly room in his oral history interview with Zuckerman, "Reminiscences of H. J. Muller," 43–44.

32. Muller discusses his disappointment at not being able to begin his fly work in Muller, "Autobiographical Notes," 4; and in Muller, "Autobiographical Data," 3.

33. The fact that Morgan was cut out of his father's will is reported in Ian Shine and Sylvia Wrobel's biography of Morgan, *Thomas Hunt Morgan, Pioneer of Genetics* (Lexington: University of Kentucky Press, 1976), 34. Sturtevant spoke of Morgan's generosity during his senior year in his unpublished 1967 lecture "Reminiscences about Morgan," given at the Marine Biology Laboratory in Woods Hole (Marine Biological Library Archives).

34. Carlson, *Genes, Radiation, and Society*, 57.

35. Muller paints a disparaging picture of Morgan's thinking in Muller, "Autobiographical Notes," 3–4.

36. Ibid., 5.

37. Examples of genes that affect more than one character include *rudimentary, bifid, shifted, notch, club, dachs, morula, bent, furrowed, deformed,* and *dichaete* (Muller, "Baur Ms.," folder 4, sec. 3, "On the Relation between Phenotype and Genotype," draft B, p. 183′, Muller MSS.)

38. H. J. Muller, "Erroneous Assumptions regarding Genes," 22–23, unpublished manuscript (1911–12), Muller MSS. See also H. J. Muller, "Some Genetic Aspects of Sex," *American Naturalists* 66 (1932): 136–138.

39. Likewise it was found that pink reverted to peach, and ebony to sooty (Muller, "Baur Ms.," folder 4, sec. 3, draft B, p. 184′, Muller MSS).

40. The idea that a recessive phenotype was caused by the loss of a factor was the basic tenet of Bateson's so-called presence-absence theory. The presence-absence theory was discussed in Bateson's "Presidential Address," given to the Zoological Section of the British Association in 1904 and further elaborated on in his 1908 lecture entitled "The Methods and Scope of Genetics." See Beatrice Bateson, *William Bateson, F.R.S. Naturalist: His Essays and Addresses Together with a Short Account of His Life* (Cambridge: Cambridge University Press, 1928), 249, 320.

41. Muller described this history in his public lecture at Cold Spring Harbor, "An Episode in Science," 13.

42. Altenburg describes his place in the lab in his interview with Carlson, Muller Collection. His dates in the lab are given in the preface of his textbook, *Genetics* (New York: Henry Holt, 1945), vi. Muller confirms that Altenburg was third in the lab in H. J. Muller to A. H. Sturtevant, Apr. 30, 1959, Muller MSS.

43. Muller, "Autobiographical Notes," 5.

44. The full theory, including the terms *genotype* and *phenotype*, was first presented in Johannsen's 1909 monograph *Elemente der Exakten Erblichkeitslehre* (Jena: G. Fischer, 1909); on pp. 124–125 he first introduced the term *gen*, which he proposed to replace the well-known *pangene*. Johannsen mistakenly credited Darwin with coining the term *pangene*, which had actually been coined by De Vries in his 1889 *Intracellular Pangenesis*.

45. To see why self-fertilization results in homozygosity, consider a plant that is heterozygous (genotype *A/a*) for some gene. In the first generation, the *A/a* genotype gives rise to *A/A*, *A/a*, and *a/a* progeny in the ratio of 1:2:1. Thus, half the progeny are homozygous, *A/A* or *a/a*, and these plants will grow up and breed true. In each succeeding generation, the number of heterozygotes is similarly cut in half.

46. Johannsen coined the term *phenotype* in his 1903 study of inheritance, "Heredity in Populations and Pure Lines" (a translation can be found in James A. Peters, ed., *Classic Papers in Genetics* [Englewood Cliffs, NJ: Prentice-Hall, 1959], 20–27). Though he came to a radically different conclusion as to the significance of regression, he nonetheless dedicated the book to Galton, "the creator of the study of exact heredity."

47. E. M. East, "The Mendelian Notation as a Description of Physiological Facts," *American Naturalist* 46 (1912): 634.

48. This description of Altenburg's experiments can be found in Muller, "Baur Ms.," folder 2, sec. 3, "On the Relation between Phenotype and Genotype," p. 179, Muller MSS.

49. Just as in the case of a single segregating trait, the progeny of a strain that differed in two closely linked genes would be expected to give a 1:2:1 ratio of types. If, for example, the two parental chromosomes were of type *Ab* and *aB*, then the ratio of the three classes of progeny would be 1 *Ab/Ab* to 2 *Ab/aB* to 1 *aB/aB*. In addition there would be a small number of crossover types—*AB/aB*, *AB/Ab*, *ab/Ab*, *ab/aB*—and an even smaller number of types *ab/AB*, *ab/ab*, *AB/AB* arising from the mating of two recombinants.

In contrast, as Carl Correns had been the first to point out, two unlinked traits would be expected to produce equal numbers of the four classes of gametes—*AB*, *Ab*, *aB*, and *ab*—among the sex cells of hybrids, which would in turn give rise to a 9:3:3:1 ratio of double dominants, to those who were dominant for one or the other trait, to double recessives in the next generation. The 9:3:3:1 ratio is most easily seen by considering Reginald Punnett's famous square, which lays out the sixteen combinations of factors that can arise from the joining of the four possible germ cells at random. A translation of Corren's paper can be found in Curt Stern and Eva R. Sherwood, *The Origin of Genetics: A Mendel Source Book* (San

Francisco: W. H. Freeman and Co., 1966), 130. Punnett's square is presented in R. C. Punnett, *Mendelism* (Cambridge: Bowes and Bowes, 1907), 45.

50. Muller, "Baur Ms.," folder 2, p. 182, Muller MSS.

51. Appropriately marked truncate males were constructed as follows. An "extreme" truncate female was crossed to a normal-winged male that was homozygous for *black*—a recessive mutation located on chromosome 2 that turns ordinarily grey-bodied flies black—as well as for *pink*, which is located on chromosome 3. Because the truncate mother contained only normal grey and red alleles, one could be sure that the truncate male progeny were heterozygous for both markers (ibid., 183–184).

52. Ibid., 184.

53. The fluctuation experiment was done by comparing grey-red males (that is, males that had inherited their first, second, and third chromosomes from their truncate mothers) to their grey-red sons, who could be inferred to contain the same set of chromosomes (ibid., 186).

54. Elof Carlson's interview with Altenburg, Muller Collection.

55. In his oral history interview with Zuckerman, "Reminiscences," p. 5, Muller said that Morgan had invited Altenburg to join in writing *The Mechanism of Mendelian Heredity*.

56. Edgar Altenburg and H. J. Muller, "The Genetic Basis of Truncate Wings—An Inconstant and Modifiable Character in Drosophila," *Genetics* 5 (1920): 1–59.

57. From the notes taken by Elof Carlson during his interview with Altenburg on Nov. 28, 1966, in Houston, Muller Collection.

58. Altenburg and Muller were the first to identify the action of modifying genes (B. Glass, "H. J. Muller: A Definitive Biography," *Quarterly Review of Biology* 56 [1981]: 447).

59. Altenburg and Muller, "Truncate Wings," 51.

60. H. J. Muller to Edgar Altenburg, Mar. 5, 1946, Muller MSS.

## 11. *OENOTHERA* RECONSIDERED

1. In a letter to Elof Carlson, Julian Huxley fondly recalled his time with Muller in Texas, describing their mutual fascination with a colony of leaf-cutting ants (Julian Huxley to Elof Carlson, Jan. 6, 1972, Herman J. Muller Collection, Cold Spring Harbor Laboratory Archives [hereafter Muller Collection]).

2. Hermann J. Muller, "Baur Ms.," folder 2, p. 179, in Muller Manuscripts, Lilly Library, Indiana University (hereafter Muller MSS).

3. Ibid., 189–190.

4. John S. Dexter, "The Analysis of a Case of Continuous Variation in Drosophila by a Study of Its Linkage Relations," *American Naturalist* 48 (1914): 719–721.

5. Muller first described the beaded experiments in Hermann J. Muller, "Genetic Variability, Twin Hybrids and Constant Hybrids, in a Case of Balanced Lethal Factors," *Genetics* 3 (1918): 429. See also Muller, "Baur Ms.," folder 2, p. 191, Muller MSS.

6. Later Sturtevant would show that the suppression of crossing over was actually due, not to a new genetic factor, but to an inversion of the gene order in a small segment of one copy of a pair of homologous chromosomes that prevented the two homologous chromosomes from making a proper "synaptic" connection. Muller then confirmed Sturtevant's interpretation by showing that crossing over was normal in a strain that was homozygous for the third chromosome inversion. The history of suppressors is given in Muller's "Balanced Lethal Factors," 430–431.

7. The actual construction of the $C_b'b_d/D_f'$ heterozygote is described in ibid., 429–434.

8. The cross of two heterozygotes, each of genotype $C_b'/D_f'$, gives three classes of progeny: homozygotes with deformed eyes ($D_f'/D_f'$), heterozygotes with deformed eyes ($C_b'/D_f'$), and homozygotes with normal eyes ($C_b'/C_b'$).

9. The introduction of a recessive mutation $s_s$ for spineless into the chromosome containing $D_f'$ made it possible to distinguish the homozygote ($D_f's_s/D_f's_s$) from the heterozygotes ($C_b'/D_f's_s$) by the presence or absence of spinelessness.

10. For his theory, it would suffice to assume that there was a factor in the $D_f'$-containing chromosome that was lethal when homozygous, as Muller pointed out in "Balanced Lethal Factors," 434.

11. In fact, of the seventeen known dominant mutations, five of them were lethal when homozygous. To Muller, who had long believed that mutations were overwhelming deleterious in their effects, it was fitting that a dominant mutation, which was, by definition, more pronounced in its effects than the wild-type gene, would tend to be even more deleterious than a recessive mutation and was likely often to be lethal in its effect. (Ibid., 463.)

12. The balancing lethal isolated in this experiment was in fact the first example of a recessive lethal that had been isolated on a non-sex chromosome, but Muller believed that because they escaped detection recessive lethals were far more frequent than commonly believed (H. J. Muller, "An Oenothera-Like Case in Drosophila," *Proceedings of the National Academy of Sciences of the United States of America* 3 [1917]: 624). The following year Altenburg would prove him right by demonstrating that nearly one in fifty flies per generation showed a spontaneous recessive X-linked lethal, a frequency far higher than he or Muller had ever suspected.

13. If a new lethal arose in the grandparental generation and was inherited by two flies that were chosen to serve as the parents of a new generation, the bottle containing those two parents would contain an unusually low percentage of normals, and progeny from this bottle would therefore be chosen over other progeny in other bottles to continue the experiment.

14. Muller believed that some or all of the early impure beaded stock carried C′ and that the balancing lethal had appeared a considerable time after the selection experiments on beaded had begun (Muller, "Balanced Lethal Factors," 451–452).

15. The recombination rate was a function of the distance between the beaded and the lethal, which turned out to be 10 centimorgans.

16. "There can be no doubt that its occurrence in *O. Lamarckiana* is intimately connected with at least part of the special manner in which this species displays its mutability," De Vries wrote in regard to the splitting of the $F_1$ hybrids (Hugo de Vries, "On Twin Hybrids," *Botanical Gazette* 44 [1907]: 406).

17. Ralph E. Cleland, *Oenothera: Cytogenetics and Evolution* (Bloomington, Ind.: Academic Press, 1972), 8.

18. De Vries, "On Twin Hybrids," 401.

19. There was one notable exception: when *O. lamarckiana* females were fertilized by *O. biennis* pollen, the progeny were uniformly of the laeta type. The problem in this case, De Vries proposed, was that the *biennis* parent had lost the power to induce the labile laeta pangene to mutate. The reciprocal cross, in which the *O. biennis* eggs were fertilized by *O. lamarckiana* pollen, gave twins, which could be explained under De Vries's scheme on the assumption that the *O. biennis* contained a labile laeta pangene that could be induced to mutate to veluntina form by *O. lamarckiana.* (Cleland, *Oenothera,* 16 n.)

20. Ibid., 33.

21. Some *Oenothera* crosses showed an additional twist that distinguished them from those of the beaded flies. Unlike the flies in which the male and female roles were for all intents and purposes interchangeable, the results of the *Oenothera* crosses often were different according to which species provided the egg and which the pollen. This too, Muller suggested, could be accounted by a form of the balanced lethal idea, only in this case the lethal factors had to be envisioned to act on the germ cells—selectively killing pollen or ova that carried one or the other gene complex—rather than on the fertilized egg (the zygote). De Vries himself had proposed a similar theory in 1911 to explain his double reciprocal crosses, but he had refused to acknowledge that a similar mechanism might have been at work in all the crosses (Muller, "Balanced Lethal Factors," 471).

22. Hugo de Vries, "Mass Mutations and Twin Hybrids of *O. grandiflora ait.*," *Botanical Gazette* 65 (1918): 415.

23. Hugo de Vries to Jacques Loeb, June 29, 1919, Jacques Loeb Papers, Manuscript Division, Library of Congress (hereafter Loeb Papers).

24. Despite the existence of the crossover-suppressing factor, C′, a gene distantly located from C′ might break its bond to the lethal mutation and cross over to the homologous chromosome, and when this happened, a long latent recessive could suddenly spring out into the open. To illustrate the idea, Muller crossed four visible recessive mutations—sepia eye ($s_e$), spineless ($s_s$), kidney-shaped eye (k), and rough eye ($r_o$) into the pure-breeding beaded strain (containing the balanced lethals $B_d$ and l and the crossover suppressor C′) to create flies of genotype $s_e$ $s_s$ k $r_o$ $B_d$ /l C′ and allowed them to interbreed. Even in the presence of the crossover suppressor C′, a small number (2 in 1,000) of crossovers on the extreme left, between $s_e$ and $s_s$, gave rise to two new recombinant chromosomes: $s_s$ k $r_o$ $B_d$ and $s_e$ l C′. Every now and then one of the rare recombinant eggs was fertilized by a sperm containing the multiply-marked beaded chromosome giving an offspring with sepia eyes, as shown in the cross illustrated here.

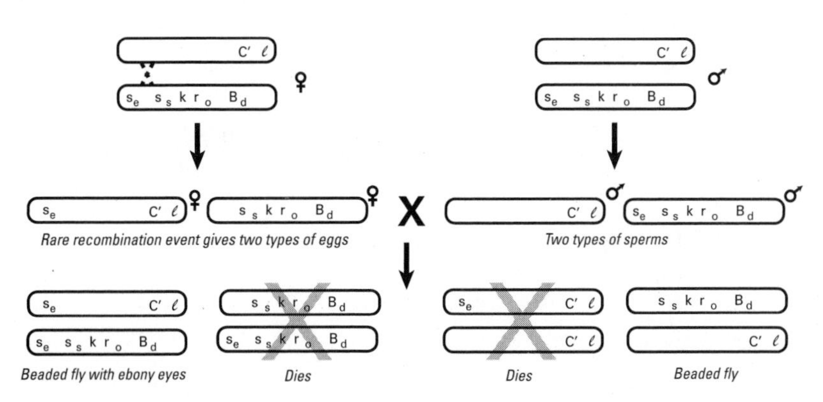


**The breaking free of recessive mutations.**

Illustration by Tamara L. Clark.

25. Hugo de Vries to Jacques Loeb, Mar. 17, 1919, Loeb Papers.

26. Cleland, *Oenothera,* 198.

27. Ibid., 216.

28. B. M. Davis, "Genetical Studies on *Oenothera.* II. Some Hybrids of *Oenothera biennis* and *O. grandiflora* that Resemble *O. lamarckiana,*" *American Naturalist* 45 (1911): 228–230.

29. B. M. Davis, "Genetical Studies on *Oenothera.* III. Further Hybrids of *Oenothera biennis* and *O. grandiflora* that Resemble *O. lamarckiana,*" *American Naturalist* 46 (1912): 377.

30. B. M. Davis, "Was Lamarck's Evening Primrose (*Oenothera lamarckiana seringe*) a Form of *Oenothera grandiflora solander*?" *Bulletin of the Torrey Botanical Club* 39 (1912): 532. See also Hugo de Vries, "The Probable Origin of *Oenothera lamarckiana* Ser.," *Botanical Gazette* 57 (1914): 359.

31. B. M. Davis, "Professor De Vries on the Probable Origin of *Oenothera lamarckiana,*" *American Naturalist* 49 (1915): 63. See also Bradley Moore Davis, "Species, Pure and Impure," *Science* 55 (1922): 109.

32. Cleland, *Oenothera,* 232–235.

33. Ibid., 239.

34. Ibid., 224–225, 232.

## 12. X-RAYS

1. H. J. Muller to Julian Huxley, Dec. 13, 1919, in Muller Manuscripts, Lilly Library, Indiana University (hereafter Muller MSS).

2. H. J. Muller, "Revelations of Biology and Their Significance," address to the Peithologian Society, Mar. 24, 1910, p. 22, Muller MSS.

3. In 1966 Sturtevant claimed credit for the suggestion that lethals could be used to measure the frequency of mutation. He mentions his crucial contribution for the first time forty years after the fact. It is ironic that Sturtevant claims credit for such an important idea in a passage purporting to show the cooperative nature of the work in the fly room (A. F. Sturtevant, *History of Genetics* [Cold Spring Harbor, N.Y.: Cold Spring Harbor Press, 2001], 50). Muller says it first occurred to him in 1912 that many, if not most, mutations would be lethals (H. J. Muller, "Genetic Variability, Twin Hybrids and Constant Hybrids, in a Case of Balanced Lethal Factors," *Genetics* 3 [1918]: 464).

4. Muller, "Balanced Lethal Factors," 464–465.

5. There is a probability of one-half that the inseminating sperm carries a Y chromosome, and likewise a probability of one-half that it carries an X chromosome.

6. The presence of a new lethal is revealed by the absence of certain non-crossover classes among the $F_1$ sons. If, for example, a new lethal arose between

vermillion ("v") and forked ("f") on a female X chromosome, one would never see vermillion-eyed sons with forked bristles, as is illustrated in the diagram below.

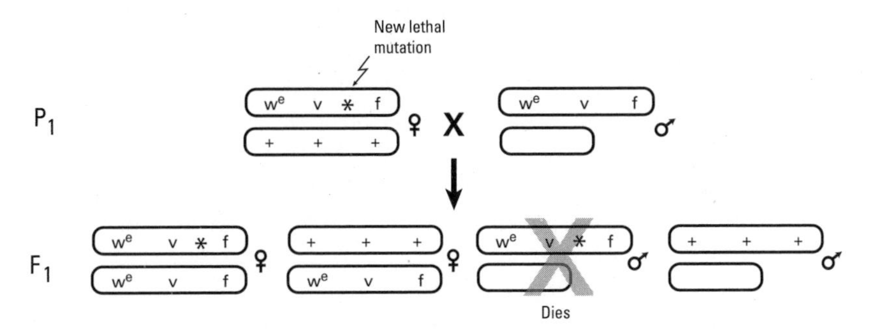


**Mapping lethals.**

Illustration by Tamara L. Clark.

7. Each of fifteen students performed ten crosses, and each of the crosses was carried through three generations. The first generation was a control cross, used to weed out any preexisting X-linked lethals in the mother. Once it was determined that the grandmother did not contain any lethals, two further generations were bred, mother and daughter. In this process four individual X chromosomes (two from the mother, and two from the granddaughter) were tested for the presence of newly created lethal mutations. Thus four X's were examined per mating, and each of fifteen students performed ten matings, so 600 X chromosomes were examined in total (H. J. Muller, "The Measurement of Gene Mutation Rate in Drosophila, Its High Variability, and Its Dependence upon Temperature," *Genetics* 13 [1928]: 306).

8. H. J. Muller, "The Problem of Genic Modification," paper read at the Fifth International Genetics Congress, Berlin, and published in *Verhandlungen des V. Internationalen Kongresses für Vererbungswissenschaft: Supplementband I der Zeitschrift für inductive Abstammungs- und Vererbungslehre* (1928): 235.

9. Muller, "Mutation Rate in Drosophila," 311–312.

10. H. J. Muller and Edgar Altenburg, "The Rate of Change of Hereditary Factors in Drosophila," *Proceedings of the Society for Experimental Biology and Medicine* 17 (1919): 12.

11. H. J. Muller to Edgar Altenburg, Sept. 10, 1919, Muller MSS.

12. Muller, "Genic Modification," 235.

13. Ibid., 318.

14. The *Bar*-containing copy of the mother's X also contained four recessive visible mutations. Ordinarily the parental type and the various recombinant types would be visible among the sons, but in this case only the parental type appeared.

15. C was for the crossover-suppressing function (after the already known crossover suppressors) and *l* for the associated lethality.

16. Elof Carlson's 1967 interview with Edgar Altenburg, in the Hermann J. Muller Collection, Cold Spring Harbor Laboratory Archives (hereafter Muller Collection).

17. A. H. Sturtevant to Tracy Sonnenborn, May 5, 1967, Muller Collection.

18. John A. Moore spoke of Sturtevant's resentment in his interview with Carlson on May 15, 1971, Muller Collection. According to Wilson's records for his graduate zoology class, Muller scored highest on the final with a 98, Sturtevant was in the middle with 80, and Bridges was near the bottom with 65. In his comments, Wilson described Muller as "A.1," Bridges as "very good," and Sturtevant as "fair." See Wilson's private journal, 1903–1928, 4th volume of four notebooks, E. B. Wilson Papers, American Philosophical Society Library.

19. H. J. Muller to Julian Huxley, Nov. 22, 1920, Muller MSS.

20. Bentley Glass, "Theophilus Shickel Painter: August 22, 1889–October 5, 1969," *Biographical Memoirs of the National Academy of Sciences* 59 (1990): 313–314.

21. Muller, "Mutation Rate in Drosophila," 322.

22. H. J. Muller, "Mutation," *Eugenics, Genetics and the Family* 1 (1923): 106–112.

23. He used the fact that a lethal arose once in each 106 chromosome generations, and the average length of a fly generation was 2 weeks. So that a lethal arose once in 212 weeks = 4 years. Assuming there were no fewer than 500 genes on each of the three large chromosomes, this implied that each individual factor must, on average, undergo mutation once every 2,000 years. The calculation appeared in Muller and Altenburg, "Rate of Change of Hereditary Factors," 4–5.

24. Muller, "Mutation," 108, 110–111.

25. H. J. Muller, "Further Changes in the White-Eye Series of Drosophila and Their Bearing on the Manner of Occurrence of Mutation," *Journal of Experimental Zoology* 31 (1920): 443–473. See also R. R. Hyde, "Two New Members of a Sex-linked Multiple (Sextuple) Allelomorph System," *Genetics* 1 (1916): 535.

26. To Muller and Altenburg, however, the evidence for the facts of linear linkage—that the sum rule (AB + BC = AC) held for any three closely spaced genes and that the relative lengths of the chromosomes as determined by the crossover frequencies matched the relative lengths of the cytological chromosomes—had al-

ready clinched the case for the chromosome theory long before the discovery of the nondisjunctional flies.

27. W. Bateson, "The Progress of Genetics since the Rediscovery of Mendel's Papers," *Progressus Rei Botanicae* 1 (1907): 408; cited by William Coleman, "Bateson and Chromosome: Conservative Thought in Science," *Centaurus* 15 (1970): 255.

28. William Bateson to Beatrice Bateson, Dec. 26, 1921, in William Bateson Manuscripts, microfilm, American Philosophical Society Library, Philadelphia (hereafter Bateson MSS), reel F, no. 35.

29. Beatrice Bateson to William Bateson, Jan. 1, 1922, Bateson MSS, reel F, no. 35.

30. William Bateson, "Evolutionary Faith and Modern Doubts," *Science* (1922): 56.

31. A. G. Cock, "Bateson's Two Toronto Addresses, 1921: 1. Chromosomal Skepticism," *Journal of Heredity* 80 (1989): 93.

32. H. J. Muller, "Variation due to Change in the Individual Gene," *American Naturalist* 56 (1922): 35.

33. H. J. Muller, "The Gene as the Basis of Life," *Proceedings of the International Congress of Plant Sciences* 1 (1929): 918; presented before the International Congress of Plant Sciences, Section of Genetics, symposium "The Gene," Ithaca, N.Y., Aug. 19, 1926.

34. Although the original idea appears already in his 1921 lecture, Muller fleshed out the model in 1937 (H. J. Muller, "Physics in the Attack on the Fundamental Problems of Genetics," *Scientific Monthly* 44 [1937]: 212).

35. Thirty years later, James Watson, guided by Muller's ideas about the gene, which he had absorbed during two year-long courses at the Indiana University in Bloomington, would set out to find a molecule with the properties that Muller said a gene must have.

36. Muller, "Variation Due to Change," 49.

37. H. J. Muller, "The Gene Material as the Initiator and the Organizing Basis of Life," in A. Brink and E. D. Styles, eds., *Heritage from Mendel* (Madison: Wisconsin University Press, 1966), 441.

38. H. J. Muller to Jessie Jacobs, June 17, 1922, Muller Collection.

39. Ibid.

40. Jessie Jacobs to Carlos Offerman, June 2, 1936, Muller Collection.

41. H. J. Muller to Jessie Jacobs, June 28, 1922, and Muller to Jacobs, July 6, 1922, Muller Collection.

42. H. J. Muller to Jessie Jacobs, Aug. 3, 1922, Muller Collection.

43. Letters from H. J. Muller to Jessie Jacobs dated July 22, 1922, and Aug. 10, 1922, Muller Collection.

44. "Most of the scientific workers lost about 30 pounds in weight" (H. J. Muller,

"Observations of Biological Science in Russia," *Scientific Monthly* 16 [1923]: 541). Dobzhansky also describes the summer of 1920 when he was starving (Theodosius Grigorievich Dobzhansky, *Dobzhansky's Genetics of Natural Populations,* chaps. 1–43, ed. R. C. Lewontin [New York: Columbia University Press, 1981], pp. 8–10).

45. Muller describes this work in an article he wrote upon his return, "Biological Science in Russia," 540. Two of Serebrovsky's students, Agol and Levitt, would win Rockefeller Fellowships to work with Muller in Texas in 1930.

46. William Bateson to Beatrice Bateson, Sept. 9, 1925, Bateson MSS, reel F, no. 35.

47. N. I. Vavilov to William Bateson, May 10, 1922, Bateson MSS, reel F, no. 43. The fact that Vavilov lived in his office can be found in Th. Dobzhansky, "N. I. Vavilov: A Martyr of Genetics, 1887–1942," *Journal of Heredity* 38 (1947): 229.

48. With his "Law of Parallel Variations," Vavilov believed he could determine the geographic origin of certain cereals (Muller, "Biological Science in Russia," 549).

49. Letter from H. J. Muller to O. Mohr, Mar. 24, 1925, Muller MSS.

50. Harold H. Plough, "Radium Radiations and Crossing Over," *American Naturalist* 58 (1924): 85–87.

51. J. W. Mavor, "An Effect of X Rays on the Linkage of Mendelian Characters in the First Chromosome of Drosophila," *Genetics* 8 (1923): 355.

52. H. J. Muller to Edgar Altenburg, Oct. 22, 1924, Muller MSS. Letter must be misdated because he proposed that Edgar visit not just him and Jessie, but also David, who was not yet born.

53. Ibid.

54. H. J. Muller to Edgar Altenburg, Nov. 12, 1924, Muller MSS.

55. Following Mavor, a radiologist at Union College in Schenectady, who had found that a machine that produced X-rays accelerated the rate of exchange in the X chromosomes and decreased the rate of exchange in the 2nd chromosome (Mavor, "Effect of X Rays," 355).

56. Robert P. Wagner and James F. Crow, "The Other Fly Room: J. T. Patterson and Texas Genetics," *Genetics* 157 (2001): 1.

57. Muller, "Mutation Rate in Drosophila," 343–344.

58. Ibid., 344.

59. When males were X-rayed, only those chromosomes derived from the male parent were mutated, and likewise only the chromosomes derived from the female parent showed new lethals when the female was X-rayed.

60. Male flies whose sperm had been X-rayed were mated in individual vials to $C\ell B$-females and grown up for two generations. As can be seen by considering the fig-

ure illustrating the CℓB-method for detecting lethals, any nonlethal mutation will appear in half of the $F_2$ generation. See also Muller, "Genic Modification," 242.

61. James F. Crow, "Seventy Years Ago: Mutation Becomes Experimental," *Genetics* 147 (1997): 1491.

62. H. J. Muller, "The Method of Evolution," *Scientific Monthly* 29 (1929): 492.

63. Elof Axel Carlson, *Genes, Radiation, and Society: The Life and Work of H. J. Muller* (Ithaca: Cornell University Press, 1981), 148. Dobzhansky confirms that Morgan was highly skeptical of Muller's X-ray results and openly derisive of him when the discovery was first announced ("Reminiscences of Theodosius Dobzhansky" [1962–1963], 281–282, interview, Columbia University Oral History Research Office Collection).

64. H. J. Muller to Hartman, Oct. 28, 1927, Muller Collection.

65. Curt Stern, "Domain of Genetics," *Genetics* 78 (1974): 29.

## 13. TRIUMPH OF THE MODERN GENE

1. H. J. Muller, "Effects of X-radiation on Genes and Chromosomes," *Anatomical Record* 37 (1927): 174.

2. He found the first translocation in December 1926 and reported on it in detail at meetings in Nashville, Tennessee, the following year (Theophilus Shickel Painter and Hermann J. Muller, "Parallel Cytology and Genetics of Induced Translocations and Deletions in Drosophila," *Journal of Heredity* 20 [1929]: 288). He also discusses the history of the transposition work in H. J. Muller, "The Production of Mutations by X-Rays," *Proceedings of the National Academy of Sciences* 14 (1928): 714.

3. Painter and Muller, "Parallel Cytology and Genetics," 288.

4. Muller, "Mutations by X-Rays," 720. See also H. J. Muller and Edgar Altenburg, "Chromosome Translocations Produced by X-Rays in Drosophila," *Anatomical Record* 41 (1928): 100.

5. H. J. Muller to Edgar Altenburg, July 14, 1928, in Muller Manuscripts, Lilly Library, Indiana University (hereafter Muller MSS).

6. "Reminiscences of Theodosius Dobzhansky" (1962–1963), p. 295, interview, Columbia University Oral History Research Office Collection.

7. In his letter to Calvin Bridges of Oct. 24, 1928, Muller says that Edgar and Painter had been working on translocation for nearly a year (Hermann J. Muller Collection, Cold Spring Harbor Laboratory Archives; hereafter Muller Collection).

8. B. Glass, "Theophilus Shickel Painter: August 22, 1889–October 5, 1969," *Biographical Memoirs of the National Academy of Sciences* 59 (1990): 319.

9. H. J. Muller to Curt Stern, Oct. 3, 1929, Curt Stern Papers, American Philosophical Society.

10. This case, which was called the "Star-Curly" transposition, is described in Painter and Muller, "Parallel Cytology and Genetics," 298.

11. H. J. Muller and T. S. Painter, "The Cytological Expression of Changes in Gene Alignment Produced by X-Rays in Drosophila," *American Naturalist* 63 (1929): 198.

12. Ibid., 193.

13. Painter and Muller, "Parallel Cytology and Genetics," 298.

14. Hermann Muller to Jessie Muller, Aug. 29, 1930, Muller Collection.

15. David Muller, personal communication, Feb. 10, 2004.

16. H. J. Muller to Carlos Offerman, July 2, 1931, Muller Collection.

17. Jessie Muller to Carlos Offerman, June 26, 1931, Muller Collection.

18. H. J. Muller to Jessie Muller, Jan. 22, 1933, Muller Collection.

19. H. J. Muller to Jessie Muller, Feb. 21, 1935, Muller Collection.

20. The suicide note is found in the letters of Edgar Altenburg, Muller MSS.

21. He left around 8 a.m. on Sunday, Jan. 10, 1932 (*Daily Texan,* Jan. 12, 1932).

22. E. A. Carlson, *Genes, Radiation, and Society: The Life and Work of H. J. Muller* (Ithaca: Cornell University Press, 1981), 179.

23. H. J. Muller, "The Dominance of Economics over Eugenics," *Scientific Monthly* 37 (1933): 40–47.

24. R. C. Cook, "The Eugenics Congress," *Journal of Heredity* 23 (1932): 355–360.

25. Curt Stern, "The Domain of Genetics," *Genetics* 78 (1974): 30.

26. A. H. Sturtevant, "Unequal Crossing Over at the Bar Locus," *Genetics* 10 (1925): 117, 145.

27. To illustrate the process, he represented a chromosome by a string of genes (or a linear segment containing multiple genes), A B C D. If, then, this string broke between genes B and C, it would produce two fragments, A B and C D. If likewise a second chromosome, L M N O, broke between M and N into fragments L M and N O, then the chromosome fragments might rejoin at their broken ends, forming two new chromosomes A B N O and C D N O (H. J. Muller, "Further Studies on the Nature and Causes of Gene Mutations," *Proceedings of the Sixth International Congress of Genetics,* 1 [1932]: 217–218).

28. Ibid., 246.

29. H. J. Muller to Edgar Altenburg, Mar. 5, 1932, Muller Collection.

30. He'd arranged to spend his Guggenheim year in the lab of the Russian émigré Timoféef Ressovsky (H. J. Muller to Edgar Altenburg, Sept. 24, 1932, Muller MSS).

31. H. J. Muller to Jessie Muller, Nov. 8, 1932, Muller Collection.

32. Vavilov made a brief stopover in Austin during the winter of 1929–30, hoping to persuade Muller to spend a sabbatical year in Russia. David Muller remembers Vavilov saying "Okay" and David not knowing what he meant because David was less fluent in colloquial English than Vavilov (David Muller, personal communication, Feb. 9, 2004). Dobzhansky confirms he was in the United States that year ("Reminiscences of Theodosius Dobzhansky," p. 172). Mark Papovsky also mentions the visit (Mark Papovsky, *The Vavilov Affair* [Hamden: Archon Books, 1984], 35). See also Carlos Offerman to Elof Carlson, Jan. 1982, Muller Collection.

33. Popovsky, *The Vavilov Affair,* 33–35. See also David Joravsky, *The Lysenko Affair* (Cambridge: Harvard University Press, 1970), 30.

34. Valery N. Soyfer, *Lysenko and the Tragedy of Soviet Science,* trans. Leo and Rebecca Gruliow (New Brunswick: Rutgers University Press, 1994), 45, 48.

35. "Reminiscences of Theodosius Dobzhansky," p. 164.

36. Joravsky, *The Lysenko Affair,* 77.

37. Zhores A. Medvedev, *The Rise and Fall of T. D. Lysenko,* trans. I. M. Lerner (New York: Columbia University Press, 1969), 40.

38. H. J. Muller to Jessie Muller, Feb. 21, 1933, Muller Collection.

39. Ibid.

40. H. J. Muller to Jessie Muller, Mar. 21, 1933, Muller Collection.

41. H. J. Muller to Jessie Muller, May 11, 1933, Muller Collection.

42. H. J. Muller to Jessie Muller, Jan. 22, 1933, Muller Collection.

43. H. J. Muller to David Muller, Sept. 4 and Oct. 31, 1934, Muller Collection. Also David Muller, personal communication.

44. H. J. Muller to Jessie Muller, Dec. 9, 1934, Muller Collection.

45. Embassy of the USSR to Jessie J. Muller, Apr. 2, 1935. See also H. J. Muller to Jessie Muller, Feb. 5, 1935, Muller Collection.

46. H. J. Muller to Jessie Jacobs, Apr. 27, 1935, Muller Collection.

47. In the 1927–28 harvest, 46,000 square miles (a region approximately four times the size of the state of Massachusetts) of winter wheat had been lost. Data from Joravsky, *Lysenko Affair,* 61.

48. Six months after introducing his method of treating winter wheat, he switched to advocating cold treatment of spring rather than winter varieties, ignoring the fact that the new claim about the improvement of spring varieties was fundamentally different from the original claim, which amounted to a hereditary transformation of winter into spring wheat (Soyfer, *Tragedy of Soviet Science,* 21).

49. Ibid., 20.

50. Medvedev, *Rise and Fall,* 15; see also Soyfer, *Tragedy of Soviet Science,* 27–31.

51. Joravsky gives several examples of the genre in *Lysenko Affair,* 92.

52. Article appeared in *Pravda* on Feb. 15, 1936 (Medvedev, *Rise and Fall,* 15–17). The scene of the speech was described in Soyfer, *Tragedy of Soviet Science,* 61. Three months later Lysenko was made an academician on his way to taking over for Vavilov as head of the All-Union Academy of Agricultural Science several years later (Popovsky, *The Vavilov Affair,* 65).

53. Kerkis was also distressed by the fact that his wife, Natasha, had begun an affair with Carlos and had become "sick and estranged" (H. J. Muller to Jessie Muller, June 16, 1935, Muller Collection).

54. Prokofyeva had made use of so-called salivary gland maps, which had been discovered by Muller's Austin colleague, Theophilus Painter, in 1931. Fearing that Morgan et al. would try to exploit the new finding before he and the Texas group had a chance to capitalize on the new method of studying the salivary gland chromosomes, Painter delayed publishing the result until December 1933, almost two year after he'd first discovered it (B. Glass, "Theophilus Shickel Painter: August 22, 1889–October 5, 1969," *Biographical Memoirs of the National Academy of Sciences* 55 [1990]: 321).

55. H. J. Muller to Edgar Altenburg, Dec. 15, 1936, Muller MSS.

56. Muller, "Further Studies," 217–218.

57. H. J. Muller, A. A. Prokofyeva-Belgovskaja, and K. V. Kossikov, "Unequal Crossing-Over in the Bar Mutant as a Result of Duplication of a Minute Chromosome Section," *Comptes Rendus de L'Académie des Sciences de L'URRS* 1 (1936), reprinted in H. J. Muller, *Studies in Genetics* (Bloomington: Indiana University Press, 1962), 474.

58. Ibid.

59. H. J. Muller, "Bar Duplication," *Science* 83: 530.

60. H. J. Muller to J. Alexander, June 27, 1946, Muller MSS.

61. H. J. Muller to Edgar Altenburg, Mar. 30, 1936, Muller MSS.

62. If the repeated section were inserted at an arbitrary point to the left or right of the original section, as Bridges assumed, the unequal crossing over between two chromosomes would result in a recombinant chromosome that lacked the region between the repeated sections and would not be a true revertant.

63. Muller, "Bar Duplication," 529.

64. In fact, Bridges was famous for stealing other men's wives as well as their ideas. His behavior had become so problematical that he was given strict warnings before he was allowed to attend a summer conference at Cold Spring Harbor (letters between Calvin Bridges and Milislav Demerec, Mar. 21 and Mar. 26, 1934, Demerec Papers, American Philosophical Society).

65. Letter from H. J. Muller to Edgar Altenburg, Mar. 30, 1936, Muller Collection.

66. *Daily Texan,* Apr. 28, 1936.

67. Valery N. Soyfer, "Tragic History of the VII International Congress of Genetics," *Genetics* 165 (2003): 1–9.

68. H. J. Muller to David Muller, Dec. 1, 1936, Muller Collection.

69. Conway Zirkle, ed., *Death of a Science in Russia: The Fate of Genetics as Described in Pravda and Elsewhere* (Philadelphia: University of Pennsylvania Press, 1949), 2–3.

70. Ibid., 4–5.

71. Throughout the country, peasants were provided with a pair of scissors and tweezers and told to cut open sepals and spread pollen in fields of normally self-pollinating wheat. As a public relations stunt intra-varietal crossing was brilliant, both cost effective and photogenic, but there was little, if any, credible scientific evidence that it had any effect. N. I. Vavilov, "The Report Read by N. I. Vavilov," *Imperial Bureau of Pastures and Forage Crops: Aberystwyth, Great Britain,* Mimeographed Translation no. 1/1938, Fourth Session of the Academy of Agricultural Science, Moscow, U.S.S.R., December 1936, issued Jan. 1938, p. 4–5, Cold Spring Harbor Laboratory Archives.

72. Even *Pravda* acknowledged that vernalization wasn't working (Soyfer, *Tragedy of Soviet Science,* 83).

73. The cultivation of highly inbred homozygous lines was an essential part of the strategy of American corn farmers, who found that hybrids made from two highly homozygous lines were remarkably vigorous and high yielding.

74. Joravsky, *Lysenko Affair,* 379 n. 129.

75. Soyfer, *Tragedy of Soviet Science,* 84.

76. Ibid., 88. See also Raissa L. Berg, *Acquired Traits: Memoirs of a Geneticist from the Soviet Union,* trans. David Lowe (New York: Viking, 1988), 36.

77. T. D. Lysenko, "The Report Read by T. D. Lysenko," *Imperial Bureau,* 9–10, Cold Spring Harbor Laboratory Archives.

78. Soyfer, *Tragedy of Soviet Science,* 86.

79. Ibid., 87.

80. Lysenko, "The Report Read by T. D. Lysenko," 15–18, Cold Spring Harbor Laboratory Archives.

81. H. J. Muller, "Basis of the Theory of the Gene: The Experimental Evidence concerning the Properties of the Gene," summary of address given at session of Lenin Academy of Agricultural Science, Dec. 23, 1936, typescript in Muller MSS.

82. Popovsky, *The Vavilov Affair,* 82. Popovsky references a transcript of the session he found in government archives.

83. H. J. Muller to Julian Huxley, Mar. 9, 1937, Muller MSS.

84. Soyfer, *Tragedy of Soviet Science,* 103.

85. In the end, however, the Academy never published the final version, which was eventually published in England instead. See H. J. Muller to Julian Huxley, Mar. 9, 1937, Muller MSS; Carlson, *Life and Work of H. J. Muller,* 234.

86. Carlson, *Life and Work of H. J. Muller,* 239.

87. H. J. Muller to Carlos Offerman, May 14, 1937, Muller Collection. See also H. J. Muller to Julian Huxley, Feb. 24, 1937, Muller MSS.

88. H. J. Muller to Edgar Altenburg, July 15, 1937, Muller Collection.

89. H. J. Muller to Carlos Offerman, Aug. 20, 1937. See also H. J. Muller to David Muller, July 27, 1937, Muller Collection.

90. H. J. Muller to Edgar Altenburg, Sept. 16, 1937, Muller Collection. See also H. J. Muller to Carlos Offerman, Aug. 11, 1937, Muller Collection; H. J. Muller to Julian Huxley, Oct. 22, 1937, Muller MSS.

91. Vavilov's and Muller's final meeting in Leningrad was described by Raissa Berg, who had worked in Muller's lab (Berg, *Acquired Traits,* 39).

92. H. J. Muller to Jessie Offerman, Sept. 27, 1937, Muller Collection.

93. H. J. Muller to Carlos Offerman, Sept. 28, 1937, Muller Collection.

94. H. J. Muller to Edgar Altenburg, Oct. 25, 1937, Muller MSS.

95. H. J. Muller to Edgar Altenburg, Dec. 6, 1937, Muller Collection. See also H. J. Muller to N. I. Vavilov, Feb. 21, 1938, Muller MSS.

96. H. J. Muller to Edgar Altenburg, Oct. 19, 1937, Muller MSS.

97. H. J. Muller to Edgar Altenburg, Oct. 18, 1938, Muller Collection.

98. H. J. Muller to Edgar Altenburg, Jan. 30, 1939, Muller MSS.

99. H. J. Muller to Edgar Altenburg, Mar. 23, 1939, and May 20, 1939, Muller Collection.

100. N. I. Vavilov to H. J. Muller, Dec. 18, 1938, Muller MSS.

101. H. J. Muller to Cyril Darlington, Aug. 4, 1939, Muller MSS.

102. Letter from H. J. Muller to Carlos and Jessie Offerman, Aug. 16, 1939, Muller Collection.

103. H. J. Muller to Edgar Altenburg, Oct. 20, 1939, Muller Collection.

104. H. J. Muller to David Muller, May 23, 1940, Muller Collection.

105. Ada Griesmaier to Edgar Altenburg, July 27, 1940, Muller Collection.

106. Thea Muller, "With Yeast and Flies in Portugal," typescript, 1973, Muller MSS.

107. H. J. Muller to Edgar Altenburg, Oct. 1940, from Amherst (last 2 pages only), Muller Collection.

108. Carlson, *Life and Work of H. J. Muller,* 274.
109. H. J. Muller to Edgar Altenburg, May 6, 1941, Muller MSS.
110. Edgar Altenburg to H. J. Muller, Mar. 9, 1941, Muller MSS.
111. Ibid.
112. The theory of "genic balance" was one of Muller's most profound and original insights, to which he would return over and over again in the course of his life. It had first arisen in connection with sex determination in *Drosophila* in 1912. At that point, Muller had suggested that it was not simply a matter of whether an individual possessed one or two X chromosomes that determined its sex but rather the ratio of the number of X's to the number of sets of autosomes (the non-sex-linked chromosomes). In particular, Muller had argued in an unpublished 1912 lecture that flies possessing a 1:1 ratio of X's to autosomes (two X's to two sets of autosomes) were female, while those possessing a 1:2 ratio of X's to autosomes (a single X chromosome to two sets of autosomes) were male (H. J. Muller, "Erroneous Assumptions Regarding Genes," 20, unpublished manuscript [1911–1912], Muller MSS). The theory was spectacularly confirmed by Bridges, who discovered that flies with a 2:3 ratio of X's to autosomes (containing two X's and three sets of autosomes) were intersexes, sometimes containing both an ovary and a testis. In what was by then a well-established pattern, Bridges smoothly glossed over the genesis of the idea in the announcement of his new experimental finding, writing in 1921, "A significant new conclusion proved by the intersexes is that sex in *D. melanogaster* is determined by a balance between the genes contained in the X-chromosome and those contained in the autosomes" (Calvin B. Bridges, "Triploid Intersexes in Drosophila Melanogaster," *Science* 54 [1921]: 253). The fact that the "significant new conclusion" had been reached by Muller nearly ten years prior to the publication of the paper was not mentioned.
113. Carlson, *Life and Work of H. J. Muller,* 280.
114. H. J. Muller to Jessie Offerman, May 17, 1944, Muller Collection.
115. H. J. Muller to Jessie Offerman, Aug. 5, 1945, Muller Collection.
116. Popovsky, *The Vavilov Affair,* 1984.
117. Ibid., 130, 145, 155, 175, 185, 186–191.
118. Edgar Altenburg to H. J. Muller, Jan. 19, 1946, Muller MSS.
119. Edgar Altenburg to H. J. Muller, Feb. 13, 1946, Muller MSS.
120. Edgar Altenburg to H. J. Muller, Feb. 19, 1946, Muller MSS.
121. H. J. Muller to Edgar Altenburg, Mar. 1, 1946, Muller MSS.
122. Edgar Altenburg to H. J. Muller, Mar. 4, 1946, Muller MSS.

123. H. J. Muller, "Thomas Hunt Morgan, 1866–1945," *Science* 103 (1946): 550.

124. H. J. Muller to Thea Muller, Dec. 9, 1946, Muller Collection.

## EPILOGUE

1. H. J. Muller, "The Gene," *Proceedings of the Royal Society,* series B, 134 (1947): 23 n.

2. H. J. Muller, "The Gene as the Basis of Life," *Proceedings of the International Congress of Plant Sciences* 1 (1929): 918; originally presented as an address in Ithaca, New York, on Aug. 19, 1926.

3. H. J. Muller, "The Development of the Gene Theory," in Leslie Clarence Dunn, ed., *Genetics in the 20th Century: Essays on the Progress of Genetics during Its First 50 Years* (New York: Macmillan, 1951), 97.

4. H. J. Muller, "Physics in the Attack on the Fundamental Problems of Genetics," *Scientific Monthly* 44 (1937): 212.

5. Linus Pauling and Max Delbrück, "The Nature of the Intermolecular Forces Operative in Biological Processes," *Science* 92 (1940): 78.

6. Robert Olby, *The Path to the Double Helix* (Seattle: University of Washington Press, 1974), 120.

7. James D. Watson, *The Double Helix* (New York: Atheneum, 1968), 56.

8. Ibid., 52.

9. Brenda Maddox, *Rosalind Franklin: The Dark Lady of DNA* (New York: HarperCollins, 2002), 165.

10. In the MRC report, Franklin gave the exact parameters of the repeating unit of the DNA crystal as well as noting that it was "face-centred monoclinic." From these facts Crick instantly deduced that there were two chains that ran in opposite directions (Horace Freeland Judson, *The Eighth Day of Creation: Makers of the Revolution in Biology* [Plainview: Cold Spring Harbor Laboratory Press, 1996], 142–143).

11. Olby, *The Path,* 215.

12. For Watson and Crick's differing accounts of this, see Watson, *The Double Helix,* 193–196; and Olby, *The Path,* 411–412.

13. J. D. Watson and F. H. C. Crick, "Molecular Structure of Nucleic Acid," *Nature* 171 (1953): 737–738.

14. Victor A. McKusick and Frank H. Ruddle, "The Status of the Gene Map of the Human Chromosomes," *Science* 196 (1977): 402.

15. It was impossible to distinguish between the case in which the two dominant alleles (A B) were on one member of a parental pair and the two recessives (a b)

were on the other member, and the case in which one dominant allele and one recessive were on each member of the parental pair, giving (A b) and (a B).

16. T. Grodzicker, J. Williams, P. Sharp, and J. Sambrook, "Physical Mapping of Temperature-Sensitive Mutations of Adenoviruses," *Cold Spring Harbor Symposium on Quantitative Biology* 39 (1975): 439–446.

17. Roger Lewin, "Jumping Genes Help Trace Inherited Diseases," *Science* 211 (1981): 690.

18. Botstein et al. compute that a gene must be within 10 million base pairs of an RFLP to stay linked—at that distance there was a 90 percent chance that the gene would not undergo recombination. Thus they estimated that 150 RFLPs, one every 20 million bases to span the 3 billion bases of the haploid genome, would be needed. See David Botstein, Raymond L. White, Mark Skolnick, and Ronald W. Davis, "Construction of a Genetic Linkage Map in Man Using Restriction Fragment Length Polymorphisms," *American Journal of Human Genetics* 32 (1980): 319–320.

19. David Botstein, "1989 Allen Award Address: The American Society of Human Genetics Annual Meeting, Baltimore," *American Journal of Human Genetics* 47 (1989): 889–890.

20. Victor A. McKusick, "Mapping and Sequencing the Human Genome," *New England Journal of Medicine* 320 (1989): 910–915. See also Roger Lewin, "National Academy Looks at Human Genome Project, Sees Progress," *Science* 235 (1987): 747.

21. Nine years later the estimate of the frequency of the lesser allele was downgraded to 1 percent.

22. Example with three SNPs: Consider a block of DNA that contains three SNPs as pictured below:

```
————X————————X————X——————————
    SNP1         SNP2   SNP3
```

Let A and B stand for the two possible alleles at each location. Then there are 8 (= $2^3$) possible patterns, which can be written out as follows:

```
Type1:————A————————A————A——————————
Type2:————A————————A————B——————————
Type3:————A————————B————A——————————
Type4:————B————————A————A——————————
Type5:————A————————B————B——————————
Type6:————B————————A————B——————————
Type7:————B————————B————A——————————
Type8:————B————————B————B——————————
```

Because neighboring SNPs do not associate at random, only a subset of these patterns would appear in significant numbers. For example, in a randomly chosen group, 80 to 90 percent of individuals would be found to carry chromosomes containing one of only 3 or 4 of these 8 possible types. These common patterns are the ancestral segments (haplotypes) passed down through the generations.

23. The International HapMap Consortium, "A Haplotype Map of the Human Genome," *Nature* 437 (2005): 1299–1320.

24. H. J. Muller, "Human Values in Relation to Evolution," *Science* 127 (1958): 629.

# ACKNOWLEDGMENTS

WRITING THIS BOOK has been a long journey that began when I read Elof Carlson's remarkable, kaleidoscopic biography of Hermann J. Muller. E.C.'s love for his subject jumped off of every page of his book. Over the years, I have benefited greatly from his encyclopedic knowledge of the history of genetics as well as from his flashes of insight, one of which led to the title of this book.

Both Helen and David Muller have been very generous with me, and through many hours of conversation with them I gained a much deeper understanding of their father. I am particularly indebted to David Muller for allowing me to stay at his home for several weeks and sharing with me the experience of reading his parents' correspondence. The time I spent with him at the edge of the Chihuahuan desert reading letters was one of the high points of my life as a writer.

My literary agent, Stuart Bernstein, has been a constant source of encouragement and guidance, literary and otherwise, for many years. I am also deeply indebted to Ruth Cowan, who from early on in my research seemed to understand what kind of book I might be aiming toward and helped me shape the story as it emerged.

I thank my editor, Michael Fisher, first for his enthusiasm about doing a book with me, then for sticking through the endless iterations of this one, and lastly for his fine editing. I am also indebted to Nomi Pierce, who first introduced me to the world of evolutionary biology, got me excited about science writing, and has been immensely generous and supportive ever since. Also, I am grateful to one of the last true renaissance men, Alex Star, for publishing my first science writing. Gerhard Hochschild and Jim Crow each read the book thoroughly at two different stages, and made many useful suggestions on both readings. In addition, G.H.'s translations of the writings of several nineteenth-century cell biologists were invaluable. Also, I thank Michael Arnold, Tony Schwartz, Mike Greenberg, Kai Zinn, Elof Carlson, Jan Witkowski, Jim Watson, Matt Meselson, David Haig, David Altshuler, and Alex Gann for taking the time to read and critique this manuscript and, in Alex's case, for serving as my guide and emissary at Cold Spring Harbor.

Also, I wish to thank librarians everywhere for being an exceptionally selfless and dedicated group. More particularly, I thank Valerie-Ann Lutz at the American Philosophical Society, who was always responsive and ready to provide assistance, and Ben Steinberg of the Brookline Public Library, who has aided my research efforts since my first foray into science writing in 1998. Thanks also to Ronnie Broadfoot and Mary Sears of the Ernst Mayr Library, who were unfailingly helpful to me at all stages in this project. Saundra Taylor at the Lilly library helped me locate Muller letters and materials that would otherwise never have been found. The staff at the Marine Biology Laboratory Library, including Heidi Nelson, Elinor Uhlinger, and Joanne Westburg, helped me in various ways. Ludmila Pollock welcomed me at the Cold Spring Harbor Archives, and Clare Bunce provided me with valuable assistance. Amanda York Focke, archivist at Rice University's Fondren Library, came to my rescue twice. I am grateful to Tamara Clark for her expert artwork, Wendy Nelson for her excellent copyediting, and Susan Fels for her fine work on the index; to Alice Baxter for lending me a copy of her thesis on E. B. Wilson; to Harriet Zuckerman for allowing me permission to read and quote from her interview of Hermann Muller; and to David Harmin for straightening out my statistical reasoning at a critical time.

I thank both Kai Zinn and Pamela Bjorkman for their stimulating company over the years and for their hospitality when I visited the Cal Tech Archives. Thanks also to Niza and Richard Davidson for putting me up when I visited the American Philosophical Society. Also, I am grateful to the hugely humorous and magnanimous Bob Schneider, proprietor of the Woods Hole Inn, who three times lowered my rent and finally waived it altogether in order to encourage me to stay on at the Marine Biological Laboratory Library and finish the work that needed to be done. I would also like to thank the American Philosophical Society, the Lilly Library, and the MBL Library for their financial support.

Lastly, I wish to thank my father and mother, who always encouraged me in my academic and other pursuits, and my brother and sister, with whom I feel a deep connection, both genetic and otherwise. Finally, it is with special pleasure that I acknowledge my debt to my two sons, Eli and Daniel, who provided a living laboratory in which to test my ideas about inheritance and the relative importance of nature and nurture.

# INDEX